D1806775

Boston Studies in the Philosophy and History of Science

Volume 337

Series Editors
Alisa Bokulich, Boston University, Boston, MA, USA
Jürgen Renn, Max Planck Institute for the History of Science, Berlin, Germany
Michela Massimi, University of Edinburgh, Edinburgh, UK

Managing Editor
Lindy Divarci, Max Planck Institute for the History of Science, Berlin, Berlin, Germany

Editorial Board Members
Theodore Arabatzis, University of Athens, Athens, Greece
Heather E. Douglas, University of Waterloo, Ontario, Canada
Jean Gayon, Université Paris 1, Paris, France
Thomas F. Glick, Boston University, Boston, USA
Hubert Goenner, University of Goettingen, Göttingen, Niedersachsen, Germany
John Heilbron, University of California, Berkeley, UK
Diana Kormos-Buchwald, California Institute of Technology, Pasadena, CA, USA
Christoph Lehner, Max Planck Institute for the History of Science, Berlin, Germany
Peter McLaughlin, Universität Heidelberg, Heidelberg, Baden-Württemberg, Germany
Agustı Nieto-Galan, Universitat Autònoma de Barcelona, Bellaterra (Cerdanyola del V.), Spain
Nuccio Ordine, Universitá della Calabria, RENDE, Cosenza, Italy
Ana Simões, Universidade de Lisboa, Lisboa, Portugal
John J. Stachel, Boston University, Brookline, MA, USA
Baichun Zhang, Chinese Academy of Science, Beijing, China
Sylvan S. Schweber, Harvard University, Cambridge, USA

The series *Boston Studies in the Philosophy and History of Science* was conceived in the broadest framework of interdisciplinary and international concerns. Natural scientists, mathematicians, social scientists and philosophers have contributed to the series, as have historians and sociologists of science, linguists, psychologists, physicians, and literary critics.

The series has been able to include works by authors from many other countries around the world.

The editors believe that the history and philosophy of science should itself be scientific, self-consciously critical, humane as well as rational, sceptical and undogmatic while also receptive to discussion of first principles. One of the aims of Boston Studies, therefore, is to develop collaboration among scientists, historians and philosophers.

Boston Studies in the Philosophy and History of Science looks into and reflects on interactions between epistemological and historical dimensions in an effort to understand the scientific enterprise from every viewpoint.

More information about this series at http://www.springer.com/series/5710

Vincenzo De Risi

Editor

Leibniz and the Structure of Sciences

Modern Perspectives on the History of Logic, Mathematics, Epistemology

 Springer

Editor
Vincenzo De Risi
Labratoire SPHère - CNRS
Paris, France

Max Planck Institute for the History
of Science
Berlin, Germany

ISSN 0068-0346   ISSN 2214-7942   (electronic)
Boston Studies in the Philosophy and History of Science
ISBN 978-3-030-25571-8   ISBN 978-3-030-25572-5   (eBook)
https://doi.org/10.1007/978-3-030-25572-5

© Springer Nature Switzerland AG 2019
This work is subject to copyright. All rights are reserved by the Publisher, whether the whole or part of
the material is concerned, specifically the rights of translation, reprinting, reuse of illustrations, recitation,
broadcasting, reproduction on microfilms or in any other physical way, and transmission or information
storage and retrieval, electronic adaptation, computer software, or by similar or dissimilar methodology
now known or hereafter developed.
The use of general descriptive names, registered names, trademarks, service marks, etc. in this publication
does not imply, even in the absence of a specific statement, that such names are exempt from the relevant
protective laws and regulations and therefore free for general use.
The publisher, the authors, and the editors are safe to assume that the advice and information in this book
are believed to be true and accurate at the date of publication. Neither the publisher nor the authors or
the editors give a warranty, expressed or implied, with respect to the material contained herein or for any
errors or omissions that may have been made. The publisher remains neutral with regard to jurisdictional
claims in published maps and institutional affiliations.

This Springer imprint is published by the registered company Springer Nature Switzerland AG.
The registered company address is: Gewerbestrasse 11, 6330 Cham, Switzerland

Foreword

The present volume brings together eight essays on Gottfried Wilhelm Leibniz's work on logic, mathematics, and epistemology. The essays have been arranged in broadly systematic order. The book begins with two studies on various aspects of Leibniz's logic and its connection with other exact sciences (the contribution by Marko Malink and Anubav Vasudevan on the notion of logical coincidence and the one by Massimo Mugnai on Leibniz's mereology). It continues with an essay on infinity and cardinality (by Richard Arthur) which succeeds in bridging the gap between pure logic and mathematics and with further contributions on the foundations of geometry (by Vincenzo De Risi on continuity in elementary geometry and David Rabouin and Valérie Debuiche on the possibility of non-Euclidean geometry in Leibniz's thought). The geometrical perspective is combined with considerations on the calculus in an essay on Leibniz's treatment of curves (by Davide Crippa), and a foray is made into the realm of mathematical physics with a paper on Leibniz's calculus of variations (by Jürgen Jost). The final essay (by Nabeel Hamid) draws on what these preceding investigations in the history of mathematics have established to offer some new views on Leibniz's epistemology of the physical sciences.

The book can serve to open up an interesting line of approach to many aspects of the work that Leibniz carried out in the exact sciences. The history of epistemology is transversally addressed in all the contributions, with a close connection being maintained throughout to Leibniz's actual scientific practice as a working mathematician. Several essays deal with the history of science as such and enter, in considerable depth, into Leibniz's technical achievements in logic, mathematics, or mathematical physics.

The title of the book, *Leibniz and the Structure of Sciences*, stresses the "structural" approach which was characteristic of Leibniz's own epistemological and scientific research and which may well represent the main step forward that Leibniz took as compared to several of his contemporaries. The structure of Leibnizian science, however, is also illustrated by the internal connection among the essays published in this volume and their complex relations of grounding: so that logic seems to ground mereology, mereology grounds the theory of cardinality and mathematics as a whole, the latter grounds geometry and then analysis, and analysis forms the foundation of mathematical physics. In the course of the systematic exposition of the structures of these disciplines, however, the reader will come to realize that the connections between them are more complicated still and that, for instance, Leibniz's epistemology of geometry had repercussions on his conceptions of whole and part and these latter on his conception of logic itself. Leibniz's scientific practices (as opposed to his epistemological views) further enrich the picture, and the result is a complex relation of mutual grounding among various disciplines – a relation that Leibniz himself would probably have called a peculiar form of harmony – which we ourselves can look upon as a remarkable structure of the foundations of sciences.

The subtitle of the book—*Modern Perspectives on the History of Logic, Mathematics, Epistemology*—underlines the fact that several essays in the collection offer some insightful comparison with *modern* treatments of the subject. Thus, the essay by Malink and Vasudevan deals with Leibniz's logical system from the perspective of modern logic and Boolean algebras; Mugnai reads Leibniz's definitions of whole and part through the spectacles of contemporary mereology; Arthur compares Leibniz's and Cantor's views on the infinite; De Risi discusses Leibniz's notion of continuity in relation to Dedekind completeness; Rabouin and Debuiche deal with Leibniz and non-Euclidean geometry; and Jost looks at Leibniz's calculus of variations with the tools of today's mathematics. These *modern* readings of Leibniz do not, indeed, form the center of these essays, which are not written *in order* to draw such comparisons; but the authors nonetheless discovered that, in some cases, a modern perspective was apt to shed light on the actual force, structure, or limits of Leibniz's endeavors. Though aware of the fact that a historiographical approach making use of modern cognitive tools to interpret early modern scientific results may easily result in blatant anachronism, we think, and hope, that we have avoided such a pitfall and have provided, on the contrary, some tools with which to better appreciate the actual historical dimension of Leibniz's researches. In the past, indeed, Leibniz has often been anachronistically addressed as the unknowing father of modern logic, modern geometry, topology, nonstandard analysis, relativity theory, and so forth. That is to say, no really attentive comparison between his views and these further developments has really been undertaken—but rather an analogy or similarity has been uncritically assumed to exist. By confronting Leibniz with Cantor or Dedekind, the present volume wants to rectify such anachronisms, and to point up *both* the similarity *and* the distance between Leibniz's project and the modern results. We hope that, in this way, the exact extent of Leibniz's scientific innovations will be better grasped. At the same time, the present volume may offer materials for reflection not just to Leibniz specialists but also to historians of science and epistemology, and philosophers of science, working on modern authors and problems.

Several of the essays in this volume emerged from the renewed scholarly work around Leibniz that was fostered by the celebrations of Leibniz's tercentenary in 2016. In particular, the editor of this volume, appointed Leibniz Chair at the University of Leipzig in the Winter Semester 2016–2017, contributed to organizing, in collaboration with the Max Planck Institute for Mathematics in the Sciences and the Max Planck Institute for the History of Science, a Summer School on Leibniz that was held in Leipzig and Hannover on July 7–16, 2016, and a conference on "Leibniz and the Sciences" that took place in Leipzig on November 14–16, 2016. We thank the Leibniz Program of the University of Leipzig and the Max Planck Institutes for their generous support and their contributions to the advancement of Leibniz studies.

We also thank Andrew Arana, François Duchesneau, Daniel Garber, Eberhard Knobloch, Jeffrey McDonough, John Mumma, and Justin Smith for having contributed to improving the essays in the present volume.

Contents

Contributors

Richard T. W. Arthur McMaster University, Hamilton, ON, Canada

Davide Crippa Institute of philosophy, Centre for Science, Technology, and Society Studies, Czech Academy of Sciences, Prague, Czech Republic

Valérie Debuiche Aix Marseille Univ, CNRS, Centre Gilles-Gaston Granger, Aix-en-Provence, France

Nabeel Hamid Concordia University, Montréal, Canada

Jürgen Jost Max Planck Institute for Mathematics in the Sciences, Leipzig, Germany

Marko Malink New York University, New York, NY, USA

Massimo Mugnai Scuola Normale Superiore, Pisa, Italy

David Rabouin Université de Paris, Laboratoire SPHERE, UMR 7219, CNRS, Paris, France

Vincenzo De Risi Laboratoire SPHère, Paris, France
Max Planck Institute for the History of Science, Berlin, Germany

Anubav Vasudevan University of Chicago, Chicago, IL, USA

Chapter 1
Leibniz on the Logic of Conceptual Containment and Coincidence

Marko Malink and Anubav Vasudevan

Abstract In a series of early essays written around 1679, Leibniz sets out to explore the logic of conceptual containment. In his more mature logical writings from the mid-1680s, however, his focus shifts away from the logic of containment to that of coincidence, or mutual containment. This shift in emphasis is indicative of the fact that Leibniz's logic has its roots in two distinct theoretical frameworks: (i) the traditional theory of the categorical syllogism based on rules of inference such as Barbara, and (ii) equational systems of arithmetic and geometry based on the rule of substitution of equals. While syllogistic reasoning is naturally modeled in a logic of containment, substitutional reasoning of the sort performed in arithmetic and geometry is more naturally modeled in a logic of coincidence. In this paper, we argue that Leibniz's logic of conceptual containment and his logic of coincidence can in fact be viewed as two alternative axiomatizations, in different but equally expressive languages, of one and the same logical theory. Thus, far from being incoherent, the varying syllogistic and equational themes that run throughout Leibniz's logical writings complement one another and fit together harmoniously.

1.1 Conceptual Containment and Coincidence

Among the notions that figure most prominently in Leibniz's logical and metaphysical writings is that of conceptual containment. It is well known that Leibniz subscribed to a containment theory of truth, according to which the truth of any proposition consists in one term's being conceptually contained in another. This theory applies most directly to simple predicative propositions of the form *A is B*. In

M. Malink (✉)
New York University, New York, NY, USA
e-mail: mm7761@nyu.edu

A. Vasudevan
University of Chicago, Chicago, IL, USA
e-mail: anubav@uchicago.edu

© Springer Nature Switzerland AG 2019
V. De Risi (eds.), *Leibniz and the Structure of Sciences*, Boston Studies in
the Philosophy and History of Science 337,
https://doi.org/10.1007/978-3-030-25572-5_1

Leibniz's view, whenever such a proposition is true, its truth consists in the predicate term, *B*, being conceptually contained in the subject term, *A*. Famously, this is so not only for necessary propositions such as *God is wise*, but also for contingent propositions such as *Caesar is just*. Thus, Leibniz writes:[1]

> The predicate or consequent always inheres in the subject or antecedent, and the nature of truth in general or the connection between the terms of a statement, consists in this very thing, as Aristotle also observed ... This is true for every affirmative truth, universal or particular, necessary or contingent, and in both an intrinsic and extrinsic denomination. And here lies hidden a wonderful secret. (*Principia logico-metaphysica*, A VI.4 1644)

Leibniz's containment theory of truth implies that every proposition expresses the conceptual containment of one term in another.[2] Or, as Leibniz puts it, 'in every proposition the predicate is said to inhere in the subject'.[3] Given his commitment to this theory, it is perhaps not surprising that in several of his logical writings Leibniz undertakes to develop a calculus in which conceptual containment figures as the sole primitive relation between terms. For example, in an essay titled *De calculo analytico generali*, written around 1679, he develops a calculus in which every simple proposition is of the form *A is B*.[4] Similar containment calculi appear in a number of essays written by Leibniz in the late 1670s, including *De characteristica logica*, *Propositiones primitivae*, and *Specimen calculi universalis*.[5]

By contrast, in Leibniz's more mature logical writings from the mid-1680s there is a notable shift in focus away from the relation of conceptual containment to that of coincidence, or mutual containment, between terms. In these later writings, Leibniz develops calculi in which the primitive propositions are not of the form *A is B* but instead of the form *A coincides with B* (or, equivalently, *A is the same as B*). The canonical version of such a coincidence calculus is developed by Leibniz in his 1686 treatise *Generales inquisitiones de analysi notionum et veritatum*. In this treatise, Leibniz regards propositions of the form *A coincides with B* as 'the most

---

[1] In what follows, we adopt the following abbreviations for editions of Leibniz's writings:

A   *Gottfried Wilhelm Leibniz: Sämtliche Schriften und Briefe.* Ed. by the Berlin-Brandenburgische Akademie der Wissenschaften and the Akademie der Wissenschaften zu Göttingen, Berlin: de Gruyter. Cited by series, volume, and page.

C   *Opuscules et fragments inédits de Leibniz.* Ed. by L. Couturat, Paris: Félix Alcan, 1903.

GM *Leibnizens mathematische Schriften.* Ed. by C. I. Gerhardt, Halle: H. W. Schmidt, 1849–63. Cited by volume and page.

GP *Die philosophischen Schriften von Gottfried Wilhelm Leibniz.* Ed. by C. I. Gerhardt, Berlin: Weidmann, 1875–90. Cited by volume and page.

In addition, we sometimes use the abbreviation '*GI*' for Leibniz's essay *Generales inquisitiones de analysi notionum et veritatum* (A VI.4 739–88). This essay consists of a series of numbered sections, which we designate by references such as '*GI* §16' and '*GI* §72'.

[2] See Adams 1994: 57–63; similarly, Parkinson 1965: 6 and 33.

[3] A VI.4 223; see also A VI.4 218 n. 1, 551, *GI* §132.

[4] See A VI.4 148–50.

[5] A VI.4 119–20, 140–5, and 280–8; see also A VI.4 274–9.

simple' kind of proposition.[6] Propositions expressing containment are then defined in terms of coincidence. Specifically, Leibniz defines *A is B* as the proposition *A coincides with AB*, where *AB* is a composite term formed from the terms *A* and *B*.[7] By means of this definition, Leibniz is able to derive the laws of conceptual containment as theorems of his coincidence calculus. Thus, in the *Generales inquisitiones* coincidence is treated as the sole primitive relation between terms, with conceptual containment and all other relations being defined as coincidences between complex terms.[8] This same approach, aimed at developing a coincidence calculus, is adopted by Leibniz in a series of related essays written shortly after the *Generales inquisitiones*, including *Specimina calculi rationalis*, *Specimen calculi coincidentium*, and *Specimen calculi coincidentium et inexistentium*.[9]

Despite this shift in emphasis from conceptual containment to coincidence in his logical writings, Leibniz never abandons his commitment to the containment theory of truth. Indeed, he reaffirms this theory in numerous subsequent works, such as the *Discours de Metaphysique* (1686), *Principia logico-metaphysica* (around 1689), and the *Nouveaux Essais* (1704).[10] Even in the *Generales inquisitiones* (1686), Leibniz takes for granted the containment theory of truth. This can be seen, for example, from the following passage:

> Every true proposition can be proved. For since, as Aristotle says, the predicate inheres in the subject, or, the concept of the predicate is involved in the concept of the subject when this concept is perfectly understood, surely it must be possible for the truth to be shown by the analysis of terms into their values, or, those terms which they contain. (*Generales inquisitiones* §132)

Thus, Leibniz appears to persist in his commitment to the primacy of conceptual containment even in those writings in which his central aim is to develop a logic of coincidence. Accordingly, in the *Generales inquisitiones*, Leibniz undertakes to analyze the meaning of coincidence in terms of containment:

> A proposition is that which states what term is or is not contained in another. Hence, ... a proposition is also that which says whether or not a term coincides with another; for those terms which coincide are mutually contained in one another. (*Generales inquisitiones* §195)

---

[6]*GI* §157 and §163. Accordingly, the axioms of the calculus developed by Leibniz in the *Generales inquisitiones* almost all take the form of coincidence propositions; see the preliminary axiomatizations of the calculus in *GI* §§1–15, §171, and §189, as well as the final axiomatization in §198.

[7]*GI* §83, §113, A VI.4 751 n. 13, 808. For other definitions of containment in terms of coincidence, see *GI* §16, §17, §189.4, and §198.9.

[8]As Castañeda points out, in the *Generales inquisitiones* Leibniz develops 'a strict equational calculus in which all propositions are about the coincidence of terms' (Castañeda 1976: 483). See also Schupp 1993: 156.

[9]A VI.4 807–14, 816–22, 830–45; see also A VI.4 845–55 and C 421–3. Prior to the *Generales inquisitiones*, no systematic attempt is made by Leibniz to develop a calculus of coincidence. As far as we can see, the earliest indication of Leibniz's desire to develop such a calculus appears in an essay written around 1685 titled *Ad Vossii Aristarchum* (see A VI.4 622–4).

[10]See A VI.4 1539–41, 1644, and VI.6 396–8.

In spite of this characterization of coincidence as mutual containment, Leibniz opts to formulate the calculus developed in the *Generales inquisitiones* in a language in which coincidence, and not containment, figures as the sole primitive relation between terms. Indeed, it is striking that the passage just quoted appears immediately before Leibniz's final axiomatization of his coincidence calculus in §198, in which containment is defined in terms of coincidence rather than the other way around.[11]

Thus, we are left with an exegetical puzzle. Given Leibniz's unwavering commitment to the containment theory of truth, why does he choose to treat, not containment, but coincidence as the sole primitive relation between terms in his mature logical writings? If, as Leibniz insists, every proposition expresses a containment between terms, why does he nonetheless maintain that 'it will be best to reduce propositions from predication and from being-in-something to coincidence'?[12]

If containment and coincidence were straightforwardly interdefinable in the object language of Leibniz's calculus, the choice to treat the latter relation as primitive would be a mere notational preference of only marginal interest. But this is not the case. For, while containment is readily definable in the language of Leibniz's coincidence calculus by means of the formula *A coincides with AB*, there is no obvious way of defining coincidence in the language of Leibniz's containment calculus. Now, this is not to say that coincidence cannot be characterized in terms of containment. Indeed, as we have seen, Leibniz characterizes coincidence as mutual containment, appealing to the law that *A coincides with B* just in case both *A is B* and *B is A*.[13] This law of antisymmetry, however, does not constitute a definition of coincidence in the object language of Leibniz's containment calculus. For, crucially, this language does not include any primitive propositional operators for forming complex propositions from simpler ones. In particular, the language does not include any primitive operators for forming conditionals, disjunctions, or conjunctions.[14] Consequently, while both *A is B* and *B is A* are well-formed propositions of the

---

[11]See *GI* §198.9.

[12]A VI.4 622.

[13]*GI* §30, §88, and §195; see also A VI.4 813.

[14]In some of his earlier essays, Leibniz seems to employ an object language which includes primitive propositional operators for implication and conjunction (see, e.g., A VI.4 127–31, 142–3, 146–50). In his mature logical writings, however, Leibniz does not include such propositional operators in the object language of his calculi (see, e.g., the coincidence calculi formulated at *GI* §§198–200, A VI.4 816–22, 830–55). In the *Generales inquisitiones*, for example, Leibniz formulates his calculus in an object language in which every proposition is of the form *A coincides with B* (see Castañeda 1976: 483–4, Malink and Vasudevan 2016: 689–96). This decision to abstain from the use of primitive propositional operators is not an arbitrary choice on Leibniz's part, but is an essential part of his broader theoretical ambition to reduce propositional logic to a categorical logic of terms and, in particular, to 'reduce hypothetical to categorical propositions' (A VI.4 992; see also *GI* §75, §137, VI.4 811 n. 6, 862–3). For Leibniz's commitment to the Peripatetic program of reducing propositional to categorical logic, see Barnes 1983: 281 and Malink and Vasudevan 2018: §1.

language of Leibniz's containment calculus, this language does not include any primitive operator '&' by which to form their conjunction, (*A is B*) & (*B is A*).[15] Absent such a conjunction operator, or any other propositional operators that might be used to define conjunction, Leibniz's characterization of coincidence as mutual containment is not directly expressible in the object language of his containment calculus. Hence, there is no obvious way for Leibniz to provide an explicit definition of coincidence in terms of containment.

There is, however, an alternative, more indirect way for Leibniz to reduce the logic of coincidence to that of containment without exceeding the syntactic bounds of his calculus. Instead of taking the law of antisymmetry to provide an explicit definition of coincidence, one can simply appeal to the fact that this law licenses the following introduction and elimination rules for coincidence:

$$\frac{A \; is \; B \qquad B \; is \; A}{A \; coincides \; with \; B}$$

$$\frac{A \; coincides \; with \; B}{A \; is \; B} \qquad\qquad \frac{A \; coincides \; with \; B}{B \; is \; A}$$

If the object language of a containment calculus in which every proposition is of the form *A is B* is extended so as to include new primitive propositions of the form *A coincides with B*, these introduction and elimination rules would allow us to derive laws of coincidence from the underlying laws of containment.[16] In this way, coincidence could be viewed as a purely nominal relation the meaning of which is determined by its introduction and elimination rules in the calculus.[17] Thus, by adopting this approach, Leibniz would be able to integrate coincidence into his containment calculus without thereby abandoning his commitment to the logical and metaphysical primacy of conceptual containment.

---

[15] *Pace* Lenzen (1984a: 194–5, 1987: 3), who adopts such a conjunctive definition of coincidence in his reconstruction of Leibniz's containment calculus.

[16] This strategy for deriving laws of coincidence from underlying laws of containment is adopted, for example, by Schröder (1890: 184–5). There are, in fact, a number of striking parallels between Leibniz's containment calculus and Schröder's algebraic system of logic (see nn. 24, 34, and 38 below).

[17] This presupposes that the introduction and elimination rules for coincidence are conservative in the sense that they do not allow us to derive any new theorems of the form *A is B* not already derivable in the underlying containment calculus. For, if these rules were not conservative, the newly introduced concept of coincidence would have, as Brandom puts it, 'substantive content ... that is not already implicit in the contents of the other concepts being employed' (Brandom 1994: 127, 2000: 71). If, on the other hand, coincidence is introduced by means of conservative introduction and elimination rules, then 'no new content is really involved' (Brandom 1994: 127, 2000: 71). In this way, the connective of coincidence can be introduced into a pure containment calculus without the need to rely on 'an antecedent idea of the independent meaning of the connective' (Belnap 1962: 134; see also Dummett 1991: 247). This condition of conservativity is satisfied for all the versions of Leibniz's containment calculus discussed in this paper.

In what follows, we show that, by means of this approach, all the laws of Leibniz's coincidence calculus can be derived from the laws of his containment calculus. We thus argue that the shifting emphasis on containment and coincidence that is manifest in Leibniz's logical writings does not betray any ambivalence or incoherence on his part. Instead, Leibniz's logic of conceptual containment and his logic of coincidence can be viewed as two alternative axiomatizations of a single, unified logical theory. To show this, we will gradually build up a version of Leibniz's containment calculus that is strong enough to derive all the principles of the mature coincidence calculus developed in the *Generales inquisitiones*. We begin by considering the most characteristic principle of Leibniz's coincidence calculus, namely, the rule of substitution.

1.2   The Rule of Substitution

The first principle listed by Leibniz in his final axiomatization of the coincidence calculus presented in the *Generales inquisitiones* is the rule of substitution of coincidents. Leibniz formulates this rule as follows:[18]

> 1st. Coincidents can be substituted for one another. (*Generales inquisitiones* §198.1)

This principle licenses the substitution *salva veritate* of one coincident term for another in any proposition of Leibniz's calculus. In the *Generales inquisitiones*, Leibniz writes '$A = B$' to indicate that the terms A and B coincide. Using this notation, the rule of substitution can be formulated as follows, where the turnstile '$\vdash$' indicates derivability in Leibniz's calculus:

> RULE OF SUBSTITUTION: For any terms A and B and any proposition φ, if φ^* is the result of substituting A for an occurrence of B, or vice versa, in φ, then:

$$A = B, \varphi \vdash \varphi^*$$

While this rule of substitution plays a prominent role in Leibniz's coincidence calculus, it does not play any role in the various containment calculi developed in his earlier logical writings. One reason for this is that the language of Leibniz's containment calculi does not include any propositions of the form $A = B$. If, on the other hand, this language were to be extended by adding coincidence as a new primitive relation between terms, the rule of substitution might then be derivable by means of the introduction and elimination rules for coincidence. In order to provide such a derivation of the rule of substitution, it suffices to establish that the underlying containment calculus licenses free substitution under the relation of

---

[18]The rule of substitution also appears as the first principle listed by Leibniz in many subsequent versions of the coincidence calculus (see, e.g., A VI.4 810, 816, 831, 846). For further statements of the rule of substitution, see *GI* §9, §19, A VI.4 626, 746.

mutual containment. If we write '$A \supset B$' to indicate that A contains B, the requisite rule of substitution under mutual containment can be formulated as follows:[19]

> RULE OF SUBSTITUTION FOR CONTAINMENT: For any terms A, B, C, and D, if $C^* \supset D^*$ is the result of substituting A for an occurrence of B in the proposition $C \supset D$, then:

$$A \supset B, B \supset A, C \supset D \vdash C^* \supset D^*$$

Unlike the rule of substitution of coincidents, this rule of substitution for containment is expressible in the language of Leibniz's containment calculi. Nevertheless, Leibniz does not include the latter rule among the principles of any of these calculi. This is perhaps not surprising since rules for substitutional reasoning appear more natural when formulated for equivalence relations such as coincidence than they do when formulated for non-symmetric relations such as containment. Thus, were Leibniz to posit the above rule of substitution for containment as a principle of his containment calculus, this would seem rather ad hoc and unmotivated.

At the same time, however, Leibniz clearly took the rule of substitution for containment to be valid. In fact, he often observes the validity of this rule in his essays on the logic of containment in the late 1670s. Crucially, however, in these early essays Leibniz does not simply posit substitution for containment as a primitive rule of inference, but instead presents it as a theorem to be derived from more fundamental principles of containment.[20] For example, in his essay *Elementa ad calculum condendum*, written around 1679, Leibniz derives the rule of substitution for containment as follows:

> There is only one fundamental inference: A is B and B is C, therefore A is C. Hence, if A is B and B is A, the one can be substituted in place of the other *salva veritate*. Take the proposition B is C; it will then be possible to substitute A is C, since A is B. Next, take the proposition D is B; it will then be possible to substitute D is A, since B is A. Evidently, since A is B, A can be substituted in place of a subject B. But also since B is A, it will be possible to substitute A in place of a predicate B. (*Elementa ad calculum condendum*, A VI.4 154)

In the first sentence of this passage, Leibniz states the following law of containment:

$$A \supset B, B \supset C \vdash A \supset C$$

This is the most basic principle of Leibniz's logic of containment, and it appears as a primitive rule of inference in all the versions of his containment calculi.[21] Since Leibniz identifies propositions of the form $A \supset B$ with the universal affirmative propositions appearing in Aristotelian syllogisms, this rule amounts to

---

[19]It is easy to verify that, given the introduction and elimination rules for coincidence, this rule of substitution for containment implies the general rule of substitution for the extended language of Leibniz's containment calculus, i.e.: $A = B, \varphi \vdash \varphi^*$, where φ is any proposition of the form $C \supset D$ or $C = D$.

[20]See, e.g., A VI.4 154, 275, 284, 294.

[21]See, e.g., A VI.4 144, 154, 275, 281, 293, 497, 551.

the traditional syllogistic mood Barbara.[22] In the above passage, Leibniz appeals to the rule of Barbara to show that, if $A \supset B$ and $B \supset A$, then either of the terms A and B 'can be substituted in place of the other *salva veritate*'. His proof of this claim proceeds by considering two cases: first, the case in which B occurs as the subject term of a containment proposition, as in $B \supset C$; and, second, the case in which B occurs as the predicate term, as in $D \supset B$. In each of these cases, Leibniz justifies the substitution of A for B by an application of Barbara, using the premises $A \supset B$ and $B \supset A$, respectively. Obviously, the same argument shows that B can be substituted for A when the latter appears as either the subject or the predicate term of a containment proposition.[23]

It is clear from this argument that Leibniz aims to justify the rule of substitution for containment by appeal to what he takes to be the more fundamental principle of Barbara.[24] In fact, the rule of Barbara can itself be viewed as a partial rule of substitution whereby the subject term of a proposition $A \supset B$ can be replaced by any term that contains it, and the predicate term by any term it contains. For this reason, Leibniz refers to Barbara as a rule of 'one-sided substitution' (*substitutio unilateralis*).[25] Thus, in Leibniz's containment calculi, Barbara takes the place of substitution as the most prominent and characteristic rule of inference. This is diametrically opposed to the approach adopted by Leibniz in his mature logical writings, in which he derives Barbara as a theorem in his coincidence calculus by means of a straightforward substitutional proof.[26]

In the passage just quoted, Leibniz establishes the rule of substitution for the specific case in which the substituted term occurs as either the subject or the predicate of the containment proposition in which the substitution takes place. If these were the only two syntactic contexts in which a term could occur in the language of Leibniz's calculus, the proof given by Leibniz in the passage based on the rule of Barbara would succeed in establishing the rule of substitution for containment in full generality. In richer syntactic settings, however, in which these are not the only two contexts in which a term can occur in a proposition, Barbara alone does not suffice to establish the rule of substitution. This is because the latter rule, unlike Barbara, is sensitive to the variety of syntactic contexts in which a term may occur.

---

[22]See, e.g., A VI.4 280–1, *GI* §47, §124, §129, §§190–1.

[23]See A VI.4 282.

[24]For a similar proof of the rule of substitution based on Barbara in a containment calculus, see Schröder 1890: 186.

[25]A VI.4 809; cf. A VI.4 143–5, 154, 275, 672.

[26]In his essay *Principia calculi rationalis*, for example, Leibniz provides the following substitutional proof of Barbara based on his definition of $A \supset B$ as $A = AB$ (C 229–30). Suppose $A = AB$ and $B = BC$ (i.e., $A \supset B$ and $B \supset C$). Substituting BC for B in the former proposition yields $A = ABC$. Substituting AB for A in this last proposition yields $A = AC$ (i.e., $A \supset C$). For similar substitutional proofs of Barbara, see *GI* §19 and A VI.4 813. A substitutional proof of Barbara along these lines is also given by Jevons (1869: 29–30).

To illustrate this last point, consider the syntax of the containment calculus developed by Leibniz in the *Elementa ad calculum condendum*. While the only propositions in the language of this calculus are those of the form $A \supset B$, Leibniz allows the terms A and B to be syntactically complex. In particular, he admits complex terms which are constructed out of simpler ones by the logical operation of composition. In the language of Leibniz's calculus, composition is expressed by concatenating the expressions signifying the terms to be composed. For example, the complex term *rational animal* is the result of composing the simpler terms *rational* and *animal*. More generally, if A and B are terms, then AB is the term which results from composing them. One of the basic laws posited by Leibniz governing the logic of such composite terms is $AB \supset A$.[27] In a proposition of this form, the term B does not occur as either the subject or the predicate of the proposition, but rather as a component of the composite subject term AB. Hence, Leibniz's proof of substitution based on the rule of Barbara does not license substitutions for the term B in the proposition $AB \supset A$. For example, given the premises $B \supset C$ and $C \supset B$, Barbara alone does not allow us to infer $AC \supset A$ from $AB \supset A$. In order to justify such substitutions for components of composite terms, one must appeal to additional principles pertaining to composite terms in the containment calculus. In sum, Leibniz's proof of substitution based on the rule of Barbara is only complete for a simple language of containment in which terms have no syntactic complexity in the sense that no term can occur as a constituent of another. For richer languages in which terms do exhibit such syntactic complexity, however, this proof fails to establish the rule of substitution in full generality.

These considerations illustrate the fact that, as the syntactic complexity of terms increases, the rule of substitution undergoes a corresponding increase in deductive power, since it thereby comes to license substitutions into a wider variety of syntactic contexts. Hence, in order to derive the rule of substitution for containment in increasingly complex syntactic settings, the underlying logic of containment needs to be correspondingly stronger. Exactly which additional principles are needed will depend on the specific kind of syntactic complexity exhibited by the terms of the language. In the *Elementa ad calculum condendum*, the only syntactic operation for forming complex terms is that of composition; in his mature logical writings, however, Leibniz posits additional syntactic operations for forming complex terms. Specifically, the language of the calculus developed by Leibniz in the *Generales inquisitiones* includes the following three operations for forming complex terms:

COMPOSITION: If A and B are terms, then AB is a composite term. For example, *rational animal* is the composite of *rational* and *animal*.

PRIVATION: If A is a term, then *non-A* is a privative term. For example, *non-animal* is the privative of *animal*.

---

[27]In the *Elementa ad calculum condendum*, this law is stated at A VI.4 154. For other statements of this law, see, e.g., A VI.4 148, 149, 150, 151, 274, 280, 292, 813.

PROPOSITIONAL TERMS: If φ is a proposition, then $\ulcorner \varphi \urcorner$ is a propositional term. For example, $\ulcorner Man\ is\ animal \urcorner$ is the propositional term generated from the proposition *Man is animal*.

In what follows, we discuss each of these three operations in turn.

1.3   Composition in the Containment Calculus

In a series of essays written in the late 1670s, Leibniz explores the logic of containment for a language which includes composite terms of the form AB. The syntax of the language employed by Leibniz in these essays can be characterized as follows:

Definition 1 Given a nonempty set of primitive expressions referred to as simple terms, the terms and propositions of the language $\mathcal{L}_c$ are defined as follows:

(1) Every simple term is a term.
(2) If A and B are terms, then AB is a term.
(3) If A and B are terms, then $A \supset B$ is a proposition.

Given this definition of the language $\mathcal{L}_c$, what logical principles regarding composite terms are needed to derive the rule of substitution for containment? While Leibniz does not explicitly address this question, he does describe a calculus for the language $\mathcal{L}_c$ which is strong enough to derive the rule of substitution. This containment calculus is presented in an essay titled *De calculo analytico generali*, written around 1679. In the concluding sections of the essay, Leibniz enumerates the principles of the calculus as follows:

Axioms:

(1) *Every A is A*
(2) *Every AB is A*
(3) If *A is B*, then *A is B*.
    If *A is B* and *B is C*, then *A is B*. ...
(4) If *A is B* and *Every B is C*, then *A is C*. ...
(7) If *A is B* and the same *A is C*, then the same *A is BC*.[28]

(*De calculo analytico generali*, A VI.4 149–50)

While the principles stated by Leibniz in (1), (2), (4), and (7) pertain specifically to the relation of containment, the two principles stated in (3) describe more general, structural features of the relation of derivability in the calculus. The first of these two principles asserts that the relation of derivability is reflexive, i.e.: $\varphi \vdash \varphi$. The second

---

[28]We have omitted items (5) and (6) from this list since they do not contain principles of the calculus, but theorems which Leibniz derives from the principles stated in (1)–(4).

asserts a rule of weakening to the effect that $\varphi, \psi \vdash \varphi$.[29] Throughout his logical writings, Leibniz also tacitly assumes the structural rule of cut, which licenses the construction of complex derivations through the consecutive application of rules of inference.[30] Given the structural rules of cut and weakening, it follows that the derivability relation is monotonic.[31] Hence, Leibniz's containment calculus obeys the standard structural rules of reflexivity, monotonicity, and cut that are typically assumed in the definition of a calculus in modern logic:

Definition 2 A calculus in a language $\mathcal{L}$ is a relation $\vdash$ between sets of propositions of $\mathcal{L}$ and propositions of $\mathcal{L}$, such that:

Reflexivity: $\{\varphi\} \vdash \varphi$
Monotonicity: If $\Gamma \vdash \varphi$, then $\Gamma \cup \Delta \vdash \varphi$.
Cut: If $\Gamma \cup \{\varphi\} \vdash \psi$ and $\Delta \vdash \varphi$, then $\Gamma \cup \Delta \vdash \psi$.

Here, $\Gamma \vdash \varphi$ indicates that the set of propositions Γ stands in the relation $\vdash$ to the proposition φ. In what follows, we use $\vdash \varphi$ as shorthand for $\emptyset \vdash \varphi$, and $\psi_1, \ldots, \psi_n \vdash \varphi$ as shorthand for $\{\psi_1, \ldots, \psi_n\} \vdash \varphi$.

In addition to the structural principles stated by Leibniz in (3), his list of axioms includes a number of principles pertaining specifically to containment and composition. Taken together, these principles define the following containment calculus, $\vdash_c$:

Definition 3 The calculus $\vdash_c$ is the smallest calculus in the language $\mathcal{L}_c$ such that:

(C1) $\vdash_c A \supset A$
(C2) $\vdash_c AB \supset A$
(C3) $\vdash_c AB \supset B$

---

[29] In item (3), Leibniz asserts only a restricted version of weakening in which the propositions φ and ψ have a term in common. This restriction, however, is likely inadvertent, since Leibniz asserts the rule of weakening in full generality immediately before the list of principles quoted above: '*A is B* and *C is D, therefore A is B*' (A VI.4 149).

[30] This is clear, for example, from the way in which Leibniz derives the theorem $A = B, B = C, C = D \vdash A = D$ in his coincidence calculus by two consecutive applications of $A = B, B = C \vdash A = C$ (A VI.4 831). In this derivation, Leibniz employs the following instance of cut: if $A = C, C = D \vdash A = D$ and $A = B, B = C \vdash A = C$, then $A = B, B = C, C = D \vdash A = D$. See also the application of cut at A VI.4 837.

[31] Monotonicity asserts that, if $\Gamma \vdash \varphi$, then $\Gamma \cup \Delta \vdash \varphi$. The structural rules of cut and weakening imply this principle for the case in which Γ and Δ are finite sets of propositions. To see this, we first establish monotonicity for the case in which Γ is empty and Δ is the singleton set $\{\psi\}$, i.e.: if $\vdash \varphi$, then $\psi \vdash \varphi$. By weakening, we have $\varphi, \psi \vdash \varphi$. Hence, given $\vdash \varphi$, it follows by cut that $\psi \vdash \varphi$. Next, we establish monotonicity for the case in which Γ is non-empty and Δ is the singleton set $\{\psi\}$, i.e.: if $\Gamma \vdash \varphi$ and $\Gamma \neq \emptyset$, then $\Gamma \cup \{\psi\} \vdash \varphi$. Let χ be any proposition in Γ. By weakening, we have $\chi, \psi \vdash \chi$. Hence, given $\Gamma \vdash \varphi$, it follows by cut that $\Gamma \cup \{\psi\} \vdash \varphi$. The finite version of monotonicity, in which Γ and Δ are any finite sets of propositions, follows immediately from these two cases. Thus, Leibniz's containment calculus is finitely monotonic. For the sake of simplicity, we will also assume that the calculus is monotonic under the addition of infinitely many premises, although this assumption plays no essential role in what follows and can be freely omitted.

(C4) $A \supset B, B \supset C \vdash_c A \supset C$
(C5) $A \supset B, A \supset C \vdash_c A \supset BC$

Of these principles, C1, C2, C4, and C5 are explicitly stated by Leibniz in items (1), (2), (4), and (7) of his list of axioms.[32] While C3 does not appear in this list, Leibniz does take this principle for granted in his calculus.[33] For example, he appeals to C3 in his proof of the main theorem established in the *De calculo analytico generali*, which appears in item (8) of Leibniz's list:

> (8) If *A* is *C* and *B* is *D*, then *AB* is *CD*. (*De calculo analytico generali*, A VI.4 150)

Leibniz's proof of this theorem proceeds as follows:

Theorem 4 $A \supset C, B \supset D \vdash_c AB \supset CD$

Proof By C2, $\vdash_c AB \supset A$. Hence, by C4, $A \supset C \vdash_c AB \supset C$. Now, by C3, $\vdash_c AB \supset B$. Hence, by C4, $B \supset D \vdash_c AB \supset D$. Thus, by C5, we have $A \supset C, B \supset D \vdash_c AB \supset CD$.   □

This theorem can be used as a lemma to prove the rule of substitution for containment in the calculus $\vdash_c$. In particular, the theorem allows us to establish the following preliminary to the rule of substitution:

Theorem 5 *For any terms A, B, C of $\mathcal{L}_c$, if C^* is the result of substituting B for an occurrence of A in C, then both:*

$$A \supset B, B \supset A \vdash_c C \supset C^*$$

$$A \supset B, B \supset A \vdash_c C^* \supset C$$

Proof First, suppose that A is the term C. Then $C \supset C^*$ is the proposition $A \supset B$, and $C^* \supset C$ is the proposition $B \supset A$. Hence, the claim follows by the reflexivity and monotonicity of $\vdash_c$.

Now, suppose that A is a proper subterm of C. Then C is a term of the form DE, where either D or E contains the occurrence of A for which B is substituted to obtain C^*. First, suppose that D contains this occurrence of A. Then C^* is the term D^*E, where D^* is the result of substituting B for an occurrence of A in D. For induction, we assume that the claim holds for substitutions in D, so that both:

$$A \supset B, B \supset A \vdash_c D \supset D^*$$

$$A \supset B, B \supset A \vdash_c D^* \supset D$$

---

[32] In his formulation of these principles, Leibniz takes the phrases '*A is B*' and '*Every A is B*' to be equivalent (see n. 22). This can be seen from the proof given by Leibniz in item (6) of his list, where he takes *Every AB is A* to entail *AB is A*, and, conversely, takes *AB is C* to entail *Every AB is C* (A VI.4 150).

[33] Moreover, C3 is explicitly stated by Leibniz at A VI.4 274–5, 281, *GI* §38, §39, §46.

Since, by C1, $\vdash_c E \supset E$, it follows, by Theorem 4, that both:

$$A \supset B, B \supset A \vdash_c DE \supset D^*E$$

$$A \supset B, B \supset A \vdash_c D^*E \supset DE$$

The same argument applies to the case in which E contains the occurrence of A for which B is substituted. This completes the proof. □

Given this theorem, the rule of substitution for containment in the calculus $\vdash_c$ can be established as follows:[34]

Theorem 6 *For any terms A, B, C, D of $\mathcal{L}_c$, if $C^* \supset D^*$ is the result of substituting A for an occurrence of B in the proposition $C \supset D$, then:*

$$A \supset B, B \supset A, C \supset D \vdash_c C^* \supset D^*$$

Proof Since $C^* \supset D^*$ is the result of replacing a single occurrence of A by B in $C \supset D$, either C^* is identical with C or D^* is identical with D. In the first case, by Theorem 5, we have $A \supset B, B \supset A \vdash_c D \supset D^*$, and so, by C4, it follows that $A \supset B, B \supset A, C \supset D \vdash_c C \supset D^*$. In the second case, by Theorem 5, we have $A \supset B, B \supset A \vdash_c C^* \supset C$, and so, by C4, it follows that $A \supset B, B \supset A, C \supset D \vdash_c C^* \supset D$. □

Thus, the containment calculus $\vdash_c$ developed by Leibniz in the *De calculo analytico generali* is strong enough to derive the rule of substitution for containment in the language $\mathcal{L}_c$. Consequently, if this language is extended by adding a new primitive relation symbol '=' governed by the introduction and elimination rules for coincidence, the resulting calculus is strong enough to validate the general rule of substitution for coincidence. In other words, given $A = B$, the term B can be substituted for A, or vice versa, *salva veritate* in any proposition of the extended language.

The operation of composition axiomatized by the calculus $\vdash_c$ gives rise to an algebra of terms that allows for free substitution under the relation of mutual containment. It can be shown that the specific type of algebraic structure determined

---

[34] In fact, in order to establish Theorem 6, it is not necessary to appeal to Theorem 4 but only to the following weaker version of this theorem:

$$A \supset C, C \supset A, B \supset D, D \supset B \vdash_c AB \supset CD$$

Given the introduction and elimination rules for coincidence, this is equivalent to the following law asserting that coincidence is a congruence relation with respect to the operation of composition:

$$A = C, B = D \vdash_c AB = CD$$

Schröder (1890: 270) appeals to this latter law to establish a variant of Theorem 6 in his containment calculus.

by the principles of the calculus is that of a semilattice. More precisely, the calculus $\vdash_c$ is sound and complete with respect to the class of semilattices, when composition is interpreted as the meet operation and containment as the order relation in the algebra.[35] Thus, the containment calculus developed by Leibniz in the *De calculo analytico generali* constitutes an axiomatization of the theory of semilattices.

1.4 Privation in the Containment Calculus

In addition to composite terms, the calculus developed by Leibniz in the *Generales inquisitiones* also includes privative terms. In particular, Leibniz assumes that, for any term A, there is a corresponding privative term *non-A*. In what follows, we write '$\overline{A}$' to designate the privative term *non-A*.

Privative terms are largely absent from Leibniz's writings on the logic of containment in the late 1670s.[36] For example, in the *De calculo analytico generali* Leibniz deliberately excludes privative terms from consideration on the grounds that doing so will allow him to 'set aside many perplexities' (*multas perplexitates abscindemus*).[37] One complication that is raised by the introduction of privative terms into the language of Leibniz's containment calculus is that they create new syntactic contexts into which terms can be substituted. Thus, in order to validate the rule of substitution, the underlying containment calculus must now include additional principles that license the substitution of one mutually containing term for another in these new syntactic contexts. For example, if A and B contain one another, the principles of the calculus must license the substitution of B for A in propositions such as $\overline{A} \supset C$. As it turns out, the only additional principle that is needed to license such substitutions is the following law of contraposition:

$$A \supset B \vdash \overline{B} \supset \overline{A}$$

To show that this law suffices to derive the rule of substitution in a language that includes not only composite but also privative terms, we first specify the syntax of such an enriched language, $\mathcal{L}_{cp}$:

Definition 7 Given a nonempty set of primitive expressions referred to as simple terms, the terms and propositions of the language $\mathcal{L}_{cp}$ are defined as follows:

---

[35]For a proof of this claim, see Theorem 20 in the appendix below. Swoyer (1994: 14–20 and 29–30) establishes a similar completeness result for the coincidence calculus developed by Leibniz in the *Specimen calculi coincidentium et inexistentium*, written around 1686 (A VI.4 830–45).

[36]Leibniz does on a few separate occasions discuss privative terms in his earlier logical writings (e.g., A VI.4 218, 224, 253, 292–7, 622). The first systematic treatment of these terms, however, appears in the *Generales inquisitiones*.

[37]A VI.4 146.

(1) Every simple term is a term.
(2) If A and B are terms, then AB is a term.
(3) If A is a term, then $\overline{A}$ is a term.
(4) If A and B are terms, then $A \supset B$ is a proposition.

In order to establish the rule of substitution in the language $\mathcal{L}_{cp}$, the key step is to extend Theorem 5 to cover the case of privative terms. To this end, let $\vdash_{cp}$ be a calculus in $\mathcal{L}_{cp}$ obeying all the principles of $\vdash_c$ as well as the above law of contraposition. The proof of the relevant theorem then proceeds as follows:

Theorem 8 *For any terms A, B, C of $\mathcal{L}_{cp}$, if C^* is the result of substituting B for an occurrence of A in C, then both:*

$$A \supset B, B \supset A \vdash_{cp} C \supset C^*$$

$$A \supset B, B \supset A \vdash_{cp} C^* \supset C$$

Proof First, suppose that A is the term C. Then $C \supset C^*$ is the proposition $A \supset B$, and $C^* \supset C$ is the proposition $B \supset A$. Hence, the claim follows from the reflexivity and monotonicity of $\vdash_{cp}$.

Now, suppose that A is a proper subterm of C. Then either (i) C is a term of the form DE, where either D or E contains the occurrence of A for which B is substituted to obtain C^*; or (ii) C is a term of the form $\overline{D}$, where D contains the occurrence of A for which B is substituted to obtain C^*. Case (i) is treated in exactly the same way as in the proof of Theorem 5.

In case (ii), C^* is the term $\overline{D^*}$, where D^* is the result of substituting B for an occurrence of A in D. For induction, we assume that the claim holds for substitutions in D, so that both:

$$A \supset B, B \supset A \vdash_{cp} D \supset D^*$$

$$A \supset B, B \supset A \vdash_{cp} D^* \supset D$$

By contraposition, it follows that both:

$$A \supset B, B \supset A \vdash_{cp} \overline{D^*} \supset \overline{D}$$

$$A \supset B, B \supset A \vdash_{cp} \overline{D} \supset \overline{D^*}$$

This completes the proof. □

Given this theorem, the rule of substitution for the calculus $\vdash_{cp}$ follows straightforwardly:

Theorem 9 *For any terms A, B, C, D of $\mathcal{L}_{cp}$, if $C^* \supset D^*$ is the result of substituting A for an occurrence of B in the proposition $C \supset D$, then:*

$$A \supset B, B \supset A, C \supset D \vdash_{cp} C^* \supset D^*$$

Proof The proof is the same as that given for Theorem 6.   □

Thus, provided that the containment calculus $\vdash_{cp}$ validates the law of contraposition, it licenses free substitution under mutual containment into any syntactic context of the language $\mathcal{L}_{cp}$.[38] Consequently, just as before, if the language $\mathcal{L}_{cp}$ is extended by adding a new primitive relation symbol '=' governed by the introduction and elimination rules for coincidence, the resulting calculus is strong enough to validate the general rule of substitution for coincidence.

It is clear that Leibniz endorses the law of contraposition. He affirms this law on numerous occasions throughout his logical writings.[39] For example, in the *Generales inquisitiones* he writes:

> If *A is B*, then *non-B is non-A*. (*Generales inquisitiones* §93)

In addition to contraposition, Leibniz posits a number of other principles pertaining to privative terms in his mature logical writings. One of these principles is a law of double privation to the effect that a term coincides with the privative of its privative.[40] In the language $\mathcal{L}_{cp}$, this law is captured by the following two principles:

$$\vdash_{cp} \overline{\overline{A}} \supset A \qquad \vdash_{cp} A \supset \overline{\overline{A}}$$

A further principle pertaining to privative terms that plays an important role in the *Generales inquisitiones* is the following:

> If I say *AB is not*, it is the same as if I were to say *A contains non-B*. (*Generales inquisitiones* §200)

In this passage, Leibniz asserts that the proposition $A \supset \overline{B}$ is equivalent to a proposition that he expresses by the phrase '*AB is not*'. He takes this latter proposition to assert that the composite term *AB* is a 'non-being' (*non-ens*) or,

---

[38] In fact, in order to establish Theorem 9, it is not necessary to appeal to the law of contraposition but only to the following weaker version of this law:

$$A \supset B, B \supset A \vdash_{cp} \overline{B} \supset \overline{A}$$

Given the introduction and elimination rules for coincidence, this is equivalent to the following law asserting that coincidence is a congruence relation with respect to the operation of privation:

$$A = B \vdash_{cp} \overline{A} = \overline{B}$$

Schröder (1890: 306) appeals to this latter law to establish a variant of Theorem 9 in his containment calculus.

[39] See, e.g., *GI* §77, §95, §189.5, §200, A VI.4 224, 813, C 422. In some cases, Leibniz does not posit the law of contraposition as a principle of his calculus, but instead undertakes to derive it from more basic principles (see Lenzen 1986: 13–14 and 27–32, 1988: 63–4).

[40] See, e.g., *GI* §2, §96, §171.4, §189.2, §198.3, A VI.4 218, 624, 740, 807, 811, 814, 877, 931, 935, 939, C 230, 235, 421.

equivalently, that the term AB is 'false' (*falsum*).[41] In what follows, we write '$\mathbf{F}(A)$' for the proposition expressing that the term A is false. In this notation, the principle stated by Leibniz in the passage just quoted asserts that the propositions $\mathbf{F}(AB)$ and $A \supset \overline{B}$ are equivalent, i.e.:[42]

$$\mathbf{F}(AB) \dashv\vdash A \supset \overline{B}$$

Now, this equivalence is not directly expressible in the language $\mathcal{L}_{cp}$, since this language does not include any primitive means for expressing propositions of the form $\mathbf{F}(A)$. Nevertheless, Leibniz gives some indication as to how these propositions can be expressed in $\mathcal{L}_{cp}$ when he asserts that a false term is simply one which 'contains a contradiction' (*Generales inquisitiones* §57). Thus he writes:

> That term is false which contains opposite terms, A *non-A*. (*Generales inquisitiones* §194; similarly, §198.4)

> A proposition is that which states what term is or is not contained in another. Hence, a proposition can also affirm that a term is false, if it says that Y *non-Y* is contained in it. (*Generales inquisitiones* §195)

According to these passages, a term A is false just in case it contains some contradictory term of the form $B\overline{B}$. Note that this characterization of falsehood involves an existential quantification over contradictory terms. Since the language $\mathcal{L}_{cp}$ does not include any means of expressing existential quantification, Leibniz's characterization of falsehood is not directly expressible in this language.[43] Nonetheless, it can be shown to entail an explicit definition of falsehood in the language $\mathcal{L}_{cp}$, given the principles of the calculus $\vdash_{cp}$ stated so far. This is because these principles entail that a term A contains some contradictory term of the form $B\overline{B}$ just in case it contains its own privative, $\overline{A}$. This is an immediate consequence of the following two theorems:

[41] See *GI* §55, §193–4, §197, A VI.4 774 n. 47, 807, 813. In the *Generales inquisitiones*, Leibniz uses a number of phrases interchangeably to expresses the non-being, or falsehood, of a given term A, e.g.: 'A *is not*' (*GI* §§149–50, §§199–200), 'A *is not a thing*' (*GI* §§149–55, §171.8), 'A *is a non-being*' (*GI* §32b, §55, A VI.4 810 n. 5, 813, 875, 930, 935), 'A *is false*' (*GI* §55, §§58–9, §189.3, §§193–5, §197, §198.4, A VI.4 939), 'A *is impossible*' (*GI* §32b, §§33–4, §55, §128, A VI.4 749 n. 11, 807, 875, 930, 935, 939).

[42] For similar statements of this equivalence, see A VI.4 862–3 and C 237.

[43] In the *Generales inquisitiones*, Leibniz explores the possibility of expressing existential quantification in the language of his calculus by means of what he calls 'indefinite letters' (§§16–31; see Lenzen 1984b: 7–13, 2004: 47–50, Hailperin 2004: 329). However, the introduction of indefinite letters into the calculus gives rise to complications analogous to those which arise in connection with the elimination of existential quantifiers in modern quantificational logic (see, e.g., *GI* §§21–31). Leibniz was aware of some of these complications and was never entirely satisfied with the use of indefinite letters, indicating that it would be preferable to omit them from the language of his calculus (see *GI* §162 in conjunction with §128 and A VI.4 766 n. 35; cf. Schupp 1993: 153, 168, 182–3). In the absence of indefinite letters, the language of Leibniz's calculus lacks the resources to express existential quantification.

Theorem 10 $A \supset \overline{A} \vdash_{\mathrm{cp}} A \supset A\overline{A}$

Proof This follows by C1 and C5.   □

Theorem 11 $A \supset B\overline{B} \vdash_{\mathrm{cp}} A \supset \overline{A}$

Proof By C2 and C4, we have $A \supset B\overline{B} \vdash_{\mathrm{cp}} A \supset B$. So, by contraposition:

$$A \supset B\overline{B} \vdash_{\mathrm{cp}} \overline{B} \supset \overline{A}$$

But, by C3 and C4:

$$A \supset B\overline{B} \vdash_{\mathrm{cp}} A \supset \overline{B}$$

Hence, by C4, $A \supset B\overline{B} \vdash_{\mathrm{cp}} A \supset \overline{A}$.   □

Since, according to Leibniz's characterization of falsehood, a false term is one which contains some contradictory term of the form $B\overline{B}$, these two theorems yield the following definition of falsehood:

$$\mathbf{F}(A) \dashv\vdash A \supset \overline{A}$$

With this definition in hand, the principle stated by Leibniz in §200 of the *Generales inquisitiones* can now be formulated in the language $\mathcal{L}_{\mathrm{cp}}$. This principle, recall, asserts that $\mathbf{F}(AB)$ is equivalent to $A \supset \overline{B}$. Since $\mathbf{F}(AB)$ is definable as $AB \supset \overline{AB}$, this principle can be formulated in $\mathcal{L}_{\mathrm{cp}}$ as follows:

$$AB \supset \overline{AB} \dashv\vdash_{\mathrm{cp}} A \supset \overline{B}$$

All told, then, the containment calculus $\vdash_{\mathrm{cp}}$ can be obtained from $\vdash_{\mathrm{c}}$ by adding to it the laws of contraposition, double privation, and Leibniz's principle asserting the equivalence of $\mathbf{F}(AB)$ and $A \supset \overline{B}$. As it turns out, the right-to-left direction of this equivalence is already derivable from contraposition and the principles of $\vdash_{\mathrm{c}}$.[44] Moreover, given this equivalence, one direction of the law of double privation, $A \supset \overline{\overline{A}}$, is already provable in the calculus.[45] Thus, without any loss of deductive power, the calculus $\vdash_{\mathrm{cp}}$ can be defined as follows:

Definition 12 The calculus $\vdash_{\mathrm{cp}}$ is the smallest calculus in the language $\mathcal{L}_{\mathrm{cp}}$ such that:

(C1) $\vdash_{\mathrm{cp}} A \supset A$
(C2) $\vdash_{\mathrm{cp}} AB \supset A$
(C3) $\vdash_{\mathrm{cp}} AB \supset B$

---

[44] See the proof of Theorem 21.i in the appendix.
[45] See Theorem 21.iii in the appendix.

(C4) $A \supset B, B \supset C \vdash_{cp} A \supset C$

(C5) $A \supset B, A \supset C \vdash_{cp} A \supset BC$

(CP1) $A \supset B \vdash_{cp} \overline{B} \supset \overline{A}$

(CP2) $\vdash_{cp} \overline{\overline{A}} \supset A$

(CP3) $AB \supset \overline{AB} \vdash_{cp} A \supset \overline{B}$

It should be acknowledged that Leibniz does not formulate this definition of the containment calculus $\vdash_{cp}$ in any of his logical writings. Nonetheless, this calculus captures all the laws of conceptual containment endorsed by Leibniz in the *Generales inquisitiones* that pertain to the operations of composition and privation.

These two operations, as axiomatized by the calculus $\vdash_{cp}$, give rise to an algebra of terms that allows for free substitution under the relation of mutual containment. It can be shown that the specific type of algebraic structure determined by the principles of the calculus is that of a Boolean algebra. More precisely, the calculus $\vdash_{cp}$ is sound and complete with respect to the class of Boolean algebras, when composition is interpreted as the meet operation, privation as the complement operation, and containment as the order relation in the algebra.[46] Thus, the containment calculus $\vdash_{cp}$ constitutes an axiomatization of the theory of Boolean algebras.

1.5 Propositional Terms in the Containment Calculus

In addition to composition and privation, the language of the calculus developed by Leibniz in the *Generales inquisitiones* includes a third operation for forming complex terms. By means of this operation, a new term can be generated from any given proposition, in accordance with Leibniz's view that 'every proposition can be conceived of as a term'.[47] Leibniz characterizes this operation as follows:

> If the proposition *A is B* is considered as a term, as we have explained that it can be, there arises an abstract term, namely *A's being B*. And if from the proposition *A is B* the proposition *C is D* follows, then from this there comes about a new proposition of the following kind: *A's being B is* (or *contains*) *C's being D*; or, in other words, *the B-ness of A contains the D-ness of C*, or *the B-ness of A is the D-ness of C*. (*Generales inquisitiones* §138)

---

[46]For a proof of this claim, see Theorem 25 in the appendix. The Boolean completeness of Leibniz's containment calculus was first established by Lenzen (1984a: 200–2), albeit with respect to a somewhat different axiomatization of the calculus than that given in Definition 12.

[47]See *GI* §75, §109, §197. When a proposition is conceived of as a term, Leibniz describes the proposition as giving rise to a 'new term' (*terminus novus*, §197 and §198.7; *contra* Parkinson 1966: 86 n. 2).

According to this passage, a proposition such as $A \supset B$ gives rise to a new abstract term, which Leibniz signifies by the phrases '*A's being B*' or '*the B-ness of A*'.[48] In what follows, we refer to such terms as propositional terms.[49] We use corner quotes to denote the operation mapping a proposition to the corresponding propositional term. Thus, $\ulcorner A \supset B \urcorner$ is the propositional term generated from the proposition $A \supset B$. In this notation, the complex proposition mentioned by Leibniz in the passage just quoted is written:

$$\ulcorner A \supset B \urcorner \supset \ulcorner C \supset D \urcorner$$

If the language $\mathcal{L}_{cp}$ is extended so as to include propositional terms, the result is the following language, $\mathcal{L}_{\supset}$:

Definition 13 Given a nonempty set of primitive expressions referred to as simple terms, the terms and propositions of the language $\mathcal{L}_{\supset}$ are defined as follows:

(1) Every simple term is a term.
(2) If A and B are terms, then AB is a term.
(3) If A is a term, then $\overline{A}$ is a term.
(4) If A and B are terms, then $A \supset B$ is a proposition.
(5) If $A \supset B$ is a proposition, then $\ulcorner A \supset B \urcorner$ is a term.

In the *Generales inquisitiones*, Leibniz posits two main principles pertaining to propositional terms. The first characterizes the relationship between propositional terms and the propositions from which they are generated. Specifically, this principle describes what it means for one propositional term to be contained in another, and is formulated by Leibniz as follows:

> For a proposition to follow from a proposition is nothing other than for the consequent to be contained in the antecedent as a term in a term. By this method we reduce consequences to propositions, and propositions to terms. (*Generales inquisitiones* §198.8)

According to this passage, one proposition follows from another just in case the propositional term corresponding to the former is contained in the propositional term corresponding to the latter.[50] When Leibniz says that one proposition follows from another, he means that the latter is derivable from the former in his calculus.[51]

---

[48] See *GI* §§138–42 and A VI.4 740.

[49] The label 'propositional term' is borrowed from Sommers (1982: 156, 1993: 172–3), Barnes (1983: 315), and Swoyer (1995: 110–11). Leibniz does not use this label, but instead uses 'complex term' to refer to propositional terms and Boolean compounds thereof (see *GI* §61, §65, §75; cf. A VI.4 528–9).

[50] Similarly, Leibniz writes that 'whatever is said of a term which contains a term can also be said of a proposition from which another proposition follows' (§189.6). See also §138, VI4A 809, 811 n. 6, 863; cf. Swoyer 1995: 110.

[51] To justify the claim that one proposition follows (*sequitur*) from another, Leibniz often provides a derivation of the former from the latter in his calculus (e.g., §49, §§52–4, §100). Moreover, there is no indication that Leibniz entertained the possibility that a proposition might follow from another

Thus, the principle of propositional containment stated by Leibniz in the passage just quoted can be formulated as follows, where φ and ψ are any propositions:

$$\varphi \vdash \psi \quad \text{if and only if} \quad \vdash \ulcorner \varphi \urcorner \supset \ulcorner \psi \urcorner$$

In the coincidence calculus of the *Generales inquisitiones*, this biconditional entails the following stronger version of the principle, where Γ is any set of propositions:[52]

$$\Gamma \cup \{\varphi\} \vdash \psi \quad \text{if and only if} \quad \Gamma \vdash \ulcorner \varphi \urcorner \supset \ulcorner \psi \urcorner$$

In Leibniz's coincidence calculus, this principle plays a role analogous to that of a deduction theorem in modern proof systems: it allows facts concerning the inferential relations between propositions to be expressed by propositions in the language of the calculus itself. Thus, by means of this principle we can, as Leibniz puts it, 'reduce consequences to propositions'.

In order to reproduce the full deductive power of the coincidence calculus developed in the *Generales inquisitiones*, Leibniz's containment calculus must validate the stronger version of the principle of propositional containment. In the coincidence calculus, as we just noted, this stronger version of the principle can be derived from the weaker version. This derivation, however, cannot be reproduced in the containment calculus, because it relies on substitutions of one coincident term for another within a propositional term. In the coincidence calculus, the legitimacy of such substitutions is taken for granted as part of the general rule of substitution licensing the free substitution of coincident terms in any syntactic context of the language, including those that occur within propositional terms.[53] By contrast, in the containment calculus the legitimacy of such substitutions cannot be taken for granted. Consequently, rather than deriving the stronger version of propositional containment from the weaker version, we will include the stronger version among the principles of the containment calculus.

---

despite not being derivable from it in his calculus. On the contrary, he asserts the completeness of the principles of his calculus when he states that 'whatever cannot be demonstrated from these principles does not follow by virtue of logical form' (§189.7).

[52] See Theorem 4.53 in Malink and Vasudevan 2016: 738.

[53] In an earlier paper, we claimed that the validity of substitutions within propositional terms can be justified by the weak principle of propositional containment asserting that $\varphi \vdash \psi$ just in case $\vdash \ulcorner \varphi \urcorner \supset \ulcorner \psi \urcorner$ (Malink and Vasudevan 2016: 697 n. 41 and 702 n. 66). This claim, however, is not correct. In order to establish the validity of substitutions within propositional terms, one in fact needs the following two-premise version of the principle of propositional containment: $\varphi, \psi \vdash \chi$ just in case $\varphi \vdash \ulcorner \psi \urcorner \supset \ulcorner \chi \urcorner$. Now, there is independent textual evidence that Leibniz endorses this two-premise version of the principle (see Malink and Vasudevan 2018: §3). Thus, Leibniz is committed to the validity of substitutions within propositional terms. This comports well with the fact that, in his various statements of the rule of substitution in the *Generales inquisitiones* and elsewhere, Leibniz gives no indication that this rule is to be restricted to syntactic contexts that do not occur within propositional terms (see, e.g., *GI* §198.1, A VI.4 746, 810, 816, 831, 846).

As it turns out, once the stronger version of the principle has been posited, this suffices to derive the rule of substitution for the language $\mathcal{L}_\supset$. The key step in the proof of this rule is to extend Theorems 5 and 8 to cover the case of propositional terms. To this end, let the unadorned turnstile '$\vdash$' designate a calculus in the language $\mathcal{L}_\supset$ that obeys all the principles of $\vdash_{cp}$ as well as the stronger version of propositional containment. The proof of the relevant theorem then proceeds as follows:

Theorem 14 *For any terms A, B, C of $\mathcal{L}_\supset$, if C^* is the result of substituting B for an occurrence of A in C, then both:*

$$A \supset B, B \supset A \vdash C \supset C^*$$

$$A \supset B, B \supset A \vdash C^* \supset C$$

Proof First, suppose that A is the term C. Then $C \supset C^*$ is the proposition $A \supset B$ and $C^* \supset C$ is the proposition $B \supset A$. Hence, the claim follows from the reflexivity and monotonicity of $\vdash$.

Now, suppose that A is a proper subterm of C. Then one of the following three cases holds: (i) C is a term of the form DE, where either D or E contains the occurrence of A for which B is substituted to obtain C^*; or (ii) C is a term of the form $\overline{D}$, where D contains the occurrence of A for which B is substituted to obtain C^*; or (iii) C is a term of the form $\ulcorner D \supset E \urcorner$, where either D or E contains the occurrence of A for which B is substituted to obtain C^*. Cases (i) and (ii) are treated in exactly the same way as in the proofs of Theorems 5 and 8, respectively.

In case (iii), first suppose that D contains the occurrence of A for which B is substituted. Then C^* is the term $\ulcorner D^* \supset E \urcorner$, where D^* is the result of substituting B for an occurrence of A in D. For induction, we assume that the claim holds for substitutions in D, so that both:

$$A \supset B, B \supset A \vdash D \supset D^*$$

$$A \supset B, B \supset A \vdash D^* \supset D$$

Given C4, the first of these claims implies:

$$A \supset B, B \supset A, D^* \supset E \vdash D \supset E$$

Hence, by the stronger version of propositional containment, we have:

$$A \supset B, B \supset A \vdash \ulcorner D^* \supset E \urcorner \supset \ulcorner D \supset E \urcorner$$

Likewise, given C4, the second claim implies:

$$A \supset B, B \supset A, D \supset E \vdash D^* \supset E$$

So, again, by the stronger version of propositional containment, we have:

$$A \supset B, B \supset A \vdash \ulcorner D \supset E \urcorner \supset \ulcorner D^* \supset E \urcorner$$

The same argument applies to the case in which E contains the occurrence of A for which B is substituted. This completes the proof.   □

Given this theorem, the rule of substitution for containment in the calculus $\vdash$ follows straightforwardly:

Theorem 15 *For any terms A, B, C, D of $\mathcal{L}_{\supset}$, if $C^* \supset D^*$ is the result of substituting A for an occurrence of B in the proposition $C \supset D$, then:*

$$A \supset B, B \supset A, C \supset D \vdash C^* \supset D^*$$

Proof The proof is the same as that given for Theorem 6.   □

Thus, provided that the containment calculus $\vdash$ validates the stronger version of propositional containment, it licenses free substitution under mutual containment into any syntactic context of the language $\mathcal{L}_{\supset}$. Consequently, just as before, if the language $\mathcal{L}_{\supset}$ is extended by adding a new primitive relation symbol '=' governed by the introduction and elimination rules for coincidence, the resulting calculus is strong enough to validate the general rule of substitution for coincidence.

In addition to the principle of propositional containment, there is one more principle pertaining to propositional terms that is posited by Leibniz in the *Generales inquisitiones*. Leibniz formulates this principle as follows:[54]

> If B is a proposition, *non-B* is the same as B *is false*, or, *B's being false*. (*Generales inquisitiones* §32a, A VI.4 753 n. 18)

In this passage, Leibniz describes the effect of applying the operation of privation to propositional terms.[55] Specifically, he asserts that, if B is a propositional term, then its privative, *non-B*, coincides with the propositional term generated from the proposition B *is false*. In our notation, this principle states that the privative term $\overline{\ulcorner \varphi \urcorner}$ coincides with the propositional term $\ulcorner \mathbf{F}(\ulcorner \varphi \urcorner) \urcorner$ generated from the proposition $\mathbf{F}(\ulcorner \varphi \urcorner)$. Since, as we have seen, $\mathbf{F}(A)$ is definable as $A \supset \overline{A}$, the principle states that $\overline{\ulcorner \varphi \urcorner}$ coincides with the propositional term $\ulcorner \ulcorner \varphi \urcorner \supset \overline{\ulcorner \varphi \urcorner} \urcorner$. In the language $\mathcal{L}_{\supset}$, this coincidence is captured by the following two laws:

$$\vdash \overline{\ulcorner \varphi \urcorner} \supset \ulcorner \ulcorner \varphi \urcorner \supset \overline{\ulcorner \varphi \urcorner} \urcorner \qquad\qquad \vdash \ulcorner \ulcorner \varphi \urcorner \supset \overline{\ulcorner \varphi \urcorner} \urcorner \supset \overline{\ulcorner \varphi \urcorner}$$

---

[54] For a similar statement of this principle, see A VI.4 809.

[55] The fact that Leibniz applies the term-operation of privation to B indicates that B is a term. Moreover, the fact that he characterizes B as a proposition strongly suggests that B is, more specifically, a propositional term.

In what follows, we refer to these laws collectively as the principle of propositional privation.

The principle of propositional privation has a number of important consequences. For one, this principle allows us to simulate in Leibniz's calculus the propositional operation of classical conjunction. To see this, let φ and ψ be any propositions of $\mathcal{L}_\supset$, and let $\varphi \,\&\, \psi$ be the following proposition:

$$\mathbf{F}\left(\ulcorner\ulcorner\varphi\urcorner \supset \overline{\ulcorner\psi\urcorner}\,\urcorner\right)$$

The principles of Leibniz's calculus then imply that the operation &, defined in this way, satisfies the classical laws of conjunction-introduction and -elimination. Specifically, if $\vdash$ is a calculus in $\mathcal{L}_\supset$ that obeys all the principles of $\vdash_{cp}$ as well as the principles of propositional containment and propositional privation, we have:[56]

$$\varphi, \psi \vdash \varphi \,\&\, \psi$$

$$\varphi \,\&\, \psi \vdash \varphi$$

$$\varphi \,\&\, \psi \vdash \psi$$

Hence, although the language of Leibniz's containment calculus lacks any primitive propositional operators for forming conjunctions or other complex propositions, the calculus is strong enough to reproduce classical conjunctive reasoning.

The availability of classical conjunction in Leibniz's containment calculus allows us to solve the problem, discussed at the beginning of the paper, of how to define coincidence in a language in which every proposition is of the form $A \supset B$. For, given the above definition of conjunction, Leibniz's characterization of coincidence as mutual containment is expressible in the language $\mathcal{L}_\supset$ by the proposition $(A \supset B) \,\&\, (B \supset A)$, i.e.:

$$\mathbf{F}\left(\ulcorner\ulcorner A \supset B\urcorner \supset \overline{\ulcorner B \supset A\urcorner}\,\urcorner\right)$$

It is an immediate consequence of the classical laws of conjunction stated above that the relation of coincidence defined in this way satisfies the introduction and elimination rules for coincidence. Thus, unlike the containment calculi $\vdash_c$ and $\vdash_{cp}$, in which coincidence had to be defined implicitly by means of its introduction and elimination rules, the full containment calculus $\vdash$ is strong enough to allow for an explicit definition of coincidence. This explicit definition is made possible by Leibniz's device of propositional terms and the two principles governing their operation, namely, the principles of propositional containment and propositional privation.

---

[56]See Theorem 29 in the appendix.

All told, then, the full containment calculus ⊢ can be obtained from ⊢$_{cp}$ by adding to it the principles of propositional containment and propositional privation. As it turns out, the principle of propositional containment can be somewhat weakened in the resulting calculus ⊢ without any loss of deductive power. Specifically, the strong version of this principle stated above can be replaced with the following weaker version, in which the set of premises Γ consists of a single proposition:[57]

$$\varphi, \psi \vdash \chi \quad \text{if and only if} \quad \varphi \vdash \ulcorner \psi \urcorner \supset \ulcorner \chi \urcorner$$

Thus, positing only this weaker version of propositional containment, the calculus ⊢ can be defined as follows:

Definition 16 The calculus ⊢ is the smallest calculus in the language $\mathcal{L}_\supset$ such that:

(C1) $\vdash A \supset A$
(C2) $\vdash AB \supset A$
(C3) $\vdash AB \supset B$
(C4) $A \supset B, B \supset C \vdash A \supset C$
(C5) $A \supset B, A \supset C \vdash A \supset BC$
(CP1) $A \supset B \vdash \overline{B} \supset \overline{A}$
(CP2) $\vdash \overline{\overline{A}} \supset A$
(CP3) $AB \supset \overline{AB} \vdash A \supset \overline{B}$
(PT1) $\vdash \overline{\ulcorner \varphi \urcorner} \supset \ulcorner \overline{\ulcorner \varphi \urcorner} \supset \overline{\ulcorner \varphi \urcorner} \urcorner$
(PT2) $\vdash \ulcorner \ulcorner \varphi \urcorner \supset \overline{\ulcorner \varphi \urcorner} \urcorner \supset \overline{\ulcorner \varphi \urcorner}$
(PT3) $\varphi, \psi \vdash \chi$ iff $\varphi \vdash \ulcorner \psi \urcorner \supset \ulcorner \chi \urcorner$

As in the case of the calculi ⊢$_c$ and ⊢$_{cp}$, the full containment calculus ⊢ gives rise to an algebra of terms that allows for free substitution under the relation of mutual containment. The specific type of algebraic structure determined by the principles of the calculus ⊢ is what we call an auto-Boolean algebra. An auto-Boolean algebra is a Boolean algebra equipped with an additional binary operation ∘ defined as follows:[58]

---

[57] This weaker version of the principle of propositional containment suffices to establish the classical laws of conjunction in the calculus ⊢ (see the proof of Theorem 29 in the appendix). Given these laws, it can be shown that this weaker version of the principle entails the strong version for those cases in which Γ is a finite set of propositions. For, given the rule of conjunction-introduction, all the propositions in the finite set Γ can be conjoined into a single proposition, to which the weaker version of the principle can be applied. This conjunctive proposition can then be unpacked again by the rule of conjunction-elimination. The extension to cases in which Γ is infinite follows from monotonicity and the fact that a proposition is derivable from an infinite set of premises in the calculus ⊢ just in case it is derivable from some finite subset of this set.

[58] Up to a change of signature, auto-Boolean algebras just are what are known as simple monadic algebras (cf. Halmos 1962: 40–8; Goldblatt 2006: 14). These are Boolean algebras equipped with an additional unary operation f defined by:

$$x \circ y = \begin{cases} 1 & \text{if } x = y \\ 0 & \text{otherwise} \end{cases}$$

Here, 1 and 0 are the top and bottom elements of the Boolean algebra, respectively. It can be shown that the calculus $\vdash$ is sound and complete with respect to the class of auto-Boolean algebras, when containment is interpreted as the order relation in the algebra, composition as the meet operation, privation as the complement operation, and the operation mapping A and B to $\ulcorner A \supset B \urcorner$ as the auto-Boolean operation mapping x and y to the element $x \circ (x \wedge y)$.[59] In this auto-Boolean semantics of the language $\mathcal{L}_\supset$, a propositional term $\ulcorner \varphi \urcorner$ designates the top element of the Boolean algebra if the proposition φ is true, and the bottom element of the algebra if φ is false.

We have now completed our survey of the fundamental laws of conceptual containment posited by Leibniz in his logical writings. These laws guarantee that the relation of containment is a preorder, i.e., that it is reflexive and transitive. Moreover, they specify how this relation interacts with the three operations of composition, privation, and forming propositional terms. Taken together, these laws constitute the containment calculus $\vdash$ in the language $\mathcal{L}_\supset$. This calculus, it turns out, has all the expressive and deductive power of the coincidence calculus developed by Leibniz in the *Generales inquisitiones*.[60] More precisely, if the relation of coincidence is expressed in the language $\mathcal{L}_\supset$ by the formula $(A \supset B) \& (B \supset A)$, where the operation $\&$ is defined in the manner just described, then all the theorems of Leibniz's coincidence calculus are derivable in the containment calculus $\vdash$.[61] Conversely, if containment is reduced to coincidence by defining $A \supset B$ as $A = AB$, then all the theorems of the containment calculus $\vdash$ are derivable in Leibniz's coincidence calculus.[62] Thus, the containment calculus $\vdash$ and the coincidence calculus developed by Leibniz in the *Generales inquisitiones* can be viewed as alternative axiomatizations, in different but equally expressive languages, of one and the same logical theory.

$$f(x) = \begin{cases} 0 & \text{if } x = 0 \\ 1 & \text{otherwise} \end{cases}$$

An auto-Boolean algebra is a simple monadic algebra in which the operation f is defined by $f(x) = (x \circ 0) \circ 0$. Conversely, a simple monadic algebra is an auto-Boolean algebra in which the operation $\circ$ is defined by $x \circ y = (f((x' \wedge y) \vee (y' \wedge x)))'$.

[59] For a proof of this claim, see Theorem 44 in the appendix.

[60] This coincidence calculus is specified in Definition 36 below. For a more detailed exposition of this calculus, see Malink and Vasudevan 2016: 696–706.

[61] For a proof of this claim, see Theorem 37 below.

[62] This can be seen by comparing the principles of the containment calculus $\vdash$ listed in Definition 16 with the laws of the coincidence calculus stated in Theorems 4.15–4.26 and 4.53 in Malink and Vasudevan 2016: 725–38.

1.6 The Two Roots of Leibniz's Logic

While Leibniz's containment calculus and his coincidence calculus are equal in their expressive and deductive power, they also differ in a number of important respects. As we have seen, one such difference concerns the way in which substitutional reasoning is implemented in the two calculi. In the coincidence calculus, a rule of substitution licensing the free substitution of one coincident term for another is posited as a primitive rule of inference. Indeed, this rule is usually stated by Leibniz as the first principle of his coincidence calculus, highlighting the fact that substitution is the characteristic mode of reasoning in this calculus.[63] By contrast, in the containment calculus, substitutional reasoning is not implemented by means of a primitive rule of inference, but is licensed by a derived rule of substitution for containment. The derivation of this rule proceeds by reducing substitutions of mutually containing terms to applications of the syllogistic mood Barbara. Thus, whereas substitutional reasoning is characteristic of the coincidence calculus, the characteristic mode of reasoning in the containment calculus is Barbara. As Leibniz puts it in his derivation of the rule of substitution for containment in the *Elementa ad calculum condendum*, 'there is only one fundamental inference: *A is B* and *B is C, therefore A is C*'.[64]

In his logical writings from the late 1670s Leibniz gives priority to conceptual containment over coincidence, while in his mature logical writings from the mid-1680s he tends to adopt the opposite approach. According to which of these two approaches he adopts, he treats either the rule of Barbara or that of substitution as the most basic mode of reasoning. This shifting emphasis between containment and Barbara, on the one hand, and coincidence and substitution, on the other, reflects two distinct approaches to the study of logic that play a formative role in Leibniz's thought.

The first of these approaches is the syllogistic approach to logic which originated with Aristotle's theory of the categorical syllogism and constituted the dominant paradigm in logic throughout the scholastic and early modern periods. As is well known, Leibniz had a fascination with the traditional theory of the syllogism that began in his early youth with his study of the works of scholastic logicians:[65]

> When I was not yet twelve years old, I filled pages with remarkable exercises in logic; I sought to exceed the subtleties of the scholastics, and neither Zabarella nor Ruvio nor Toledo caused me much delay. (*Guilielmi Pacidii initia et specimina scientiae generalis*, A VI.4 494)

---

[63] See n. 18 above.

[64] A VI.4 154.

[65] See also Leibniz's autobiographical note published in Pertz 1847: 167–8. Cf. Couturat 1901: 33–5.

Leibniz's mastery of the technicalities of syllogistic logic is already apparent in one of his earliest philosophical works, the *Dissertatio de arte combinatoria*.[66] Over the span of his career, Leibniz wrote numerous essays on the theory of the categorical syllogism, and his admiration for this theory did not wane even in his later years. Thus, for example, in the *Nouveaux Essais* he writes:

> I hold the invention of the syllogistic form to be one of the most beautiful inventions of the human mind, and indeed one of the most notable. It is a sort of universal mathematics, the importance of which is too little known. (*Nouveaux Essais*, A VI.6 478)

The traditional theory of the categorical syllogism thus had a profound influence on the shape of Leibniz's logic. As we have noted, Leibniz took the universal affirmative propositions that appear in syllogisms of the form Barbara to express the conceptual containment of one term in another.[67] Thus, the syllogistic approach to logic is especially pronounced in those of his logical writings in which priority is given to the relation of conceptual containment and the rule of Barbara.

On the other hand, those writings in which Leibniz gives priority to coincidence and the rule of substitution are informed by a different approach to logic. This second approach derives, not from Leibniz's study of the traditional theory of the syllogism, but instead from his investigations into the foundations of mathematics. In particular, Leibniz's emphasis on substitutional reasoning in his coincidence calculus is motivated by the central role that such reasoning plays in equational systems of arithmetic and geometry. Since their development in Greek antiquity, both arithmetic and geometry have made essential use of the technique of substituting equals by equals.[68] Despite this fact, ancient systems of mathematics did not include among their principles any general rule of substitution that would provide a direct justification of this technique.[69] For example, no explicit statement of a rule of substitution appears among the principles of Euclid's *Elements*.[70] Instead, Euclid only posits certain corollaries of this rule in his list of 'common notions', such as:

> CN1:   Things which are equal to the same thing are also equal to one another.
> CN2:   If equals are added to equals, the wholes are equal.
> CN3:   If equals are subtracted from equals, the remainders are equal.

In his own investigations into the foundations of geometry, Leibniz did not follow Euclid in positing these common notions as separate principles. Instead, he sought to derive them all from a single rule licensing the free substitution of equals by

---

[66] See A VI.1 179–87.

[67] See the references given in n. 22 above.

[68] See Netz 1999: 189–97. Netz argues that in Greek mathematics 'the concept of substitution *salva veritate* is the key to most proofs' (1999: 190).

[69] See Netz 1999: 191.

[70] By contrast, an explicit statement of the rule of substitution was included in some seventeenth-century axiomatizations of Euclid's *Elements* (see De Risi 2016b: 618–24).

equals.[71] For example, in his personal copy of Euclid's *Elements* Leibniz added the following marginal note immediately after the list of common notions, which are there referred to as 'axioms':[72]

> Those things are equal which are indiscernible in magnitude, or which can be substituted for one another *salva magnitudine*. From this alone all axioms are demonstrated. (quoted in De Risi 2016a: 31 n. 17)

In this note, Leibniz does not explain how such demonstrations of the common notions are to proceed. However, he supplies the desired demonstrations elsewhere, in an essay titled *Demonstratio axiomatum Euclidis*, written in 1679. Leibniz begins this essay by positing a definition of equality which licenses the free substitution of equals by equals. He then proceeds to use this rule of substitution to derive all of Euclid's common notions. For example, he derives the first of these common notions as follows:

> Definition 1. Those things are equal which are indiscernible in magnitude, or which can be substituted for one another *salva magnitudine*. ...
> Axiom 1. Those things which are equal to the same thing are equal to each other. (1) *A is equal to B* and (2) *B is equal to C*. Therefore, (3) *A is equal to C*. For, since *A is equal to B* per 1, it follows by the definition of equals that *A* can be substituted for *B* in proposition 2, from which proposition 3 will result. (*Demonstratio axiomatum Euclidis*, A VI.4 165–6)

In this passage, Leibniz formulates Euclid's first common notion as follows: *A is equal to B* and *B is equal to C*, therefore, *A is equal to C*. Rather than positing this as an indemonstrable principle, Leibniz derives it by means of the rule of substitution licensed by his definition of equality. He gives similar proofs for the remaining common notions posited by Euclid. Such substitutional proofs of the common notions appear not only in the *Demonstratio axiomatum Euclidis*, but are given by Leibniz in a number of other essays as well.[73] Thus, it is clear that Leibniz regarded the substitution of equals by equals as a primitive mode of mathematical reasoning that plays a foundational role in arithmetic and geometry.

Leibniz's substitutional proofs of Euclid's common notions had a direct impact on the shape of the coincidence calculi developed by him in his logical writings of the mid-1680s. Indeed, in many of these essays, Leibniz reproduces the same patterns of substitutional reasoning that he employed in the late 1670s in his derivations of Euclid's common notions. Consider, for example, the derivation of the transitivity of coincidence given by Leibniz in his essay *Specimen calculi coincidentium et inexistentium*, written around 1686:[74]

> Definition 1. Those things are the same or coincident of which either can be substituted for the other anywhere *salva veritate*. ...

---

[71] For Leibniz's desire to provide demonstrations of Euclid's common notions, see De Risi 2016a: 25 n. 7, 31 n. 17, 33 n. 19.

[72] See also A II.1 769, GM 7 274.

[73] See, e.g., A VI.4 165-7, 506–7, GM 5 156, GM 7 77–80.

[74] For similar substitutional proofs of the transitivity of coincidence, see A VI.4 750, 815, 816, 849.

Proposition 3. If $A = B$ and $B = C$, then $A = C$. Those things which are the same as a single third thing are the same as each other. For, if in the statement $A = B$ (which is true by hypothesis) C is substituted in place of B (by Definition 1, since $B = C$), the proposition $A = C$ will be true.[75] (*Specimen calculi coincidentium et inexistentium*, A VI.4 831)

Leibniz's reasoning in this passage follows exactly the same pattern as his proof of Euclid's first common notion quoted above. Unlike in this earlier proof, however, in the present passage the schematic letters 'A', 'B', and 'C' do not stand for mathematical objects. Instead, these letters can be taken to stand for terms, or concepts, such as *human*, *animal*, and *omniscient*.[76] On this interpretation, the symbol '=' does not express the relation of equality in magnitude between mathematical objects, but instead expresses the relation of coincidence between terms.

It is clear, then, that the basic structure of the coincidence calculi developed by Leibniz in his later logical writings is modeled closely on his equational axiomatization of Euclid's common notions. Thus, it is these later logical writings which display most clearly the influence that Leibniz's work in the foundations of mathematics had on the shape of his logical theory. At the time when Leibniz provided his axiomatization of Euclid's common notions in 1679, his logical calculi had not yet taken the form of a coincidence calculus based on the rule of substitution. Instead, his logical writings from this earlier period still embody the traditional syllogistic approach to the subject, and thus treat as primitive the relation of conceptual containment and the rule of Barbara. At the same time, through his work in the foundations of mathematics, Leibniz was made acutely aware of the power of substitutional reasoning. This realization would have naturally led him to attempt to validate substitutional reasoning in his logic of conceptual containment. As we have seen, Leibniz succeeds in deriving a rule of substitution for the elementary case in which the terms of his containment calculus have no internal syntactic structure. The derivation of the rule of substitution, however, becomes more intricate as the syntactic structure of the terms increases in complexity. With the introduction of each new operation for forming complex terms, substitutions must be licensed into a wider variety of syntactic contexts. By the time the language of Leibniz's calculus reaches its full syntactic complexity in the *Generales inquisitiones*, Leibniz forgoes any attempt to derive the rule of substitution, preferring instead to develop

---

[75] In our translation of this passage, we print '=' in place of Leibniz's symbol '∞', the former being the symbol that Leibniz uses in the *Generales inquisitiones* to express the relation of coincidence between terms.

[76] The coincidence calculus developed by Leibniz in the *Specimen calculi coincidentium et inexistentium* is intended to be an abstract calculus that admits of multiple interpretations. In particular, Leibniz intends the calculus to be applicable to any algebraic structure which satisfies the laws of commutativity and idempotence, $AB = BA$ and $AA = A$ (A VI.4 834). As Leibniz explains, the law of idempotence excludes arithmetical interpretations of the calculus (A VI.4 834; see also *GI* §129, A VI.4 512 and 811). At the same time, Leibniz points out that the calculus is applicable to the domain of terms, or concepts (*notiones*), when '=' is taken to expresses the relation of coincidence, and the concatenation of letters is taken to express the operation of composition on terms (see Swoyer 1994: 7 and 18; Lenzen 2000: 79–82; Mugnai 2017: 175–6).

a coincidence calculus in which substitution is simply posited as a primitive rule of inference. In this way, Leibniz transitions from an approach to logic informed by the traditional theory of the categorical syllogism to a substitutional approach modeled on the paradigm of equational systems of mathematics.

The substitutional approach to logic exemplified in Leibniz's mature logical writings of the mid-1680s prefigures in many important respects the algebraic approach to logic which rose to prominence in the second half of the nineteenth century through the work of logicians such as Boole and Jevons. Jevons, in particular, regarded the rule of substitution as 'the one supreme rule of inference'.[77] In his treatise *The Substitution of Similars: The True Principle of Reasoning*, he writes:[78]

> That most familiar process in mathematical reasoning, of substituting one member of an equation for the other, appears to be the type of all reasoning, and we may fitly name this all-important process the substitution of equals. (Jevons 1869: 20)

Jevons notes that the idea of a logical calculus based on the rule of substitution 'can be traced back to no less a philosopher than Leibnitz'.[79] He expresses admiration for Leibniz's derivation of the laws of coincidence by means of the rule of substitution, praising in particular the substitutional derivations given by Leibniz in his essay *Non inelegans specimen demonstrandi in abstractis*.[80] This emphasis on coincidence and substitutional reasoning, Jevons maintains, 'anticipates the modern views of logic'.[81] At the same time, Jevons laments the fact that Leibniz ignores the rule of substitution in those of his logical writings in which he adopts a more traditional, syllogistic approach to the subject:

> When he [Leibniz] proceeds to explain the syllogism, as in the paper *Definitiones logicae*, he gives up substitution altogether, and falls back upon the notion of inclusion of class in class .... He proceeds to make out certain rules of the syllogism involving the distinction of subject and predicate, and in no important respect better than the old rules of the syllogism. (Jevons 1887: xix)

The essay *Definitiones logicae* referred to by Jevons in this passage was written in the 1690s, when Leibniz had already completed his major essays on the logic of coincidence.[82] Contrary to what Jevons suggests, however, the syllogistic approach adopted by Leibniz in this essay does not represent a 'falling back' into an

---

[77] Jevons 1887: 17.

[78] Similarly, Jevons writes: 'it is not difficult to show that all forms of reasoning consist in repeated employment of the universal process of the *substitution of equals*' (Jevons 1869: 23–4).

[79] Jevons 1887: xvi.

[80] Jevons 1887: xvii–xviii. The derivations of the laws of coincidence given by Leibniz in this essay follow exactly the same pattern as his derivations of these laws in the *Specimen calculi coincidentium et inexistentium* referred to above (cp. A VI.4 831–7 with A VI.4 846–50).

[81] Jevons 1887: xviii.

[82] This essay appears at GP 7 208–10. It also appears under the title *De syllogismo categorico ex inclusione exclusioneve terminorum* in a preliminary volume of the Academy edition (*Vorausedition* A VI.5 1120). The editors of this volume date the essay between 1690 and 1696.

older and obsolete way of doing logic. Nor does it indicate an abandonment by Leibniz of the more modern, algebraic approach to the subject based on the rule of substitution. Instead, starting with the *Generales inquisitiones* in 1686, Leibniz's mature work in logic is simultaneously informed by both approaches. On the one hand, the calculi developed by Leibniz in the *Generales inquisitiones* and related essays are coincidence calculi designed to take full advantage of the power of substitutional reasoning. In these essays, the relation of conceptual containment is reduced to that of coincidence because the latter is, as Leibniz puts it, 'more suitable for our calculus' (*calculo nostro est aptior*).[83] At the same time, as we have seen, throughout his mature logical writings Leibniz remains firmly committed to the conceptual containment theory of truth.[84] Indeed, just before presenting the final axiomatization of his coincidence calculus in §§198–200 of the *Generales inquisitiones*, Leibniz defines the relation of coincidence by analyzing it as mutual containment between terms (§195). This suggests that, despite Leibniz's recognition in his mature logical writings of the technical advantages that come with treating coincidence as a primitive relation in his calculus, he continues throughout to endorse the logical and metaphysical primacy of conceptual containment.

In this respect, Leibniz's views regarding the relative priority of containment and coincidence differ from those of Jevons. Instead, they are more in agreement with the position advanced by Peirce in his writings on algebraic logic. According to Peirce, the relation of containment expressed by universal affirmative propositions is more fundamental than the relation of coincidence expressed by algebraic equalities:

> There is a difference of opinion among logicians as to whether $\supset$ or $=$ is the simpler relation. But in my paper on the *Logic of Relatives*, I have strictly demonstrated that the preference must be given to $\supset$ in this respect.[85] (Peirce 1880: 21 n. 1; with notational adjustments)

Thus Peirce rejects the view of logicians, such as Jevons, who maintain that $=$ is a more fundamental relation than $\supset$. On Peirce's view, the laws that govern the relation of coincidence derive their justification, not from a primitive rule of substitution, but from underlying laws of containment such as Barbara. Peirce affirms this position in a letter to Jevons written in 1870:

> I have shown rigidly that according to admitted principles, the conception of $=$ is compounded of those of $\supset$ and $\subset$ (or $\leqq$ and $\geqq$). This being so the substitution-syllogism

$$\frac{A = B \qquad B = C}{A = C}$$

is a compound of the two

$$\frac{A \supset B \qquad B \supset C}{A \supset C} \quad \text{and} \quad \frac{A \subset B \qquad B \subset C}{A \subset C}$$

---

[83] A VI.4 812.

[84] See Section 1.1 above.

[85] The 'demonstration' referred to in this passage appears at Peirce 1873: 318 n. 1.

and the logician in analyzing inferences ought to represent it so. (Peirce 1984: 446; with notational adjustments)

The view espoused by Peirce in this passage agrees with that adopted by Leibniz in the exposition of his containment calculi, according to which coincidence is to be defined as mutual containment and the laws of coincidence are to be derived from the underlying laws of containment. While Peirce acknowledges that, in certain contexts, it may be more convenient for a logician to deal directly with propositions expressing coincidence, in his view this does not take away from the fact that, in the logical and metaphysical order of things, containment is the more fundamental relation:

> It frequently happens that it is more convenient to treat the propositions $A \supset B$ and $B \supset A$ together in their form $A = B$; but it also frequently happens that it is more convenient to treat them separately. ...In logic, our great object is to analyze all the operations of reason and reduce them to their ultimate elements; and to make a calculus of reasoning is a subsidiary object. Accordingly, it is more philosophical to use the copula $\supset$, apart from all considerations of convenience. Besides, this copula is intimately related to our natural logical and metaphysical ideas; and it is one of the chief purposes of logic to show what validity those ideas have. (Peirce 1880: 21 n. 1; with notational adjustments)

Like Peirce, Leibniz maintains that, from a logical and metaphysical perspective, the relation of containment is more fundamental than that of coincidence. At the same time, when developing his logical calculi, Leibniz often finds it more convenient to treat coincidence as a primitive relation and to define containment in terms of coincidence rather than the other way around. The apparent tension between these two approaches is indicative of the two distinct roots of Leibniz's logic: the syllogistic and the equational. Far from being incoherent, however, the two approaches fit together harmoniously within Leibniz's overarching conception of logic. For, as we have seen in this paper, Leibniz's logic of conceptual containment and his logic of coincidence are, in effect, two alternative axiomatizations of one and the same logical theory.

Appendix: Algebraic Semantics for Leibniz's Containment Calculi

In this appendix, we provide an algebraic semantics for the containment calculi $\vdash_c$, $\vdash_{cp}$, and $\vdash$ (introduced in Definitions 3, 12, and 16 above). The three main results established in this appendix are:

- The calculus $\vdash_c$ is sound and complete with respect to the class of semilattices (Theorem 20).
- The calculus $\vdash_{cp}$ is sound and complete with respect to the class of Boolean algebras (Theorem 25).
- The calculus $\vdash$ is sound and complete with respect to the class of auto-Boolean algebras (Theorem 44).

Semilattice Semantics for $\vdash_c$

In this section, we establish that the calculus $\vdash_c$ is sound and complete with respect to the class of semilattice interpretations of the language $\mathcal{L}_c$. As a preliminary step, we first note the following elementary fact about the calculus $\vdash_c$:

Theorem 17 *For any terms A, B of $\mathcal{L}_c$:*

(i) $\vdash_c A \supset AA$
(ii) $\vdash_c AB \supset BA$

Proof For (i): by C5, we have $A \supset A \vdash_c A \supset AA$. Hence, by C1, $\vdash_c A \supset AA$.

For (ii): by C2 and C3, we have $\vdash_{cp} AB \supset A$ and $\vdash_{cp} AB \supset B$. Hence, by C5, $\vdash_{cp} AB \supset BA$. □

We now introduce an algebraic semantics for the language $\mathcal{L}_c$:

Definition 18 An interpretation $(\mathfrak{A}, \mu)$ of $\mathcal{L}_c$ consists of (i) an algebraic structure $\mathfrak{A} = \langle \mathbb{A}, \wedge \rangle$, where $\mathbb{A}$ is a nonempty set and $\wedge$ is a binary operation on $\mathbb{A}$; and (ii) a function μ mapping each term of $\mathcal{L}_c$ to an element of $\mathbb{A}$ such that:

$$\mu(AB) = \mu(A) \wedge \mu(B)$$

A proposition $A \supset B$ of $\mathcal{L}_c$ is satisfied in an interpretation $(\mathfrak{A}, \mu)$ iff $\mu(A) = \mu(A) \wedge \mu(B)$. We write $(\mathfrak{A}, \mu) \models A \supset B$ to indicate that $(\mathfrak{A}, \mu)$ satisfies $A \supset B$.

Definition 19 An interpretation $(\langle \mathbb{A}, \wedge \rangle, \mu)$ of $\mathcal{L}_c$ is a semilattice interpretation if $\langle \mathbb{A}, \wedge \rangle$ is a semilattice. If Γ is a set of propositions of $\mathcal{L}_c$ and φ a proposition of $\mathcal{L}_c$, we write $\Gamma \models_{sl} \varphi$ to indicate that, for any semilattice interpretation $(\mathfrak{A}, \mu)$: if $(\mathfrak{A}, \mu) \models \psi$ for all $\psi \in \Gamma$, then $(\mathfrak{A}, \mu) \models \varphi$.

The following completeness proof employs the standard technique introduced by Birkhoff (1935) in his proof of the algebraic completeness of equational calculi.

Theorem 20 *For any set of propositions Γ of $\mathcal{L}_c$ and any proposition φ of $\mathcal{L}_c$: $\Gamma \vdash_c \varphi$ iff $\Gamma \models_{sl} \varphi$.*

Proof The left-to-right direction follows straightforwardly from the fact that the principles C1–C5 (as well as the general properties of calculi stated in Definition 2) are satisfied in every semilattice interpretation.

For the right-to-left direction, let Γ be a set of propositions of $\mathcal{L}_c$, and let $\equiv$ be the binary relation between terms of $\mathcal{L}_c$ defined by:

$$A \equiv B \quad \text{iff} \quad \Gamma \vdash_c A \supset B \quad \text{and} \quad \Gamma \vdash_c B \supset A$$

Clearly, if $A \equiv B$, then $B \equiv A$. Also, by C1, we have $A \equiv A$. By C4, we have: if $A \equiv B$ and $B \equiv C$, then $A \equiv C$. Hence, $\equiv$ is an equivalence relation on the set of all terms. Moreover, by Theorem 4, we have: if $A \equiv B$ and $C \equiv D$, then

$AC \equiv BD$. Thus, $\equiv$ is a congruence relation on the algebra of terms formed by the operation of composition.

Now, let μ be the function mapping each term to its equivalence class under $\equiv$, and let $\mathbb{T}$ be the set $\{\mu(A) : A \text{ is a term of } \mathcal{L}_c\}$. Since $\equiv$ is a congruence relation with respect to the operation of composition, there is a binary operation $\wedge$ on $\mathbb{T}$ such that, for any terms A and B:

$$\mu(AB) = \mu(A) \wedge \mu(B)$$

This equation implies:

$$(\mu(A) \wedge \mu(B)) \wedge \mu(C) = \mu(AB) \wedge \mu(C)$$
$$= \mu(ABC)$$
$$= \mu(A) \wedge \mu(BC) \quad = \quad \mu(A) \wedge (\mu(B) \wedge \mu(C))$$

In other words, the operation $\wedge$ is associative. Now, by C2 and Theorem 17.i, we have $\vdash_c AA \supset A$ and $\vdash_c A \supset AA$. It follows that $\mu(AA) = \mu(A)$. Hence, $\wedge$ is idempotent. Moreover, by Theorem 17.ii, we have $\vdash_c AB \supset BA$ and $\vdash_c BA \supset AB$. It follows that $\mu(AB) = \mu(BA)$. Hence, $\wedge$ is commutative. All told, then, $\wedge$ is an idempotent, associative, commutative operation on $\mathbb{T}$. Hence, $\mathfrak{T} = \langle \mathbb{T}, \wedge \rangle$ is a semilattice, and $(\mathfrak{T}, \mu)$ is a semilattice interpretation.

We now show that:

$$(\mathfrak{T}, \mu) \models A \supset B \text{ iff } \Gamma \vdash_c A \supset B$$

First, suppose that $(\mathfrak{T}, \mu) \models A \supset B$. Then, $\mu(A) = \mu(A) \wedge \mu(B) = \mu(AB)$, i.e., $A \equiv AB$. This implies that $\Gamma \vdash_c A \supset AB$, and so, since $\vdash_c AB \supset B$, it follows, by C4, that $\Gamma \vdash_c A \supset B$. Next, suppose that $\Gamma \vdash_c A \supset B$. Then, since $\vdash_c A \supset A$, by C5, we have $\Gamma \vdash_c A \supset AB$. But, since, by C2, $\vdash_c AB \supset A$, it follows that $A \equiv AB$, i.e., $\mu(A) = \mu(AB) = \mu(A) \wedge \mu(B)$. But this means that $(\mathfrak{T}, \mu) \models A \supset B$.

Now, suppose $\Gamma \nvdash_c \varphi$. Then, $(\mathfrak{T}, \mu) \nmodels \varphi$. But since $\Gamma \vdash_c \psi$ for all $\psi \in \Gamma$, it follows that $(\mathfrak{T}, \mu) \models \psi$ for all $\psi \in \Gamma$. Hence, since $(\mathfrak{T}, \mu)$ is a semilattice interpretation, $\Gamma \nmodels_{sl} \varphi$.   □

Boolean Semantics for $\vdash_{cp}$

In this section, we establish that the calculus $\vdash_{cp}$ is sound and complete with respect to the class of Boolean interpretations of the language $\mathcal{L}_{cp}$. As a preliminary step, we first note a few elementary facts about the calculus $\vdash_{cp}$.

Theorem 21 *For any terms A, B of $\mathcal{L}_{cp}$:*

(i) $A \supset \overline{B} \vdash_{cp} AB \supset \overline{AB}$

(ii) $A \supset \overline{B} \vdash_{cp} B \supset \overline{A}$

(iii) $\vdash_{cp} A \supset \overline{\overline{A}}$

(iv) $A \supset B \dashv\vdash_{cp} A\overline{B} \supset \overline{A\overline{B}}$

(v) $A \supset \overline{A} \vdash_{cp} A \supset B$

(vi) $\vdash_{cp} A\overline{A} \supset B$

Proof For (i): by C3 and CP1, we have $\vdash_{cp} \overline{B} \supset \overline{AB}$. Also, by C2 and C4, $A \supset \overline{B} \vdash_{cp} AB \supset \overline{B}$. Hence, by C4, $A \supset \overline{B} \vdash_{cp} AB \supset \overline{AB}$.

For (ii): by Theorem 17.ii and CP1, we have $\vdash_{cp} \overline{AB} \supset \overline{BA}$. Now, by (i) above and C4, we have $A \supset \overline{B} \vdash_{cp} BA \supset \overline{BA}$. Hence, by CP3, $A \supset \overline{B} \vdash_{cp} B \supset \overline{A}$.

For (iii): by (ii) above, we have $\overline{A} \supset \overline{A} \vdash_{cp} A \supset \overline{\overline{A}}$. Hence, by C1, $\vdash_{cp} A \supset \overline{\overline{A}}$.

For (iv): by CP3 and (i) above, $A \supset \overline{\overline{B}} \dashv\vdash_{cp} A\overline{B} \supset \overline{A\overline{B}}$. But by (iii) above, CP2, and C4, we have $A \supset \overline{\overline{B}} \dashv\vdash_{cp} A \supset B$. Hence, $A \supset B \dashv\vdash_{cp} A\overline{B} \supset \overline{A\overline{B}}$.

For (v): by C2 and CP1, we have both $\vdash_{cp} A\overline{B} \supset A$ and $\vdash_{cp} \overline{A} \supset \overline{A\overline{B}}$. So, by C4, $A \supset \overline{A} \vdash_{cp} A\overline{B} \supset \overline{A\overline{B}}$. Hence, by (iv) above, $A \supset \overline{A} \vdash_{cp} A \supset B$.

For (vi): by (iv) above, $A \supset A \vdash_{cp} A\overline{A} \supset \overline{A\overline{A}}$. So, by C1, $\vdash_{cp} A\overline{A} \supset \overline{A\overline{A}}$. Hence, by (v) above, $\vdash_{cp} A\overline{A} \supset B$. □

Theorem 22 *For any terms A, B, C of $\mathcal{L}_{cp}$:*

(i) $A\overline{B} \supset C\overline{C}, C\overline{C} \supset A\overline{B} \vdash_{cp} A \supset B$

(ii) $A \supset B \vdash_{cp} A\overline{B} \supset C\overline{C}$

(iii) $A \supset B \vdash_{cp} C\overline{C} \supset A\overline{B}$

Proof For (i): by Theorem 21.vi, we have $\vdash_{cp} C\overline{C} \supset \overline{A\overline{B}}$. So, by C4, $A\overline{B} \supset C\overline{C} \vdash_{cp} A\overline{B} \supset \overline{A\overline{B}}$. But, by Theorem 21.iv, this implies $A\overline{B} \supset C\overline{C} \vdash_{cp} A \supset B$. Hence, $A\overline{B} \supset C\overline{C}, C\overline{C} \supset A\overline{B} \vdash_{cp} A \supset B$.

For (ii): by Theorem 21.iv, we have $A \supset B \vdash_{cp} A\overline{B} \supset \overline{A\overline{B}}$. Hence, by Theorem 21.v, $A \supset B \vdash_{cp} A\overline{B} \supset C\overline{C}$.

Lastly, (iii) follows by Theorem 21.vi. □

We now introduce an algebraic semantics for the language $\mathcal{L}_{cp}$:

Definition 23 An interpretation $(\mathfrak{A}, \mu)$ of $\mathcal{L}_{cp}$ consists of (i) an algebraic structure $\mathfrak{A} = \langle \mathbb{A}, \wedge, ' \rangle$, where $\mathbb{A}$ is a nonempty set, $\wedge$ is a binary operation on $\mathbb{A}$, and $'$ is a unary operation on $\mathbb{A}$; and (ii) a function μ mapping each term of $\mathcal{L}_{cp}$ to an element of $\mathbb{A}$ such that:

$$\mu(AB) = \mu(A) \wedge \mu(B)$$
$$\mu\left(\overline{A}\right) = \mu(A)'$$

A proposition $A \supset B$ of $\mathcal{L}_{cp}$ is satisfied in an interpretation $(\mathfrak{A}, \mu)$ iff $\mu(A) = \mu(A) \wedge \mu(B)$. We write $(\mathfrak{A}, \mu) \models A \supset B$ to indicate that $(\mathfrak{A}, \mu)$ satisfies $A \supset B$.

Definition 24 An interpretation $(\langle \mathbb{A}, \wedge, ' \rangle, \mu)$ of $\mathcal{L}_{cp}$ is a Boolean interpretation if $\langle \mathbb{A}, \wedge, ' \rangle$ is a Boolean algebra. If Γ is a set of propositions of $\mathcal{L}_{cp}$ and φ a proposition of $\mathcal{L}_{cp}$, we write $\Gamma \models_{ba} \varphi$ to indicate that, for any Boolean interpretation $(\mathfrak{A}, \mu)$: if $(\mathfrak{A}, \mu) \models \psi$ for all $\psi \in \Gamma$, then $(\mathfrak{A}, \mu) \models \varphi$.

The following completeness proof relies on a concise axiomatization of the theory of Boolean algebras discovered by Byrne (1946). In the language $\mathcal{L}_{cp}$, the main principle in this axiomatization is given by Theorem 22.

Theorem 25 *For any set of propositions Γ of $\mathcal{L}_{cp}$ and any proposition φ of $\mathcal{L}_{cp}$:* $\Gamma \vdash_{cp} \varphi$ *iff* $\Gamma \models_{ba} \varphi$.

Proof The left-to-right direction follows straightforwardly from the fact that the principles C1–C5 and CP1–CP3 are satisfied in every Boolean interpretation.

For the right-to-left direction, let Γ be a set of propositions of $\mathcal{L}_{cp}$, and let $\equiv$ be the binary relation between terms of $\mathcal{L}_{cp}$ defined by:

$$A \equiv B \quad \text{iff} \quad \Gamma \vdash_{cp} A \supset B \quad \text{and} \quad \Gamma \vdash_{cp} B \supset A$$

Following the reasoning in the proof of Theorem 20, $\equiv$ is an equivalence relation on the set of terms of $\mathcal{L}_{cp}$. Moreover, by Theorem 4 and CP1, $\equiv$ is a congruence relation on the algebra of terms formed by the operations of composition and privation.

Now, let μ be the function mapping each term to its equivalence class under $\equiv$, and let $\mathbb{T}$ be the set $\{\mu(A) : A \text{ is a term of } \mathcal{L}_{cp}\}$. Since $\equiv$ is a congruence relation with respect to the operations of composition and privation, there is a binary operation $\wedge$ on $\mathbb{T}$ and a unary operation $'$ on $\mathbb{T}$ such that, for any terms A and B:

$$\mu(AB) = \mu(A) \wedge \mu(B)$$
$$\mu\left(\overline{A}\right) = \mu(A)'$$

The operation $\wedge$ is associative and commutative (see the proof of Theorem 20). Moreover, by Theorem 22, we have for any terms A, B, C of $\mathcal{L}_{cp}$:

$$\mu(A) \wedge \mu(B)' = \mu(C) \wedge \mu(C)' \quad \text{iff} \quad \mu(A) = \mu(A) \wedge \mu(B)$$

Byrne (1946: 269–271) has shown that this latter condition, in conjunction with the associativity and commutativity of $\wedge$, implies that $\mathfrak{T} = \langle \mathbb{T}, \wedge, ' \rangle$ is a Boolean algebra. Hence $(\mathfrak{T}, \mu)$ is a Boolean interpretation.

Following the reasoning in the proof of Theorem 20, we have:

$$(\mathfrak{T}, \mu) \models A \supset B \quad \text{iff} \quad \Gamma \vdash_{cp} A \supset B$$

Now, suppose $\Gamma \nvdash_{cp} \varphi$. Then, $(\mathfrak{T}, \mu) \not\models \varphi$. But since $\Gamma \vdash_{cp} \psi$ for all $\psi \in \Gamma$, it follows that $(\mathfrak{T}, \mu) \models \psi$ for all $\psi \in \Gamma$. Hence, since $(\mathfrak{T}, \mu)$ is a Boolean interpretation, $\Gamma \not\models_{ba} \varphi$. $\qquad\qquad\square$

Auto-Boolean Semantics for ⊢

In this section, we establish that the containment calculus ⊢ is sound and complete with respect to the class of auto-Boolean interpretations of the language $\mathcal{L}_\supset$. The proof of completeness will rely on the auto-Boolean completeness of the coincidence calculus developed by Leibniz's in the *Generales inquisitiones*. We begin by establishing some preliminary facts about the containment calculus ⊢.

Definition 26 For any term A of $\mathcal{L}_\supset$, we write $\mathbf{F}(A)$ for the proposition $A \supset \overline{A}$.

Theorem 27 *For any propositions φ, ψ, χ of $\mathcal{L}_\supset$:*

(i) *If $\varphi, \psi \vdash \chi$, then $\varphi, \mathbf{F}(\ulcorner \chi \urcorner) \vdash \mathbf{F}(\ulcorner \psi \urcorner)$*

(ii) *$\varphi \vdash \psi$ iff $\vdash \ulcorner \varphi \urcorner \supset \ulcorner \psi \urcorner$*

(iii) *If $\varphi \vdash \psi$, then $\mathbf{F}(\ulcorner \psi \urcorner) \vdash \mathbf{F}(\ulcorner \varphi \urcorner)$*

Proof For (i): suppose that $\varphi, \psi \vdash \chi$. By PT3, we have $\varphi \vdash \ulcorner \psi \urcorner \supset \ulcorner \chi \urcorner$. Hence, by CP1, $\varphi \vdash \overline{\ulcorner \chi \urcorner} \supset \overline{\ulcorner \psi \urcorner}$. By C4, PT1, and PT2, it follows that $\varphi \vdash \ulcorner \mathbf{F}(\ulcorner \chi \urcorner) \urcorner \supset \ulcorner \mathbf{F}(\ulcorner \psi \urcorner) \urcorner$. Hence, by PT3, $\varphi, \mathbf{F}(\ulcorner \chi \urcorner) \vdash \mathbf{F}(\ulcorner \psi \urcorner)$.

For (ii): by PT3, $A \supset A, \varphi \vdash \psi$ iff $A \supset A \vdash \ulcorner \varphi \urcorner \supset \ulcorner \psi \urcorner$. Hence, by C1, $\varphi \vdash \psi$ iff $\vdash \ulcorner \varphi \urcorner \supset \ulcorner \psi \urcorner$.

For (iii): suppose that $\varphi \vdash \psi$. Then, $A \supset A, \varphi \vdash \psi$. By (i) above, we have $A \supset A, \mathbf{F}(\ulcorner \psi \urcorner) \vdash \mathbf{F}(\ulcorner \varphi \urcorner)$. Hence, by C1, $\mathbf{F}(\ulcorner \psi \urcorner) \vdash \mathbf{F}(\ulcorner \varphi \urcorner)$. $\qquad\square$

Theorem 28 *For any proposition φ of $\mathcal{L}_\supset$:*

(i) $\mathbf{F}(\ulcorner \mathbf{F}(\ulcorner \varphi \urcorner) \urcorner) \vdash \varphi$

(ii) $\varphi \vdash \mathbf{F}(\ulcorner \mathbf{F}(\ulcorner \varphi \urcorner) \urcorner)$

Proof For (i): by PT2, we have $\vdash \ulcorner \mathbf{F}(\ulcorner \mathbf{F}(\ulcorner \varphi \urcorner) \urcorner) \urcorner \supset \overline{\ulcorner \mathbf{F}(\ulcorner \varphi \urcorner) \urcorner}$. Moreover, by PT1 and CP1, we have $\vdash \overline{\ulcorner \mathbf{F}(\ulcorner \varphi \urcorner) \urcorner} \supset \overline{\overline{\ulcorner \varphi \urcorner}}$. Hence, by C4, $\vdash \ulcorner \mathbf{F}(\ulcorner \mathbf{F}(\ulcorner \varphi \urcorner) \urcorner) \urcorner \supset \overline{\overline{\ulcorner \varphi \urcorner}}$. By CP2 and C4, it follows that $\vdash \ulcorner \mathbf{F}(\ulcorner \mathbf{F}(\ulcorner \varphi \urcorner) \urcorner) \urcorner \supset \ulcorner \varphi \urcorner$. Hence, by Theorem 27.ii, $\mathbf{F}(\ulcorner \mathbf{F}(\ulcorner \varphi \urcorner) \urcorner) \vdash \varphi$.

For (ii): by PT1, we have $\vdash \overline{\overline{\ulcorner \mathbf{F}(\ulcorner \varphi \urcorner) \urcorner}} \supset \ulcorner \mathbf{F}(\ulcorner \mathbf{F}(\ulcorner \varphi \urcorner) \urcorner) \urcorner$. Moreover, by PT2 and CP1, we have $\vdash \overline{\overline{\ulcorner \varphi \urcorner}} \supset \overline{\ulcorner \mathbf{F}(\ulcorner \varphi \urcorner) \urcorner}$. Hence, by C4, $\vdash \overline{\overline{\ulcorner \varphi \urcorner}} \supset \ulcorner \mathbf{F}(\ulcorner \mathbf{F}(\ulcorner \varphi \urcorner) \urcorner) \urcorner$. By Theorem 21.iii and C4, it follows that $\vdash \ulcorner \varphi \urcorner \supset \ulcorner \mathbf{F}(\ulcorner \mathbf{F}(\ulcorner \varphi \urcorner) \urcorner) \urcorner$. Hence, by Theorem 27.ii, $\varphi \vdash \mathbf{F}(\ulcorner \mathbf{F}(\ulcorner \varphi \urcorner) \urcorner)$. $\qquad\square$

Theorem 29 *For any propositions φ, ψ of $\mathcal{L}_\supset$:*

(i) $\varphi, \psi \vdash \mathbf{F}\left(^\ulcorner{^\ulcorner\varphi^\urcorner} \supset \overline{^\ulcorner\psi^\urcorner}^\urcorner\right)$

(ii) $\mathbf{F}\left(^\ulcorner{^\ulcorner\varphi^\urcorner} \supset \overline{^\ulcorner\psi^\urcorner}^\urcorner\right) \vdash \varphi$

(iii) $\mathbf{F}\left(^\ulcorner{^\ulcorner\varphi^\urcorner} \supset \overline{^\ulcorner\psi^\urcorner}^\urcorner\right) \vdash \psi$

Proof For (i): since $^\ulcorner\varphi^\urcorner \supset {^\ulcorner}\mathbf{F}(^\ulcorner\psi^\urcorner)^\urcorner \vdash {^\ulcorner\varphi^\urcorner} \supset {^\ulcorner}\mathbf{F}(^\ulcorner\psi^\urcorner)^\urcorner$, by PT3 we have $\varphi, {^\ulcorner\varphi^\urcorner} \supset {^\ulcorner}\mathbf{F}(^\ulcorner\psi^\urcorner)^\urcorner \vdash \mathbf{F}(^\ulcorner\psi^\urcorner)$. By C4 and PT1, it follows that $\varphi, {^\ulcorner\varphi^\urcorner} \supset \overline{^\ulcorner\psi^\urcorner} \vdash \mathbf{F}(^\ulcorner\psi^\urcorner)$. Hence, by Theorem 27.i, $\varphi, \mathbf{F}(^\ulcorner\mathbf{F}(^\ulcorner\psi^\urcorner)^\urcorner) \vdash \mathbf{F}\left(^\ulcorner{^\ulcorner\varphi^\urcorner} \supset \overline{^\ulcorner\psi^\urcorner}^\urcorner\right)$. The desired result follows by Theorem 28.ii.

For (ii): since $\varphi, \psi \vdash \varphi$, by Theorem 27.i we have $\varphi, \mathbf{F}(^\ulcorner\varphi^\urcorner) \vdash \mathbf{F}(^\ulcorner\psi^\urcorner)$. By PT3, it follows that $\mathbf{F}(^\ulcorner\varphi^\urcorner) \vdash {^\ulcorner\varphi^\urcorner} \supset {^\ulcorner}\mathbf{F}(^\ulcorner\psi^\urcorner)^\urcorner$. Hence, by PT2 and C4, $\mathbf{F}(^\ulcorner\varphi^\urcorner) \vdash {^\ulcorner\varphi^\urcorner} \supset \overline{^\ulcorner\psi^\urcorner}$. By Theorem 27.iii, it follows that $\mathbf{F}\left(^\ulcorner{^\ulcorner\varphi^\urcorner} \supset \overline{^\ulcorner\psi^\urcorner}^\urcorner\right) \vdash \mathbf{F}(^\ulcorner\mathbf{F}(^\ulcorner\varphi^\urcorner)^\urcorner)$. The desired result follows by Theorem 28.i.

For (iii): since $\varphi, \mathbf{F}(^\ulcorner\psi^\urcorner) \vdash \mathbf{F}(^\ulcorner\psi^\urcorner)$, by PT3 we have $\mathbf{F}(^\ulcorner\psi^\urcorner) \vdash {^\ulcorner\varphi^\urcorner} \supset {^\ulcorner}\mathbf{F}(^\ulcorner\psi^\urcorner)^\urcorner$. Hence, by C4 and PT2, $\mathbf{F}(^\ulcorner\psi^\urcorner) \vdash {^\ulcorner\varphi^\urcorner} \supset \overline{^\ulcorner\psi^\urcorner}$. By Theorem 27.iii, it follows that $\mathbf{F}\left(^\ulcorner{^\ulcorner\varphi^\urcorner} \supset \overline{^\ulcorner\psi^\urcorner}^\urcorner\right) \vdash \mathbf{F}(^\ulcorner\mathbf{F}(^\ulcorner\psi^\urcorner)^\urcorner)$. The desired result follows by Theorem 28.i. □

Definition 30 For any terms A, B of $\mathcal{L}_\supset$, we write $A \approx B$ for the proposition:

$$\mathbf{F}\left(^\ulcorner{^\ulcorner A \supset B^\urcorner} \supset \overline{^\ulcorner B \supset A^\urcorner}^\urcorner\right)$$

Theorem 31 *For any terms A, B of $\mathcal{L}_\supset$:*

(i) $A \supset B \dashv\vdash A \approx AB$

(ii) $\vdash AA \approx A$

(iii) $\vdash AB \approx BA$

(iv) $\vdash \overline{\overline{A}} \approx A$

Proof For (i): by C2, we have $\vdash AB \supset A$, and so $A \supset B \vdash AB \supset A$. Moreover, by C1 and C5, we have $A \supset B \vdash A \supset AB$. Hence, by Theorem 29.i, $A \supset B \vdash A \approx AB$. For the converse, by Theorem 29.ii, we have $A \approx AB \vdash A \supset AB$. Hence, by C3 and C4, $A \approx AB \vdash A \supset B$.

For (ii): by C2, we have $\vdash AA \supset A$. Moreover, by Theorem 17.i, we have $\vdash A \supset AA$. Hence, by Theorem 29.i, $\vdash AA \approx A$.

For (iii): by Theorem 17.ii, we have both $\vdash AB \supset BA$ and $\vdash BA \supset AB$. Hence, by Theorem 29.i, $\vdash AB \approx BA$.

For (iv): by CP2, we have $\vdash \overline{\overline{A}} \supset A$. Moreover, by Theorem 21.iii, we have $\vdash A \supset \overline{\overline{A}}$. Hence, by Theorem 29.i, $\vdash \overline{\overline{A}} \approx A$. □

In the remainder of this section, we establish that the calculus $\vdash$ is complete with respect to the class of auto-Boolean algebras. This completeness proof will rely on the auto-Boolean completeness of the coincidence calculus developed by Leibniz in the *Generales inquisitiones*. The crucial step in the proof will be to show that all the theorems of the latter calculus can be reproduced in the containment calculus $\vdash$. To this end, we first specify the language $\mathcal{L}_=$ of Leibniz's coincidence calculus:

Definition 32 The terms and propositions of the language $\mathcal{L}_=$ are defined as follows:

(1) Every simple term of the language $\mathcal{L}_\supset$ is a term.
(2) If A and B are terms, then AB is a term.
(3) If A is a term, then $\overline{A}$ is a term.
(4) If A and B are terms, then $A = B$ is a proposition.
(5) If $A = B$ is a proposition, then $\ulcorner A = B \urcorner$ is a term.

The following two definitions describe how the terms and propositions of $\mathcal{L}_\supset$ can be translated into the language $\mathcal{L}_=$ and vice versa:

Definition 33 The function σ maps every term or proposition of $\mathcal{L}_\supset$ to a term or proposition of $\mathcal{L}_=$ as follows:

(1) If A is a simple term of $\mathcal{L}_\supset$, then $\sigma(A)$ is the term A.
(2) $\sigma(AB)$ is the term $\sigma(A)\sigma(B)$.
(3) $\sigma(\overline{A})$ is the term $\overline{\sigma(A)}$.
(4) $\sigma(A \supset B)$ is the proposition $\sigma(A) = \sigma(A)\sigma(B)$
(5) $\sigma(\ulcorner \varphi \urcorner)$ is the term $\ulcorner \sigma(\varphi) \urcorner$.

If Γ is a set of propositions of $\mathcal{L}_\supset$, we write $\sigma(\Gamma)$ for the set $\{\sigma(\varphi) : \varphi \in \Gamma\}$.

Definition 34 The function τ maps every term or proposition of $\mathcal{L}_=$ to a term or proposition of $\mathcal{L}_\supset$ as follows:

(1) If A is a simple term of $\mathcal{L}_\supset$, then $\tau(A)$ is the term A.
(2) $\tau(AB)$ is the term $\tau(A)\tau(B)$.
(3) $\tau(\overline{A})$ is the term $\overline{\tau(A)}$.
(4) $\tau(A = B)$ is the proposition $\tau(A) \approx \tau(B)$
(5) $\tau(\ulcorner \varphi \urcorner)$ is the term $\ulcorner \tau(\varphi) \urcorner$.

If Γ is a set of propositions of $\mathcal{L}_=$, we write $\tau(\Gamma)$ for the set $\{\tau(\varphi) : \varphi \in \Gamma\}$.

Theorem 35 *For any term A and any proposition φ of $\mathcal{L}_\supset$:*

(i) $\vdash A \supset \tau(\sigma(A))$ *and* $\vdash \tau(\sigma(A)) \supset A$
(ii) $\varphi \dashv\vdash \tau(\sigma(\varphi))$

Proof The proof proceeds by mutual induction on the structure of the terms and propositions of $\mathcal{L}_\supset$ (see Definition 13). First, suppose that A is a simple term of $\mathcal{L}_\supset$. Then $\tau(\sigma(A))$ just is the term A, and so the claim follows by C1.

Next, suppose that the claim holds for the terms A and B. By Theorem 4, we have:

$$\vdash AB \supset \tau(\sigma(A))\tau(\sigma(B))$$
$$\vdash \tau(\sigma(A))\tau(\sigma(B)) \supset AB$$

But since $\tau(\sigma(A))\tau(\sigma(B))$ is the term $\tau(\sigma(AB))$, the claim holds for the term AB. Moreover, by CP1, we have:

$$\vdash \overline{A} \supset \overline{\tau(\sigma(A))}$$
$$\vdash \overline{\tau(\sigma(A))} \supset \overline{A}$$

But since $\overline{\tau(\sigma(A))}$ is the term $\tau(\sigma(\overline{A}))$, the claim holds for the term $\overline{A}$.

Next, given that the claim holds for the terms A and B, it follows by the rule of substitution for containment established in Theorem 15 that:

$$A \approx AB \dashv\vdash \tau(\sigma(A)) \approx \tau(\sigma(A))\tau(\sigma(B))$$

Hence, by Theorem 31.i, we have:

$$A \supset B \dashv\vdash \tau(\sigma(A)) \approx \tau(\sigma(A))\tau(\sigma(B))$$

But since $\tau(\sigma(A)) \approx \tau(\sigma(A))\tau(\sigma(B))$ is the proposition $\tau(\sigma(A \supset B))$, the claim holds for the proposition $A \supset B$.

Finally, suppose that the claim holds for the proposition φ. By Theorem 27.ii, we have:

$$\vdash \ulcorner\varphi\urcorner \supset \ulcorner\tau(\sigma(\varphi))\urcorner$$
$$\vdash \ulcorner\tau(\sigma(\varphi))\urcorner \supset \ulcorner\varphi\urcorner$$

But since $\ulcorner\tau(\sigma(\varphi))\urcorner$ is the term $\tau(\sigma(\ulcorner\varphi\urcorner))$, the claim holds for the term $\ulcorner\varphi\urcorner$. This completes the induction. $\qquad\square$

We now introduce the coincidence calculus, $\Vdash$, developed by Leibniz in the *Generales inquisitiones* (as reconstructed in Malink and Vasudevan 2016: 696–706):

Definition 36 The calculus $\Vdash$ is the smallest calculus in the language $\mathcal{L}_=$ such that:

(P1) $\Vdash AA = A$
(P2) $\Vdash AB = BA$
(P3) $\Vdash \overline{\overline{A}} = A$
(P4) $A\overline{B} = A\overline{B}\,\overline{A}\overline{B} \dashv\Vdash\vdash A = AB$
(P5) $\Vdash \ulcorner\overline{\varphi}\urcorner = \ulcorner\ulcorner\varphi\urcorner\urcorner = \ulcorner\varphi\ulcorner\overline{\varphi}\urcorner\urcorner$
(P6) $\varphi \Vdash \psi$ iff $\Vdash \ulcorner\varphi\urcorner = \ulcorner\varphi\urcorner\ulcorner\psi\urcorner$
(P7) $A = B, \varphi \Vdash \varphi^*$, where φ^* is the result of substituting B for an occurrence of A, or vice versa, in φ.

The following result states that all theorems of the coincidence calculus $\Vdash$ can be reproduced under the translation τ in the containment calculus $\vdash$:

Theorem 37 *For any set of propositions Γ of $\mathcal{L}_=$ and any proposition φ of $\mathcal{L}_=$: if $\Gamma \Vdash \varphi$, then $\tau(\Gamma) \vdash \tau(\varphi)$.*

Proof The proof proceeds by induction on the definition of the coincidence calculus $\Vdash$ (Definition 36). For P1–P6, it suffices to show that for any terms A, B of $\mathcal{L}_\supset$ and any propositions φ, ψ of $\mathcal{L}_\supset$:

(i) $\vdash AA \approx A$

(ii) $\vdash AB \approx BA$

(iii) $\vdash \overline{\overline{A}} \approx A$

(iv) $A\overline{B} \approx A\overline{B}\,\overline{AB}$ $\dashv\vdash$ $A \approx AB$

(v) $\vdash \ulcorner\ulcorner\varphi\urcorner \approx \ulcorner\ulcorner\varphi\urcorner \approx \ulcorner\varphi\urcorner\overline{\ulcorner\varphi\urcorner}\urcorner$

(vi) $\varphi \vdash \psi$ iff $\vdash \ulcorner\varphi\urcorner \approx \ulcorner\varphi\urcorner\ulcorner\psi\urcorner$

Claims (i)–(iii) are stated in Theorem 31.ii–iv. Claim (iv) follows by Theorems 21.iv and 31.i. For claim (v), by Theorems 31.i and 27.ii, we have:

$$\vdash \ulcorner\ulcorner\varphi\urcorner \supset \overline{\ulcorner\varphi\urcorner}\urcorner \supset \ulcorner\ulcorner\varphi\urcorner \approx \ulcorner\varphi\urcorner\overline{\ulcorner\varphi\urcorner}\urcorner$$

$$\vdash \ulcorner\ulcorner\varphi\urcorner \approx \ulcorner\varphi\urcorner\overline{\ulcorner\varphi\urcorner}\urcorner \supset \ulcorner\ulcorner\varphi\urcorner \supset \overline{\ulcorner\varphi\urcorner}\urcorner$$

Hence, by PT1, PT2, and C4, we have:

$$\vdash \overline{\ulcorner\varphi\urcorner} \supset \ulcorner\ulcorner\varphi\urcorner \approx \ulcorner\varphi\urcorner\overline{\ulcorner\varphi\urcorner}\urcorner$$

$$\vdash \ulcorner\ulcorner\varphi\urcorner \approx \ulcorner\varphi\urcorner\overline{\ulcorner\varphi\urcorner}\urcorner \supset \overline{\ulcorner\varphi\urcorner}$$

Claim (v) then follows by Theorem 29.i.

Claim (vi) follows by Theorems 27.ii and 31.i.

Finally, for P7 we must show that, for any terms A, B of $\mathcal{L}_=$ and any proposition φ of $\mathcal{L}_=$: if φ^* is the result of substituting B for an occurrence of A, or vice versa, in φ, then $\tau(A) \approx \tau(B), \tau(\varphi) \vdash \tau(\varphi^*)$. To see this, we first note that, given Definition 34, if φ^* is the result of substituting A for an occurrence of B, or vice versa, in φ, then $\tau(\varphi^*)$ is the result of substituting $\tau(A)$ for one or more occurrences of $\tau(B)$, or vice versa, in $\tau(\varphi)$. Moreover, it follows by Theorems 15 and 29.ii–iii that, if $\tau(\varphi^*)$ is the result of substituting $\tau(A)$ for one or more occurrences of $\tau(B)$, or vice versa, in $\tau(\varphi)$, then $\tau(A) \approx \tau(B), \tau(\varphi) \vdash \tau(\varphi^*)$. This completes the induction. $\qquad\square$

We now introduce an algebraic semantics for the language $\mathcal{L}_=$:

Definition 38 An interpretation $(\mathfrak{A}, \mu)$ of $\mathcal{L}_=$ consists of (i) an algebraic structure $\mathfrak{A} = \langle \mathbb{A}, \wedge, {}' \rangle$, where $\mathbb{A}$ is a nonempty set, $\wedge$ is a binary operation on $\mathbb{A}$, and $'$ is a unary operation on $\mathbb{A}$; and (ii) a function μ mapping each term of $\mathcal{L}_=$ to an element of $\mathbb{A}$ such that:

$$\mu(AB) = \mu(A) \wedge \mu(B)$$

$$\mu\left(\overline{A}\right) = \mu(A)'$$

A proposition $A = B$ of $\mathcal{L}_=$ is satisfied in an interpretation $(\mathfrak{A}, \mu)$ iff $\mu(A) = \mu(B)$. We write $(\mathfrak{A}, \mu) \Vdash A = B$ to indicate that $(\mathfrak{A}, \mu)$ satisfies $A = B$.

Definition 39 An interpretation $(\langle \mathbb{A}, \wedge, ' \rangle, \mu)$ of $\mathcal{L}_=$ is an auto-Boolean interpretation iff $\langle \mathbb{A}, \wedge, ' \rangle$ is a Boolean algebra such that, for any terms A, B of $\mathcal{L}_=$:

$$\mu\left(\ulcorner A = B \urcorner\right) = \begin{cases} 1 & \text{if } \mu(A) = \mu(B) \\ 0 & \text{otherwise} \end{cases}$$

Here, 1 and 0 are the top and bottom elements of the Boolean algebra $\langle \mathbb{A}, \wedge, ' \rangle$, respectively. If Γ is a set of propositions of $\mathcal{L}_=$ and φ a proposition of $\mathcal{L}_=$, we write $\Gamma \Vdash \varphi$ to indicate that, for any auto-Boolean interpretation $(\mathfrak{A}, \mu)$ of $\mathcal{L}_=$: if $(\mathfrak{A}, \mu) \Vdash \psi$ for all $\psi \in \Gamma$, then $(\mathfrak{A}, \mu) \Vdash \varphi$.

The following theorem asserts that the coincidence calculus $\Vdash$ is complete with respect to the class of auto-Boolean interpretations of $\mathcal{L}_=$:

Theorem 40 *For any set of propositions Γ of $\mathcal{L}_=$ and any proposition φ of $\mathcal{L}_=$: if $\Gamma \Vdash \varphi$, then $\Gamma \vdash \varphi$.*

The proof of this completeness theorem is given in Malink and Vasudevan 2016: 744–8 (Theorem 4.94).

With this completeness theorem for the coincidence calculus $\Vdash$ in hand, we are now in a position to establish the auto-Boolean completeness of the cointainment calculus $\vdash$. To this end, we first introduce an algebraic semantics for the language $\mathcal{L}_\supset$:

Definition 41 An interpretation $(\mathfrak{A}, \mu)$ of $\mathcal{L}_\supset$ consists of (i) an algebraic structure $\mathfrak{A} = \langle \mathbb{A}, \wedge, ' \rangle$, where $\mathbb{A}$ is a nonempty set, $\wedge$ is a binary operation on $\mathbb{A}$, and $'$ is a unary operation on $\mathbb{A}$; and (ii) a function μ mapping each term of $\mathcal{L}_\supset$ to an element of $\mathbb{A}$ such that:

$$\mu(AB) = \mu(A) \wedge \mu(B)$$
$$\mu\left(\overline{A}\right) = \mu(A)'$$

A proposition $A \supset B$ of $\mathcal{L}_\supset$ is satisfied in an interpretation $(\mathfrak{A}, \mu)$ iff $\mu(A) = \mu(A) \wedge \mu(B)$. We write $(\mathfrak{A}, \mu) \models A \supset B$ to indicate that $(\mathfrak{A}, \mu)$ satisfies $A \supset B$.

Definition 42 An interpretation $(\langle \mathbb{A}, \wedge, ' \rangle, \mu)$ of $\mathcal{L}_\supset$ is an auto-Boolean interpretation iff $\langle \mathbb{A}, \wedge, ' \rangle$ is a Boolean algebra such that, for any terms A, B of $\mathcal{L}_\supset$:

$$\mu\left(\ulcorner A \supset B \urcorner\right) = \begin{cases} 1 & \text{if } \mu(A) = \mu(A) \wedge \mu(B) \\ 0 & \text{otherwise} \end{cases}$$

Here, 1 and 0 are the top and bottom elements of the Boolean algebra $\langle \mathbb{A}, \wedge, ' \rangle$, respectively. If Γ is a set of propositions of $\mathcal{L}_\supset$ and φ a proposition of $\mathcal{L}_\supset$, we write $\Gamma \models \varphi$ to indicate that, for any auto-Boolean interpretation $(\mathfrak{A}, \mu)$ of $\mathcal{L}_\supset$: if $(\mathfrak{A}, \mu) \models \psi$ for all $\psi \in \Gamma$, then $(\mathfrak{A}, \mu) \models \varphi$.

Theorem 43 *For any set of propositions Γ of $\mathcal{L}_\supset$ and any proposition φ of $\mathcal{L}_\supset$: if $\Gamma \models \varphi$, then $\sigma(\Gamma) \Vdash \sigma(\varphi)$.*

Proof Let $(\mathfrak{A}, \mu) = (\langle \mathbb{A}, \wedge, ' \rangle, \mu)$ be an auto-Boolean interpretation of $\mathcal{L}_=$, and let μ^* be the function mapping each term of $\mathcal{L}_\supset$ to an element of $\mathbb{A}$ defined by:

$$\mu^*(A) = \mu(\sigma(A))$$

We then have:

$$
\begin{aligned}
\mu^*(\ulcorner A \supset B \urcorner) &= \mu(\sigma(\ulcorner A \supset B \urcorner)) \\
&= \mu(\ulcorner \sigma(A \supset B) \urcorner) \\
&= \mu(\ulcorner \sigma(A) = \sigma(A)\sigma(B) \urcorner) \\
&= \begin{cases} 1 & \text{if } \mu(\sigma(A)) = \mu(\sigma(A)\sigma(B)) \\ 0 & \text{otherwise} \end{cases} \\
&= \begin{cases} 1 & \text{if } \mu(\sigma(A)) = \mu(\sigma(A)) \wedge \mu(\sigma(B)) \\ 0 & \text{otherwise} \end{cases} \\
&= \begin{cases} 1 & \text{if } \mu^*(A) = \mu^*(A) \wedge \mu^*(B) \\ 0 & \text{otherwise} \end{cases}
\end{aligned}
$$

Hence, by Definition 42, $(\mathfrak{A}, \mu^*)$ is an auto-Boolean interpretation of $\mathcal{L}_\supset$.

Now, suppose that $\Gamma \models \varphi$ and that $(\mathfrak{A}, \mu) \Vdash \psi$ for all $\psi \in \sigma(\Gamma)$. By Definition 41 and the definition of μ^*, we have for any proposition ψ of $\mathcal{L}_\supset$: $(\mathfrak{A}, \mu) \Vdash \sigma(\psi)$ iff $(\mathfrak{A}, \mu^*) \models \psi$. Hence, $(\mathfrak{A}, \mu^*) \models \psi$ for all $\psi \in \Gamma$. But since $(\mathfrak{A}, \mu^*)$ is an auto-Boolean interpretation of $\mathcal{L}_\supset$, it follows that $(\mathfrak{A}, \mu^*) \models \varphi$. Thus, by the above biconditional, $(\mathfrak{A}, \mu) \Vdash \sigma(\varphi)$. So, if $(\mathfrak{A}, \mu) \Vdash \psi$ for all $\psi \in \sigma(\Gamma)$, then $(\mathfrak{A}, \mu) \Vdash \sigma(\varphi)$. But since $(\mathfrak{A}, \mu)$ was an arbitrary auto-Boolean interpretation of $\mathcal{L}_=$, this means that $\sigma(\Gamma) \Vdash \sigma(\varphi)$. □

The following theorem asserts that the containment calculus $\vdash$ is sound and complete with respect to the class of auto-Boolean interpretations of $\mathcal{L}_\supset$:

Theorem 44 *For any set of propositions Γ of $\mathcal{L}_\supset$ and any proposition φ of $\mathcal{L}_\supset$: $\Gamma \vdash \varphi$ iff $\Gamma \models \varphi$.*

Proof The left-to-right direction follows straightforwardly from the fact that the principles C1–C5, CP1–CP3, and PT1–PT3 are satisfied in every auto-Boolean interpretation.

For the right-to-left direction, suppose that $\Gamma \models \varphi$. Then, by Theorem 43, $\sigma(\Gamma) \Vdash \sigma(\varphi)$. By Theorem 40, $\sigma(\Gamma) \Vdash \sigma(\varphi)$. By Theorem 37, $\tau(\sigma(\Gamma)) \vdash \tau(\sigma(\varphi))$. Hence, by Theorem 35.ii, it follows that $\Gamma \vdash \varphi$. □

Bibliography

Adams, R.M. 1994. *Leibniz: Determinist, Theist, Idealist*. New York: Oxford University Press.

Barnes, J. 1983. Terms and sentences. *Proceedings of the British Academy* 69: 279–326.

Belnap, N. 1962. Tonk, plonk and plink. *Analysis* 22: 130–134.

Birkhoff, G. 1935. On the structure of abstract algebras. *Mathematical Proceedings of the Cambridge Philosophical Society* 31: 433–454.

Brandom, R.B. 1994. *Making It Explicit: Reasoning, Representing, and Discursive Commitment*. Cambridge, MA: Harvard University Press.

Brandom, R.B. 2000. *Articulating Reasons: An Introduction to Inferentialism*. Cambridge, MA: Harvard University Press.

Byrne, L. 1946. Two brief formulations of Boolean algebra. *Bulletin of the American Mathematical Society* 52: 269–272.

Castañeda, H.-N. 1976. Leibniz's syllogistico-propositional calculus. *Notre Dame Journal of Formal Logic* 17: 481–500.

Couturat, L. 1901. *La logique de Leibniz*. Paris: Félix Alcan.

De Risi, V. 2016a. *Leibniz on the Parallel Postulate and the Foundations of Geometry: The Unpublished Manuscripts*. Heidelberg: Birkhäuser.

De Risi, V. 2016b. The development of Euclidean axiomatics: The systems of principles and the foundations of mathematics in editions of the *elements* in the early modern age. *Archive for History of Exact Sciences* 70: 591–676.

Dummett, M.A.E. 1991. *The logical Basis of Metaphysics*. Cambridge, MA: Harvard University Press.

Goldblatt, R. 2006. Mathematical modal logic: A view of its evolution. In *Logic and the Modalities in the Twentieth Century*, Handbook of the History of Logic, vol. 7, ed. D.M. Gabbay and J. Woods, 1–98. Amsterdam: Elsevier.

Hailperin, T. 2004. Algebraical logic 1685–1900. In *The Rise of Modern Logic: From Leibniz to Frege*, Handbook of the History of Logic, vol. 3, ed. D.M. Gabbay and J. Woods, 323–388. Amsterdam: Elsevier.

Halmos, P.R. 1962. *Algebraic Logic*. New York: Chelsea Publishing Company.

Jevons, W.S. 1869. *The Substitution of Similars: The True Principle of Reasoning*. London: Macmillan.

Jevons, W.S. 1887. *The Principles of Science: A Treatise on Logic and Scientific Method*, 2nd ed. London: Macmillan.

Lenzen, W. 1984a. Leibniz und die Boolesche Algebra. *Studia Leibnitiana* 16: 187–203.

Lenzen, W. 1984b. „Unbestimmte Begriffe" bei Leibniz. *Studia Leibnitiana* 16: 1–26.

Lenzen, W. 1986. ‚Non est' non est ‚est non'. Zu Leibnizens Theorie der Negation. *Studia Leibnitiana* 18: 1–37.

Lenzen, W. 1987. Leibniz's calculus of strict implication. In *Initiatives in Logic*, ed. J. Srzednicki, 1–35. Dordrecht: Kluwer.

Lenzen, W. 1988. Zur Einbettung der Syllogistik in Leibnizens "Allgemeinen Kalkül". In *Leibniz: Questions de logique*, ed. A. Heinekamp, 38–71. Stuttgart: Steiner Verlag.

Lenzen, W. 2000. Guilielmi Pacidii Non plus ultra, oder: Eine Rekonstruktion des Leibnizschen Plus-Minus-Kalküls. *Logical Analysis and History of Philosophy* 3: 71–118.

Lenzen, W. 2004. Leibniz's logic. In *The Rise of Modern Logic: From Leibniz to Frege*, Handbook of the History of Logic, vol. 3, ed. D.M. Gabbay and J. Woods, 1–83. Amsterdam: Elsevier.

Malink, M., and A. Vasudevan. 2016. The logic of Leibniz's *Generales inquisitiones de analysi notionum et veritatum*. *Review of Symbolic Logic* 9: 686–751.

Malink, M., and A. Vasudevan. 2018. The peripatetic program in categorical logic: Leibniz on propositional terms. *Review of Symbolic Logic*. https://doi.org/10.1017/S1755020318000266.

Mugnai, M. 2017. Leibniz's mereology in the essays on logical calculus of 1686–90. In *Für unser Glück oder das Glück Anderer: Vorträge des X. Internationalen Leibniz-Kongresses, Band VI*, ed. W. Li, 175–194. Hildesheim: Olms Verlag.

Netz, R. 1999. *The Shaping of Deduction in Greek Mathematics: A Study in Cognitive History*. Cambridge: Cambridge University Press.

Parkinson, G.H.R. 1965. *Logic and Reality in Leibniz's Metaphysics*. Oxford: Clarendon Press.

Parkinson, G.H.R. 1966. *Gottfried Wilhelm Leibniz: Logical Papers. A Selection*. Oxford: Clarendon Press.

Peirce, C.S. 1873. Description of a notation for the logic of relatives, resulting from an amplification of the conceptions of Boole's calculus of logic. *Memoirs of the American Academy of Arts and Sciences* 9: 317–378.

Peirce, C.S. 1880. On the algebra of logic. *American Journal of Mathematics* 3: 15–57.

Peirce, C.S. 1984. *Writings of Charles S. Peirce: A Chronological Edition, Volume 2: 1867–1871*. Bloomington: Indiana University Press.

Pertz, G.H. 1847. *Leibnizens gesammelte Werke aus den Handschriften der Königlichen Bibliothek zu Hannover I.4*. Hannover: Hahn.

Schröder, E. 1890. *Vorlesungen über die Algebra der Logik (erster Band)*. Leipzig: Teubner.

Schupp, F. 1993. *Gottfried Wilhelm Leibniz: Allgemeine Untersuchungen über die Analyse der Begriffe und Wahrheiten*, 2nd ed. Hamburg: Felix Meiner.

Sommers, F. 1982. *The Logic of Natural Language*. Oxford: Clarendon Press.

Sommers, F. 1993. The world, the facts, and primary logic. *Notre Dame Journal of Formal Logic* 34: 169–182.

Swoyer, C. 1994. Leibniz's calculus of real addition. *Studia Leibnitiana* 26: 1–30.

Swoyer, C. 1995. Leibniz on intension and extension. *Noûs* 29: 96–114.

Chapter 2
Leibniz's Mereology in the Essays on Logical Calculus of 1686–1690

Massimo Mugnai

Abstract Mereology, the doctrine of the relations of part to whole and of parts to parts, has so far awoken the interest of only a small number of Leibniz's scholars. Since the publication of the pioneering paper of Hans Burkhardt and Wolfgang Degen (Topoi 9(1):3–13, 1990), entirely devoted to Leibniz's mereology, very few works have been published on the same topic. Moreover, these works tend to consider mereology in the general setting of Leibniz's metaphysics and do not pay due attention to those essays where Leibniz systematically develops a mereological calculus. In the years 1686–1690, indeed, Leibniz wrote a series of essays, concerning the so-called 'plus-minus calculus', where a very interesting mereological doctrine is developed.

While several scholars have investigated the ancient and medieval attempts to develop a more or less embryonic mereology, modern theories of parthood are less explored. In this paper my aim is to fill at least partially this gap, focusing on Leibniz's mereological ideas. Leibniz, indeed, just in the *Dissertation on Combinatorial Art* (1666) elaborated the project of constructing a very general mereological doctrine, which evolved later in a series of papers centered on the logical operation of 'real addition'. Leibniz considered the real addition as a kind of non restricted sum, capable of being applied to any sort of things and satisfying the conditions of 'idempotence', 'reflexivity' and 'transitivity'. With the real addition,

A first version of this paper was presented at the 10th International Leibniz-Congress in Hannover (2016) and was published in the sixth volume of the conference proceedings: "Für unser Glück oder das Glück Anderer" – *Vorträge des X. Internationalen Leibniz-Kongresses*, Herausg. von Wenchao Li in Verbindung mit Ute Beckmann, Sven Erdner, Esther Maria Errulat, Jürgen Herbst, Helena Iwasinski und Simona Noreik, Georg Olms Verlag, Hildesheim-Zürich-New York, 2017, Band VI, pp. 175–194. The paper grew out of several discussions with Achille Varzi, to whom I am deeply indebted for his helpful advice on questions concerning mereology. I am indebted to an anonymous referee, as well, who gave me some suggestions that contributed (I hope) to improve the paper.

M. Mugnai (✉)
Scuola Normale Superiore, Pisa, Italy

© Springer Nature Switzerland AG 2019
V. De Risi (eds.), *Leibniz and the Structure of Sciences*, Boston Studies in the Philosophy and History of Science 337,
https://doi.org/10.1007/978-3-030-25572-5_2

the relation of *containment* (reflexive, transitive and anti-symmetric), and the notion of *proper parthood*, Leibniz elaborates a quite interesting mereology. My main purpose is to offer an exhaustive analysis of the essays (written around 1690) in which Leibniz proposes his mereological calculi based on 'real addition'.

1. Even though the notions of part and whole play an important role in Leibniz's metaphysics, *Mereology*, the doctrine of the relations of part to whole and of parts to parts, has so far awoken the interest of only a small number of Leibniz scholars. In the period since the publication of the pioneering paper by Hans Burkhardt and Wolfgang Degen (1990), which is entirely devoted to Leibniz's mereology, very few works have been published on the same topic. Moreover, these works tend to consider mereology in the general setting of Leibniz's metaphysics and do not pay attention to those essays where Leibniz systematically develops a proper mereological calculus[1]

Leibniz's interest in mereology can be traced back to the year 1666, when the *Dissertation on Combinatorial Art* was published. According to the image of the world that we get from the *Dissertation*, every existing thing is thought of as a whole, which can be decomposed into lesser wholes and these latter, in turn, into lesser wholes again, till the smallest components parts (atoms or molecular aggregates of some sort) are reached. By a simple process of combination of the parts, in the reverse direction, one may then recompose the whole. The same procedure can be applied to non-material things as well as, for example, to concepts, propositions, geometrical figures and numbers. Thus, the *art of combinations* applied to mereology, the doctrine of the whole and parts, becomes in Leibniz's hand a kind of all-purpose tool capable of introducing us to the intimate secrets of nature:

> Since all things that exist, or can be conceived in thought, may be said to be made up of parts, either real or at least conceptual, whatever differs in kind must necessarily differ either in parts, and here lie the Applications of Complexions, or by a different situs, hence the application of Dispositions.[2]

The system of variations elaborated in the *Dissertation on the Combinatorial Art*, Leibniz observes, "leads the mind that yields to it almost through all infinity, and embraces at once the harmony of the world, the inner workings of things, and the series of forms."[3] Introducing the seventh specimen, devoted to the application of the art of combinations to geometrical figures, Leibniz emphatically writes:

---

[1]Cf. Burkhardt, Degen (1990); Cook (2000); Lodge (2001); Hartz (2006), esp. pp. 54–79.

[2]A VI, 1, p. 187. All passages from the *Dissertation on Combinatorial Art* are quoted from Martin Wilson's forthcoming translation for Oxford University Press.

[3]A VI, 1, p. 187.

With these complications, not only can geometry be enriched by an infinite number of new Theorems, as every complication brings into being a new, compound figure, by the contemplation of whose properties we may devise new theorems and new demonstrations; but we have also (if it is indeed true that great things are made up of little things, whether you call them atoms or molecules) a unique way of penetrating into the arcana of nature. This is because the more one has perceived the parts of a thing, the parts of its parts, and their shapes and arrangements, the more perfectly one can be said to know the thing. [. . .] when you enter upon natural history and the question of being, that is, upon the question of the real constitution of bodies, the vast portals of Physics will stand open, and the character of the elements, the origin and mixture of the qualities, the origin of mixtures, the mixing of those mixtures, and everything that formerly lay hidden in darkness will be revealed.[4]

Later in his life, Leibniz recalls that for a brief period in his youth he favored atomism.[5] This is quite in agreement with the parenthetical remark made in this passage, which shows no prejudice against atoms as the least components of natural bodies.

In the *Dissertation on the Combinatorial Art* Leibniz's atomism is, however, wedded with a metaphysical doctrine focused on the notion of *immeation* or *perichoresis*, according to which everything is related to everything on the basis of the two relations of *similarity* and *dissimilarity*. *Perichoresis* is the transliteration into English of a Greek word meaning 'circle' or the effect of revolving, which was employed in theology "to explain the relation of *coherence* of the three persons of the Trinity".[6] The Latin word *immeatio* was coined to translate *perichoresis* and according to Johann Heinrich Bisterfeld, the main source of Leibniz on this issue, it was meant to designate the varied concourse, combination and complication of relations.[7]

Thus, in the *Dissertation on the Combinatorial Art*, Leibniz deeply modifies the account of the world usually associated with the traditional atomistic doctrines, based on the aggregation and disaggregation of atoms, without any trace of intrinsic finalism and governed by mere chance. To this view he substitutes the picture of a world composed of beings connected by a net of reciprocal relations generating a mutual union and communion, analogous to that of the three persons in the holy Trinity. This picture was clearly meant to avoid the materialistic and atheistic consequences implicit in genuine atomistic theories as, for example, the one displayed by Lucretius in his poem *On the nature of things*.

It is not difficult to see the existence of a latent conflict between Leibniz's acceptance of atomism on one hand and his agreement with a metaphysics motivated by religious issues and influenced by a philosophy largely inspired by doctrines of neoplatonic origins.[8] As soon as the conflict came to light, in the years immediately fol-

---

[4]A VI, 1, pp. 187–88.

[5]On Leibniz's atomism cf. Arthur (2004), pp. 183–227; Garber (2009), pp. 62–70, 81–82.

[6]Antognazza (2009), pp. 57–83.

[7]Cf. Mugnai (1973).

[8]On the influence of Neoplatonic ideas on Leibniz's philosophy cf. Mercer (2001).

lowing the edition of the *Dissertation on the Combinatorial Art*, Leibniz decided to abandon the atomistic theory, but in a certain sense he never stopped flirting with it.

Consider, for instance, the words with which Leibniz introduces his *Monadology*:

1. The monad, of which we will be speaking here, is nothing but a simple substance, which enters into composites; simple, meaning without parts.
2. And there must be simple substances, because there are composites; for the composite is nothing but a collection, or *aggregatum*, of simples.
3. Now, in that which has no parts, neither extension, nor shape, nor divisibility is possible. And so monads are the true atoms of nature; in a word, the elements of things.[9]

In the *New System on the Nature of Substances and Their Communication* (1695), Leibniz explains as follows the reasons for assuming the existence of such a 'formal atom':

> But thinking again about this, after much meditation I saw that it is impossible to find the principles of a real unity in matter alone, or in what is only passive, since this is nothing but a collection or aggregation of parts ad infinitum. Now a multiplicity can derive its reality only from true unities which come from elsewhere, and which are quite different from <mathematical> points, <which are only the extremities of extended things, and mere modifications,> from which it is obvious that something continuous cannot be composed. So, in order to get to these real unities I had to have recourse to a formal atom <what might be called a real and animated point, or to an atom of substance, which must contain some kind of form or activity in order to make a complete being>, since a material thing cannot simultaneously be material and perfectly indivisible, or possessed of a genuine unity.[10]

Thus, when he attempts to explain his hypothesis of the monads to his contemporaries, he uses the expression 'spiritual atom' to characterize what he properly intends to denote with the word 'monad'. As formerly in the *Dissertation*, in his mature thought Leibniz attempts to reconcile an atomistic perspective with an anti-materialistic metaphysics.

2. In the years 1686–1690, Leibniz wrote a series of essays,[11] which are usually regarded as pertaining to logic, even though their topic belongs more properly to *mereology*. That these essays primarily concern mereology is clearly shown by the simple, linguistic fact that in them Leibniz does not speak of *terms*, but of *things*, of generic *wholes* and *parts*.

Usually, Leibniz employs the Latin neuter, singular or plural, mainly of pronouns to denote the 'things' that play the role of a 'part' or a 'whole', as in the following passage:

---

[9]Leibniz (1998a), p. 268.

[10]Leibniz (1998a), p. 145.

[11]*Specimen Calculi Coincidentium* (A VI, 4A, pp. 816–22); *De casibus in quibus componendo nihil novi fieri potest* (A VI, 4A, pp. 823–28); *Specimen Calculi coincidentium et inexistentium* (A VI, 4A, pp. 830–45); *Non inelegans Specimen demonstrandi in abstractis* (A VI, 4A, pp. 845–55); *De Calculo irrepetibilium* (A VI, 4A, pp. 855–58).

si quid alteri inest adiectum ei, non facit aliud ab eo [if something which is in another thing is added to it, it does not make anything different from that other].[12]

In translations for the most part the 'quid' (= 'aliquid') of the Latin text has been rendered with an expression equivalent to the English 'term'. Thus, for instance, Clarence Irving Lewis translates the above sentence as:

If the addition of any term to another does not alter that other, then the term added is in the other.[13]

An analogous translation we find in Parkinson's excellent collection of Leibniz's logical papers:

[...] if any term which is in another is added to it, it does not make anything which is different from that other.[14]

The French translation of the same text has:

[...] si un terme est dans un autre et lui est adjouté, il ne vient rien de plus.[15]

Herbert Herring, in his translation into German employs the word *Begriff* to translate the neuter that in Latin properly corresponds to the English word 'thing'.[16] This attitude is so widespread that even the editors of Leibniz's critical edition in the introductory remark to one of the essays under consideration write that Leibniz in it "tackles... the special problem of the constitution of a concept ["wendet sich Leibniz in diesem Stück dem Spezialproblem der Begriffskonstitution zu"].[17] Yet, in this essay, Leibniz clearly states that he is mainly interested in the constitution or composition of *things* [*res*].[18] This is quite clear in a short text written in the years 1687–90, where Leibniz develops some scattered remarks concerning the notions of body, containment and continuum:

What exists in [*inexistens*], i.e. what is contained, is another thing that is the immediate or constituent requisite of the container. Therefore, we are not speaking here of terms or attributes.[19]

However, as remarked by Wolfgang Lenzen, it is not a mistake to consider 'terms' or 'concepts' as the main subjects of these essays, because Leibniz's mereological calculus can be applied *even* to 'terms' (concepts).[20] In the *Elements of a calculus*, for instance, one of his first essays entirely devoted to logical matter (dated "April

---

[12]A VI, 4A, p. 837.

[13]Lewis (1960), p. 299.

[14]LP, p. 135.

[15]Leibniz (1998b), p. 416.

[16]Cf., for instance Leibniz (1992), p. 157.

[17]A VI, 4 A, p. 823.

[18]Cf. for instance A VI, 4 A, p. 823: "Et generaliter ex quotcunque rebus [...] nihil fieri potest novi" ["And in general, from any number whatsoever of things [...] nothing new can be made"]

[19]A VI, 4A, p. 1001.

[20]Lenzen (2000), pp. 79–82.

1679"), Leibniz assimilates the distinction between genus and species to that of whole and part:

> [. . .] every true universal affirmative categorical proposition simply shows [significat] some connexion between predicate and subject [. . .] This connexion is, that the predicate is said to be in the subject, or to be contained in the subject; either absolutely and regarded in itself, or at any rate in some instance, i.e. that the subject is said to contain the predicate in a stated fashion. This is to say that the concept of the subject, either in itself or with some addition, involves the concept of the predicate, and therefore that subject and predicate are related to each other either as whole and part, or as whole and coincident whole, or as part to whole. In the first two cases the proposition is a universal affirmative; so when I say 'All gold is metal' I simply mean that in the concept of gold the concept of metal is contained directly, since gold is the heaviest metal. [. . .] But in all cases, whether the subject or predicate is a part or a whole, a particular affirmative proposition always holds.[21]

> Two terms which contain each other but do not coincide are commonly called 'genus' and 'species'. These, in so far as they compose concepts or terms (which is how I regard them here) differ as part and whole, in such a way that the concept of the genus is a part and that of the species is a whole, since it is composed of genus and differentia. For example, the concept of gold and the concept of metal differ as part and whole; for in the concept of gold there is contained the concept of metal and something else e.g. the concept of the heaviest among metals. Consequently, the concept of gold is greater than the concept of metal.[22]

Leibniz's aim in the 'mereological' essays is quite ambitious and corresponds to the attempt to construct a general theory of what contains and what is contained (*de continente et contento*) capable of being applied to a variety of things, including concepts and material objects – a theory of which, as we will see, a mereology is a special case.

In the essays we are considering, Leibniz introduces as follows the relation of *containment*, which plays such a prominent role in his logic and metaphysics:

> If several things taken together coincide with one, any one of those several things is said 'to be in' or 'to be contained in' that one thing, and the one thing itself is said to be the 'container'. Conversely, if some thing is in another, it will be among several which together coincide with that other thing.[23]

> That A 'is in' L, or, that L 'contains' A, is the same as that L is assumed to be coincident with several things taken together, among which is A.[24]

According to Leibniz, the relation of *Containment* or *inherence* has the property of being:

Reflexive: (C.1) Cxx - (x contains itself).

> A is in A. Anything is in itself.[25]

---

[21]LP, pp. 18–19.

[22]LP, p. 20.

[23]LP, p. 122 (here and in the following quotations from this edition the translation has been slightly modified according to the remarks made at the beginning of the present paper); A VI, 4A, p. 846.

[24]LP, p. 132.

[25]LP, p. 133; A VI, 4A, p. 835.

But also

Transitive: (C.2) $(Cxy \,\&\, Cyz) \rightarrow Cxz$.

> A content of a content is a content of the container; i.e. if that thing in which there is another is in a third thing, that which is in it will be in that third thing; or if A is in B and B is in C, A will also be in C.[26]

And

Antisymmetric: (C.3) $(Cxy \,\&\, Cyx) \rightarrow x = y$.

> If A is in B and B is in A, then $A = B$.[27]

Leibniz thinks of the relation of *containment* in very general terms. Given two aggregates, A and B, if all the elements that are in B *are in* A, he says that A *contains* B, without raising the question of the nature or of reciprocal connections of the elements of B. This means that the elements of B may be heterogeneous not only as regards the elements of A, but even as regards themselves:

> We say that the concept of the genus is in the concept of the species, the individuals of the species in the individuals of the genus; a part in the whole, and the indivisible in the continuum - such as a point in a line, even though a point is not a part of a line. Thus, the concept of an affection or predicate is in the concept of the subject. In general, this consideration extends very widely. We also say that inexistents are contained in those things in which they are. Nor does it matter here, with regard to this general concept, how those things which are in something are related to each other or to the container.[28]

The relation of *containment* applies to concepts and to the individuals 'falling' under the concepts, as well. Employing a distinction that Leibniz himself introduced in the *New Essays*, we may say that it holds from both points of view: the *intensional* and the *extensional* one.[29] This is clearly expressed in one of the 'mereological' essays entitled *Specimen of a calculus of what coincides and what inheres*:

> 'Being quadrilateral' is in 'parallelogram', and 'being a parallelogram' is in 'rectangle' (i.e. a figure every angle of which is a right angle). Therefore 'being quadrilateral' is in 'rectangle'. These can be inverted, if instead of concepts considered in themselves we consider the individuals [singularia] comprehended under a concept; A can be a rectangle, B a parallelogram, C a quadrilateral. For all rectangles are comprehended in the number of parallelograms, and all parallelograms in the number of quadrilaterals; therefore all rectangles are contained in quadrilaterals. In the same way, all men are contained in all animals, and all animals in all corporeal substances; therefore all men are contained in corporeal substances. On the other hand, the concept of corporeal substance is in the concept

---

[26]LP, p. 126; A VI, 4A, p. 850.

[27]LP, p. 136; A VI, 4A, p. 839.

[28]LP, p. 141; A VI, 4A, pp. 832–33.

[29]Cf. *NE*, p. 487: "For when I say *Every man is an animal* I mean that all the men are included amongst all the animals; but at the same time I mean that the idea of animal is included in the idea of man. 'Animal' comprises more individuals than 'man' does but 'man' comprises more ideas or more attributes: one has more instances, the other more degrees of reality; one has the greater extension, the other the greater intension."

of animal and the concept of animal is in the concept of man; for being a man contains being an animal.[30]

If the relation of containment holds between two concepts A and B, then it holds *in the reverse order* between the *extensions* corresponding to these concepts.

To the relation of containment Leibniz associates a principle of *composition* that applies to every kind of objects and that in the essays on mereology is called *real addition*.[31] Leibniz observes that to compose a whole it is not necessary for the components of the whole to exist all "at the same time or place":

> If, as soon as several things are put together, we understand that one thing immediately originates, then those things are called *parts* and the latter *whole*. And it is not necessary that they all exist at the same time or place, but it is sufficient that they are considered at the same time. Thus, from all Roman emperors we compose simultaneously one aggregate.[32]

Around 1690, the *composition* becomes even more general and evolves into the *real addition*, which strongly resembles what is now known in contemporary mereology as *non restricted sum* or *composition*[33]:

> We must show that, once several things are given, different one by one, it is possible to compose something different from each of them. [...] In the case of addition or composition, we do not care about quality and order among things, but we care simply about position, so that two things are both considered insofar as they constitute only one thing, equivalent to the single items taken together.[34]

3. Leibniz calls this kind of addition *real* to distinguish it from the arithmetical one and employs the symbol '$\oplus$' to designate it. Whereas in the case of arithmetical sum we have $A + A = 2A$, the real addition is *idempotent*, i.e. it obeys the law according to which $A \oplus A = A$[35]

> Next, no account is taken here of repetition; i.e. *AA* is the same for us as *A*. Consequently, whenever these laws are observed, the present calculus can be applied. It is evident that this is observed in the composition of absolute concepts, where no account is taken of order or of repetition. Thus, it is the same to say *hot and bright* as to say *bright and hot*, and to speak of *hot fire* or *white milk*, with the poets, is a pleonasm; *white milk* is simply *milk*, and *rational man* - i.e. *rational animal which is rational* - is simply *rational animal*. It is the same when certain determinate things are said to exist in things: real addition of the same thing is vain repetition. When two and two are said to make four, the latter two must be different from the former. If they were the same, nothing new would result; it would be just as if, for a joke, I wanted to make six eggs out of three by first counting three eggs, then taking away one and counting the remaining two, and finally taking one away again and

---

[30]LP, p. 136; A VI, 4A, pp. 838–39. In the Gerhardt's edition this essay has no title and Parkinson, who translates from this edition, entitles it *A Study in the Calculus of Real Addition*. The edition of the Academy has *Specimen calculi coincidentium et inexistentium* (A VI, 4A, p. 830).

[31]*Adjectio realis*, cf. A VI 4A, p. 834.

[32]A VI, 4A, p. 627.

[33]Cf. Varzi (2016), pp. 56 ff; Lando (2017), pp. 151 ff.

[34]A VI, 4A, p. 858.

[35]On the differences between *real addition* and the arithmetical operation of sum in Leibniz, cf. Lenzen (1989).

counting the remaining one. But in the calculus of numbers and magnitudes, A, B or other signs do not stand for a certain thing, but for any thing of the same number of congruent parts. For any two feet are signified by 2, if a foot is the unit or measure, whence $2 + 2$ makes something new, 4, and 3 by 3 makes something new, 9; for it is presupposed that what are used are always different (though of the same magnitude).[36]

Real addition, however, as the arithmetical one, is *commutative* and *associative*. Thus, we may summarize the main properties of *real addition* as follows:

$A \oplus A = A$ (idempotence);
$A \oplus B = B \oplus A$ (commutativity);
$(A \oplus B) \oplus C = A \oplus (B \oplus C)$ (associativity).

In Leibniz's words:

> As the *speciosa generalis* is merely the representation and treatment of combinations by signs, and as various laws of combination can be discovered, the result of this is that various methods of computation arise. Here, however, no account is taken of the variation which consists in a change of order alone, and AB is the same for us as BA. Next, no account is taken here of repetition; i.e. AA is the same for us as A. Consequently, whenever these laws are observed, the present calculus can be applied.[37]

Here Leibniz characterizes "real addition" through the properties of *idempotence* and *commutativity* and is clearly thinking of a not interpreted calculus susceptible of being applied to any domain of things, in which an operation corresponding to the 'real addition' may be performed, and these two properties hold. This explains, for example, why in the essays based on 'real addition' he tends not to mention *terms* or *propositions*, referring instead to generic 'things': the letters employed in these essays stand, indeed, for any kind of things which obey the rules of calculus. Even though he does not mention explicitly the associative property, he usually employs it, taking its validity for granted.

The 'real addition' is *non-restricted*:

> Any plurality of things, such as A and B, can be taken together to compose one thing, $A \oplus B$, or, L.[38]

And even more clearly:

> [...] our general construction depends upon the second postulate, in which is contained the proposition that any term can be compounded with any term. Thus, God, soul, body, point and heat compose an aggregate of these five things.[39]

Leibniz also proves some theorems by linking the operation of sum to the relation of containment. Theorem 5 of the *Not inelegant specimen of abstract proof*, for instance, states that if x contains y and x contains z, then x contains the sum of y and z:

[36]LP, p. 142; A VI, 4A, p. 834

[37]LP, pp. 142–43; A VI, 4A, p. 834.

[38]LP, p. 132; A VI, 4A, p. 834.

[39]LP, p. 139; A VI, 4A, p. 842.

(Th. 5)$(Cxy \& Cxz) \rightarrow Cx(y \oplus z)$.

> If each of a number of things taken severally is in something, so also is that which is constituted by them.
>
> If A is in C and B is in C, then A + B (that which is constituted by A and B: def. 4) is in C.[40]

Theorem 6 of the same work states that if x contains y and w contains z, then the real addition of x and w contains the real addition of y and z:

(Th. 6)$(Cxy \& Cwz) \rightarrow C(x \oplus w)(y \oplus z)$.

> That which is constituted by contents is in that which is constituted by the containers.
>   If A is in M and B is in N, then A + B will be in M + N.[41]

In a paper on geometry entitled *Specimen geometriae luciferae* [*An example of geometry bearer of light*] that includes very interesting remarks on continuity and where the notion of *relation* is clearly distinguished from that of *function*, Leibniz quite proudly refers to the essays we are considering, giving a short summary of some results contained in them:

> Furthermore, we may demonstrate several universal conclusions about what contains and what is contained, i.e. about what exists in some other thing, and these conclusions will be useful in logic and in geometry. Of this I gave an example in an essay where I proved that, if A is in B and B in C, then A too is in C; and if A is in L and B is in L, then even the compound of A and B will be in L; and that if A is in B and B in A, then A and B coincide.[42]

Here (C.2) *transitivity* and (C.3) *antisymmetry* are mentioned, together with theorem (Th. 6) above.

The passage just quoted continues by mentioning the solution of the problem of finding a disposition of a plurality of things such that "nothing new can be composed out of them":

> I solved some problems as well, as, for instance, that of finding a plurality made of a number whatsoever of things disposed in such a way that nothing new can be composed out of them, which happens if they are the one in the other, composing a continuum as, for instance, if A is in B, B in C and C in D etc. In such a case, nothing new can be composed; and this can be shown in other different ways, as, for instance, if we have five things A, B, C, D, E and $A \oplus B$ coincides with C and A is in D and, finally, $B \oplus D$ coincides with E. From these we cannot compose anything new in whatsoever way we want combine them.[43]

This 'solution' is not present in the essay where (C.2), (C.3) and (Th.6) are stated, but it belongs to another essay, written in the same period and exclusively devoted to the investigation of the various ways in which a whole may be composed out of a plurality of things.[44]

---

[40]LP, p. 126; A VI, 4A, p. 850.

[41]LP, p. 126; A VI, 4A, p. 851.

[42]GM 7, p. 261.

[43]GM 7, p. 261.

[44]Cf. A, VI, 4A, pp. 823–28.

4. Around 1680, Leibniz begins to distinguish between the relation of *inherence* or *containment* and the *part-whole* relation. The difference is clearly stated in the essay entitled *A not inelegant Specimen of Abstract Proof*:

> Not every inexistent is a part, nor is every container a whole. For example, a square inscribed in a circle and a diameter are in the circle; the square is a part of the circle, but the diameter is not a part of it. Something must therefore be added if the concept of whole and part is to be explained accurately [...] Further, those inexistents which are not parts are not only in something, but can also be taken away. For example, the center can be removed from a circle, in such a way that all the points except the center remain. This remainder will be the locus of all points within the circle whose distance from the circumference is less than the radius; the difference between this locus and the circle is a point, namely the center. In the same way you get the locus of all points which are moved if a sphere is moved whilst two separate points on its diameter are unmoved, if you take away from the sphere the axis, i.e. the diameter which goes through the two unmoved points.[45]

We find the same claim expressed in the above mentioned essay on geometry (the *Specimen geometriae luciferae*):

> There are, however, some things that *are in*, without being parts as, for instance, the points that can be taken in a straight line or the diameter in a circle [...].[46]

To inhere are not only parts, but other things as well. A square inscribed in a circle and the side of the square, for instance, both inhere in the circle, but only the square, not the side, can be properly considered a *part* of the circle.[47] If we want to discriminate a *part* from what simply inheres, we need to take into account the property of *similarity* or that of *congruence*.[48]

We may conclude then, as one can easily infer from the following text, that the relation of *containment* or *inherence* is more general than that of *parthood*:

> But now, however, we must explain a little bit what a whole is and what a part. It is also clear that a part inheres in the whole, i.e. that soon as the whole is given, then *eo ipso* a part immediately is given as well, or, given a part together with some other parts, then *eo ipso* the whole is given, so that if the parts are given together with their positions, they differ from the whole for the name only. And when we are reasoning, we employ the name of the whole as shorthand for the [names of the] parts themselves. There are, however, things that inhere without being parts, as the points that we may consider in a straight line, the diameter that we may consider in a circle; therefore a part must be homogeneous with the whole; and, for this reason, if there are two homogeneous things, *A* and *B* and if *B* inheres in *A*, *A* will be the whole and *B* the part. And for this reason, all demonstrations that I gave in another essay about what contains and what is contained, or what exists in some other thing, can be applied to the part and the whole.[49]

---

[45]LP, pp. 122–23; A VI, 4A, pp. 846–47.

[46]GM 7, p. 274.

[47]The circle and the square, indeed, are homogeneous: they have *two* dimensions, whereas the side of the square has only one dimension, and therefore it is not homogeneous with the circle.

[48]A VI, 4A, p. 821

[49]GM 7, p. 274.

Leibniz considers *homogeneity* as the property best suited to discriminate between *simple inherence* and *parthood*. In a text collecting several definitions, for instance, he writes: "In a more strict sense, the whole is taken as being homogeneous with the parts".[50] And in an essay entitled *Initia Mathematica* (*First Elements of Mathematics*) he links again *homogeneity* and the property of *being a part* of a whole:

> We may even define as *homogeneous* those things, which agree on some feature [...].
> If a plurality is given, as *A* or *B* and one *C*, which all agree on some property and have something homogeneous in common, or rather everything that is homogeneous in them is common to *C*, then that plurality should be called *integral parts*, and *C* a *whole*.[51]

Whereas the relation of *inherence* can subsist even in cases of aggregate of completely heterogeneous things, the part-whole relation can be applied to homogeneous things only. Thus, homogeneity imposes a restriction on the relation of inherence if we want to have a part-whole relation. As Leibniz claims "a part must be homogeneous with the whole".[52] Therefore, if some things are homogeneous and are contained in another thing, not only can the former be called *parts* and the latter a *whole*, but to them one can apply the calculus holding in the case of the relationship between container and what is contained:

> [...] therefore, if two homogeneous things *A* and *B* are given, and *B* is in *A*, then *A* should be the whole and *B* a part and consequently the demonstrations that I have done elsewhere about what contains and what is contained can be applied to the whole and the part.[53]

Leibniz's definition of *homogeneity* is quite stable over time and strongly connected with that of *similarity*. In an essay written in 1687 we find a definition of *similarity* that Leibniz repeats in many other texts:

> Similar are those things, which have the same attributes, i.e. those, which belong to the same lowest species, i.e. those which, considered in themselves, cannot be distinguished the one from the other.[54]

One of Leibniz's favorite examples to illustrate similarity is that of two circles that differ only in size. If a circle drawn on paper is presented to us and then another similar to the first differing only by virtue of a small difference in the diameter, it is quite probable that we will think that the same circle has been presented to our eyes twice. We are able, indeed, to discover that the two circles are different, only if we perceive both simultaneously, not successively, the one after the other.

Leibniz defines *homogeneity* by means of *similarity*:

---

[50]C, p. 476.
[51]GM 7, p. 30.
[52]GM 7, p. 274.
[53]GM 7, p. 274.
[54]A VI, 4A, p. 872.

> *Homogeneous* are those things, which are similar or may be made similar through a transformation.[55]

Leibniz here thinks of the *transformation* as analogous to a *continuous geometric transformation* (a translation, for example, of a figure in the Euclidean plane). The transformation goes in both directions, from homogeneous to similar things and from similar things to the homogeneous ones:

> All similar things are homogeneous. All homogeneous things can be transformed into similar. And every thing that can be transformed into similar is homogeneous.[56]

> That to whom something inheres, i.e. the container. What inheres, i.e. the contained. (And these two, if they are homogeneous, i.e. if they may become similar by means of a transformation, are said whole and part).[57]

> All homogeneous things may be transformed into similar and all similar things can be made homogeneous by means of a transformation. Two straight lines are homogeneous, because they are similar; and a straight line and an arc of a circle are homogeneous, because a circle can be stretched out in such a way to form a straight line.[58]

> [...] thus, homogeneous are those things that either are similar (the homogeneity of which is manifest by itself, as that subsisting between two squares or two circles) or, at least, can be made similar by means of a transformation.[59]

However in a marginal note to an essay written in 1685–86, Leibniz alludes to the possibility of expressing the notions of *part* and *homogeneity* without appealing to similarity and says that this is exactly what he has done "somewhere [*alicubi*]", without giving further references.

We find an attempt to define homogeneity without explicitly referring to similarity in a text written very early, probably in the year 1676:

> *Homogeneous* are those things that agree, even though in different ways, in some form or nature intelligible in itself.[60]

In this case Leibniz mentions two things as an example, one white and one black, which agree in terms of some common nature, such as mass or corporeity.[61] In the *New Essays* (1702), homogeneity continues to be associated with the property of sharing the same kind or genus:

> Furthermore, one should distinguish between the *physical* (or rather real) genus and the *logical* (or ideal) genus. Things which are of the same physical genus, or which are 'homogeneous', are so to speak of the same *matter* and can often be transformed from one into the other by changing their modifications - circles and squares for instance.[62]

---

[55]GM 7, p. 30.

[56]A VI, 4A, p. 418.

[57]A VI, 4A, p. 392.

[58]A VI, 4A, p. 418.

[59]GM 7, p. 282.

[60]A VI, 3, p. 483.

[61]*Ibidem.*

[62]A VI, 6, p. 63.

Agreement is evoked with regard to homogeneity in the following passage, belonging to the years 1680–84:

> *Homogeneous* are those things that agree [*conveniunt*] on the same thing.[63]

Occasionally, Leibniz even speaks of a *basis* for homogeneity. A *basis* in such a case is something that, existing in both, the part and the whole, is the reason of their similarity:

> A basis of homogeneous things is what is similar everywhere, in the whole as in the parts. So there is something common in a day and in an hour and in every time, and the same happens in the parts of the space, in the parts of the body, in the parts of a movement, in the degrees of heat. Such a basis of homogeneous things is even the common matter of the bodies.[64]

> I call 'basis' that in virtue of which several things are homogeneous or similar and that that by means of changes determines the differences of those very things, as, for example, space as regards figures, matter as regards bodies, time as regards hours, movement as regards its parts. Part and whole are similar. All parts together are not different from the whole.[65]

> A basis id what is common to the whole and the part, what is common to homogeneous things.[66]

Thus, according to these definitions, we may conclude that, for Leibniz, two 'things' are homogeneous if there is at least a property on which they agree. This condition, however, is too weak. As we have seen from an above quoted passage, Leibniz does not consider the property of being colored shared by three different things as sufficient to determine their homogeneity. Clearly, when considering homogeneity, we must think of some more substantive property, such as mass or a specific nature, for instance being a mineral, a vegetable or an animal.

Homogeneity is clearly *reflexive*, *transitive* and *symmetrical*:

(H.1) *Hxx* (every thing is homogeneous with itself);
(H.2) (*Hxy & Hyz*) → *Hxz* (if x is homogeneous with y and y is homogeneous with z, then x is homogeneous with z);
(H.3) *Hxy* → *Hyx* (if x is homogeneous with y, then y is homogeneous with x).

In short: homogeneity determines an *equivalence relation*. Leibniz doesn't mention explicitly these properties of the relation of homogeneity, but they easily follow from the way he characterizes the relation itself (remember that Leibniz sees a strong affinity between *similarity* and *homogeneity*, and similarity determines an equivalence relation).

   5. Since the relation of *parthood* arises from that of containment when container and contained are homogeneous and homogeneity determines an equivalence relation, it follows that even the relation of parthood is reflexive. Leibniz, however,

---

[63] A VI, 4A, p. 418.

[64] A VI, 4A, p. 311.

[65] A VI, 4A, p. 278.

[66] A VI, 4A, p. 393.

does not explicitly endorse this claim and, when speaking of parthood, he seems to be rather thinking of *proper parthood*. This clearly emerges, for example, from the following passage in the *New Essays*:

> So it can truthfully be said that the whole theory of syllogism could be demonstrated from the theory 'de continente et contento', of container and contained. The latter is different from that of whole and part, for the whole is always greater than the part [...].[67]

If Leibniz does not carefully distinguish *parthood* from *proper parthood*, analogously he does not put any emphasis on the distinction between *containment* and *proper containment*; he shows little interest in *proper containment* and *parthood* and concentrates mainly on *containment* in general and on the relation of *proper parthood*.

Whereas, according to Leibniz, the relation of containment is *reflexive*, the proper part relation is not; in other words, everything contains itself and nothing is a proper part of itself.

We may, therefore, define *Parthood*, and *Proper Parthood*:

(P.1) *Pxy* (*x* is Part of *y*) $=_{df}$ *Cxy* & *Hxy* (*x* contains y and *x* is homogeneous with *y*);
(PP.1) *PPxy* (*x* is a Proper Part of *y*) $=_{df}$ *Pxy* & $\neg(x = y)$;

it follows that *Parthood* (*P*) too is *reflexive*, *transitive* and *antisymmetric* — a *partial order*:

(P.2) *Pxx*;
(P.3) (*Pxy* & *Pyz*) $\rightarrow$ *Pxz*;
(P.4) (*Pxy* & *Pyx*) $\rightarrow$ *x* = *y*.

By contrast, *Proper Parthood* (*PP*) turns out to be a *strict partial order*, as one would expect: that is, it is *non-reflexive*, *transitive* and *asymmetric*:

(PP.2) $\neg PPxx$;
(PP.3) (*PPxy* & *PPyz*) $\rightarrow$ *PPxz*;
(PP.4) *PPxy* $\rightarrow$ $\neg PPyx$.

The distinction between *Parthood* and *Proper Parthood* parallels that between *Containment* and *Proper Containment* - that is:

(PC. 1) *PCxy* (*x* properly contains *y*) $=_{df}$ *Cxy* & $\neg(x = y)$.

From this it follows that, if *x* properly contains *y* and *y is homogeneous with x*, then *y is a proper part of x* (and vice versa):

(PP. 2) *PPxy* $\leftrightarrow$ *PCxy* & *Hxy*.

In this case as well as in the case of the relation of homogeneity mentioned above, Leibniz does not draw all these conclusions from the notions of *part* and

---

[67] *NE*, p. 486.

proper part: they easily follow, however, from what he says about *containment*, *homogeneity* and what a *part* is.

At the beginning of the essays in which Leibniz develops his mereological theory, he defines the relation of *sameness* or *coincidence*:

> *The same* or *coincidents* are those things of which either can be substituted everywhere for the other *salva veritate*.[68]

In a less concise way:

> Those things are *the same* if one of them can be substituted for the other without loss of truth. Thus, if there are A and B and A enters some true proposition and if on substituting B for A in some place in this proposition we have a new proposition, which is also true; and if this always holds good in the case of any such proposition, then A and B are said to be *the same*. Conversely, if A and B are the same, the substitution which I have mentioned will hold good. The same things are also called *coincident*. Sometimes, however, A and A are called *the same*, whereas A and B, if they are the same, are called *coincident*.[69]

As it emerges from this quotation and from other analogous texts, Leibniz carefully distinguishes *identity* or *sameness* in the proper sense from *coincidence*. The sameness of A with itself is a case of identity in the proper sense, whereas if A and B are the same, with the expression $A = B$ Leibniz denotes coincidence.[70] Since Leibniz defines *coincidence* by means of the law of substitutability, we may represent it as follows:

(EQ) *EQxy* (coincides *x with y*) $=_{\text{def}} \forall \varphi\, (\varphi(x) \leftrightarrow \varphi(y))$.

Therefore, with (C.1), (C.2), (C3), or alternatively with (C.1), (C.2), (EQ) Leibniz disposes of the basic ingredients sufficient to develop a classical mereology. The same holds for (P.1), (P.2), P.3) (or alternatively for (P.1), (P.2), (EQ)).

Coincidence and *Containment* are connected:

> *Proposition* 17. If A is in B and B is in A, then $A = B$. Things which contain each other coincide.[71]

6. In the essays we are considering, besides the relations of *Containment* and *Proper Parthood*, Leibniz employs other mereological relations such as, for instance, the relation of *Communicating*, according to which two 'things' *x* and *y* are 'communicating' (*Commxy*) if they have something in common

Commxy $=_{\text{df}} \exists z (Cxz\ \&\ Cyz)$

> If some thing, *M*, is in *A*, and the same thing is in *B*, then it will be said to be common to them, and they will be said to be *communicating*.[72]

---

[68]A VI, 4A, p. 846.

[69]*Ibidem*.

[70]Cf. Kauppi (1960), pp. 71–76.

[71]LP, p. 136; A VI, 4A, p. 839.

[72]LP, p. 123; A VI, 4A, p. 847.

The relation of *Communicating* is analogous to that of *Overlapping* in contemporary mereology.[73]

Another important relation is that of *Disjointness* (implicit in that of *Communicating*):

> If some thing, *M,* is in *A*, and the same thing is in *B*, then it will be said to be *common* to them, and they will be said to be *communicantia*. If, however, they have nothing in common [. . .], they will be called *uncommunicating*.[74]

In other words, two things are *disjoint* if they have no part (be it proper or improper) in common.

The following two passages, focused on the notion of *Disjointness* – i.e. concerning things that do not communicate – are strongly reminiscent of the so-called *supplementation principles*, according to which, roughly speaking, if x is a proper part of y, then another proper part of y exists that does not overlap with x[75]:

> If *A* and *B* do not communicate and $A + B = C$, it will be not $A = C$. Otherwise, indeed, it would be $A + B = A$. Therefore (for 15), it would be either *B* in *A*, against the Hypothesis, or *B* will be Nothing, that is against the Hypothesis, as well. If things that do not communicate, taken simultaneously coincide with the container, then only one of them cannot coincide with the container.[76]

> If *A* is in *L, and another entity, N,* should be produced, in which there remains everything which is in *L* except what is also in *A* (of which nothing must remain in *N*), *A* will be said to be 'subtracted' or removed from *L*, and *N* will be called the 'remainder'.[77]

Leibniz thinks of the operation corresponding to the sign '-' as the inverse of the *real addition*. Whereas the real addition denotes "a collection made of several things all taken simultaneously",[78] the sign '-' is employed to designate that something is removed from something else. It properly denotes the operation of *subtraction*:

> Thus, if $A + B = C$ it will be $A = C - B$ and *A* is called remainder.[79]

Closely related to the operation of *subtraction* is the notion of *Nothing* that Leibniz introduces in *Axiom 2* of *A not inelegant Specimen*:

> If the same is added and subtracted, then whatever is constituted in another as a result of this coincides with Nothing. That is, *A* (however often it is added in the constitution of some thing) - *A* (however often it is subtracted from the same thing) = *Nothing* [*Nihil*].[80]

---

[73]Cf. Varzi (2016), p. 15. Usually the relation of *overlapping* is defined as $Oxy =_{\text{def.}} \exists z(Pzx \wedge Pzy)$.

[74]LP, p. 123; A VI, 4A, p. 847.

[75]For an exhaustive presentation of the supplementation principles, cf. Varzi (2016), pp. 19–36.

[76]A VI, 4A, p. 821.

[77]LP, p. 124; A VI, 4A, p. 848.

[78]A VI, 4A, p. 819.

[79]*Ibidem.*

[80]LP, p. 124; A VI, 4A, p. 848.

As Leibniz remarks, his *Nothing* does not change the things to which it is added or subtracted: "*A + Nothing = A*".[81] Therefore, we may designate the Leibnizian *Nothing* with the usual symbol '0'.

Now, the following theorem holds:

> If something is added to something else, in which it is contained, nothing new is constituted; i.e. if B is in A, then $A + B = A$.[82]

Since '$A + 0 = A$' also holds, it follows that '0' is contained in everything.[83] But this amounts to saying that there exists at least an atom, contrary to what Leibniz firmly believes. Leibniz, however is not aware of these consequences. He never states, for instance, that 'Nothing' does in fact inhere in every part, even though, as we have seen, this follows from some very natural assumptions of his calculus. Clearly, Leibniz did not consider even the possibility that *Nothing* could be an atom or something more than . . . just nothing: he is not interested in the trivial cases of inherence or parthood involving the 'null item'.

In conclusion, all this shows that Leibniz possesses (without, obviously, being aware of this) all the fundamental ingredients constituting what Achille Varzi in the entry *Mereology* of the *Stanford Encyclopedia of Philosophy* calls *core mereology*.[84]

7. As I have remarked at the beginning of this paper, in the *Dissertation on the Combinatorial Art* Leibniz attempts to undermine the materialistic and anti-finalistic tendencies implicit in any classical atomistic account. When he writes the essays on mereology that we are now considering, what he calls 'my hypothesis of the monads' is already defined in its fundamentals. In these essays, Leibniz compares the monad to the geometrical point and the body to a segment. As the point is *in* the segment without being *part* of it, a monad is *in* the body, without being *part* of it. A segment is *in* a line (or in a bigger segment) and is *a part* of it, as well. This motivates Leibniz's distinction between the two mereological approaches, the one based on the relation of *Containment* and the other based on *Parthood*.

As Leibniz distinguishes *simple containment* from *parthood*, this distinction is paralleled in his metaphysics by that of *simple aggregates* and *wholes*. An aggregate is the sum of *non-homogeneous* things, while a *whole* is the sum of *homogeneous* ones. From this it follows, for instance, that the world we are living in is a mere *aggregate*, not a whole, constituted by individual beings.

Even though during his entire life Leibniz denied the existence of atoms understood as the smallest particles of matter of 'infinite hardness' and therefore not further divisible into parts, he never stopped flirting with atomism. In a text written during the years 1686–1690, for example, he writes:

---

[81] A VI, 4A, p. 819.

[82] LP, p. 126; A VI, 4A, 851.

[83] This holds from the *extensional* point of view. On the distinction between an *extensional* and an *intensional* interpretation of Leibniz's *nihil* cf. Lenzen (2004), pp. 248–51. On the same point, see also Leibniz (2000), pp. LXX-LXXVI.

[84] Cf. Varzi (2016), p. 14.

> Those people who established the existence of atoms, saw only a part of the truth. They acknowledged, indeed, that we need to reach something that is one and indivisible in itself, as the basis of the multiplicity; but they were mistaken insofar as they sought unity in matter. And they believed it to be possible that a body exists which is truly one and an indivisible substance.[85]

Once Leibniz has converted the atoms of the classical atomistic doctrines into 'spiritual atoms' or 'soul-like' individual substances, he attributes to them the same role that the soul (form of the body) has in the Aristotelian philosophy. It is thus that the idea of a monad dominating a cluster of other monads playing the role of a body originates:

> Everywhere there are simple substances actually separated from each other by their own actions, which continually change their relations. And each outstanding simple substance or monad which forms the center of a compound substance (such as an animal, for example), and is the principle of its uniqueness, is surrounded by a mass composed of an infinity of other monads which constitute the body belonging to this central monad, corresponding to the affections by which it represents, as in a kind of center, the things which are outside of it. The body is *organic* when it forms a kind of automaton or natural machine not only as a whole but also in its smallest observable parts.[86]

As Leibniz emphasizes on many occasions, matter in itself is devoid of unity. Every existing material body belonging to our world is the result of an aggregation of soul-like atoms, that is of *monads*. The aggregate of monads, according to Leibniz, generates the corporeal body, but neither the aggregate nor each monad belonging to it are *part* (proper or improper) of the body, even though the body itself can be part of another body (of some mass of matter):

> By monad I understand a substance truly one, namely, one which is not an aggregate of substances. Matter in itself, or bulk [*moles*], which you can call primary matter, is not a substance; indeed, it is not an aggregate of substances, but something incomplete. Secondary matter, or mass [*massa*], is not a substance, but [a collection of] substances; and so not the flock but the animal, not the fish pond but the fish is one substance. Moreover, even if the body of an animal, or my organic body is composed, in turn, of innumerable substances, they are not parts of the animal or of me. But if there were no souls or something analogous to them, then there would be no I [Ego], no monads, no real unities, and therefore there would be no substantial multitudes; indeed, there would be nothing in bodies but phantasms.[87]

Monads are simple, spiritual, and therefore indivisible beings. To each monad, God associates an aggregate of other monads, which constitute the *body* of the original monad. Only God, the 'monad of all monads', being a pure spirit, is without a body. Thus, the problem arises as to what is the principle or the main cause responsible for the aggregation of the monads constituting a body. Clearly, if we have to take seriously the analogy between points and monads, the juxtaposition or mere addition of simple monads cannot generate a body (as the addition of a plurality of points

---

[85] A VI, 4A, p. 1064.
[86] Rescher (1991), pp. 227–228.
[87] AG, p. 167.

does not generate a segment or a line). How an aggregate of simple, soul-like and indivisible beings can produce an extended material body is a serious problem for Leibniz, who attempts to solve it in ways that interpreters have traditionally found implausible or inconsistent or both.

Consider, for example, that for Leibniz an absolute space does not exist, thus the notion itself of an 'aggregate' with its correlated activity of 'aggregating' or 'putting things together' cannot be understood according to its obvious spatial interpretation. Fortunately, the aim of the present paper is to investigate Leibniz's mereological doctrine and we may take for granted the constitution of the bodies out of an aggregate of monads, dispensing ourselves from inquiring into how Leibniz thinks that a given aggregate of simple monads does in fact produce a body.

Thus, let us simply assume the existence of material bodies. As we have seen, Leibniz overtly admits that a body can be part of another. Monads, instead, are not parts of a body, but they *inhere* in the body to whose coming into being they contribute:

> Therefore, we need not to say that the indivisible substance enters the composition of the body as a part, but rather that it enters as an essential, internal requisite.[88]

Simple, indivisible substances (monads) are situated *in* the body, but they don't have a precise recognizable location in it. We may only say that the monads constituting a body have a spatial situation by means of the body; they are entirely located where the body is, but we cannot determine in what portion of the body they are. Here, however, the analogy between point and line on one hand and monad and body on the other seems to fail. We are indeed able to find the exact situation of a point in a line, whereas we cannot do the same in the case of a monad in a body.

To briefly sum up the basic ingredients of Leibniz's ontology, we have:

1. simple monads;
2. clusters of simple monads;
3. simple monads associated with a cluster of other simple monads.[89]

Leibniz calls *corporeal substances* the items corresponding to (3).

In a draft of a letter addressed to Thomas Burnett (1699), Leibniz observes concerning the relationship between *matter, force* and *corporeal substances*:

> In bodies I distinguish corporeal substance from matter, and I distinguish primary from secondary matter. Secondary matter is an aggregate or composite of several corporeal substances, as a flock is composed of several animals. But each animal and each plant is also a corporeal substance, having in itself a principle of unity which makes it truly a substance and not an aggregate. And this principle of unity is that which one calls soul, or it

---

[88] A VI, 4B, p. 1669.

[89] Cf. Leibniz's letter to De Volder (AG, p. 177): "Therefore I distinguish: (1) the primitive entelechy or soul; (2) the matter, namely, the primary matter or primitive passive power; (3) the monad made up of these two things; (4) the mass [massa] or secondary matter, or the organic machine in which innumerable subordinate monads come together; and (5) the animal, that is, the corporeal substance, which the dominating monad in the machine makes one."

is something analogous to soul. But, besides the principle of unity, corporeal substance has its mass or its secondary matter, which is, again, an aggregate of other smaller corporeal substances—and that goes to infinity. However, primitive matter, or matter taken in itself is what we conceive in bodies when we set aside all the principles of unity, that is, it is what is passive, from which arise two qualities: resistance, and tardiness or inertia [resistentia et restitantia vel inertia]. That is to say, a body gives way to another rather than allowing itself to be penetrated, but it does not give way without difficulty and without weakening the total motion of the body pushing it. Thus one can say that matter in itself, besides extension, contains a primitive, passive power. But the principle of unity contains the primitive active power, or the primitive force, which can never be destroyed and always persists in the exact order of its internal modifications, which represent those outside it.[90]

As I pointed out, it is quite puzzling how a concrete, material body emerges from an aggregate of simple, soul-like monads, but once we assume that the mysterious transformation has taken place, we may represent the structure of the Leibnizian (existing) world as follows.

Corporeal substances are the building blocks of two main types of aggregates: *without* and *with a dominant monad*. Simple aggregates, that is aggregate *without a dominant monad*, are collections of corporeal substances lacking in unity, like a heap of stones. Of this nature are all inorganic bodies, as for instance a piece of metal, which receives its unity from our perceptions. Aggregates with a dominant monad, instead, are all organic bodies, i.e. all the aggregates possessing a unity in themselves, as in the case of animals and plants. Thus, the blade of a knife, which appears to us as perfectly smooth and without holes and fractures, is an aggregate of an infinity of other aggregates (of corporeal substances), and receives its unity by means of our imagination - a faculty that plays an essential role in our perceiving activity. A fish, a man and a tree, instead, receive their unity by means of a monad that dominates their bodies. Therefore, all organic beings are 'animated' and die - i.e. they cease to be *one* being, breaking down into a multitude of other corporeal substances - when the dominant monad detaches itself from the aggregate subordinate to it.

A peculiar feature of Leibniz's world is that organic bodies are the fundamental constituents of any other body, *inorganic* bodies included. As Leibniz claims on many occasions, if we could investigate the fine-grained structure of a slab of marble with the help of a very powerful microscope, we would discover a 'whole world of creatures':

66. From this one sees that there is a whole world of creatures - of organisms, animals, entelechies, and souls - even in the least piece of matter.

67. Every bit of matter can be conceived as a garden full of plants or a pond full of fish. But each branch of the plant, each member of the animal, each drop of its bodily fluids, is also such a garden or such a pond.[91]

It follows that 'every bit of matter' is made up of an infinity of items or parts. Therefore, Leibniz's mereology is *atomless*. This, however, seems at odds with the

---

[90] AG, pp. 289–90.
[91] Rescher (1991), pp. 227–228.

fact that, if we interpret as the *nullset* the *Nothing* of the calculus of 'real addition', we seem to be forced to conclude that *Nothing* is a mereological atom. I do not think, however, that we are committed to this conclusion. To interpret the Leibnizian *Nothing* as the *null-set* is a mistake: it amounts to attributing to Leibniz ideas and theories of a logician of the twentieth century.

Because of the presence of an immaterial principle that gives unity to bodies and to the elementary aggregates composing bodies, Leibniz's mereology is evocative of some kind of hylomorphism, which probably has its roots in a remote influence of Aristotelian philosophy. From this point of view, Leibniz's mereological account presents some analogies with recent hylomorphic theories in the field of mereology, such as those proposed by Kathrin Koslicki or Thomas Sattig, for instance.[92] Leibniz's hylomorphism, however, is quite peculiar; in the last analysis, *matter* and *form*, i.e. the body and the dominant monad, are made of the same ingredients, namely monads. As Leibniz emphasizes on various occasions, simple substances are the only true substances, and strictly speaking everything else (i.e. aggregates) exist only thanks to our activity of perceiving.

Bibliography[93]

Antognazza, M.R. 2009. Bisterfeld and "immeatio": origins of a key concept in the early modern doctrine of universal harmony. In *Spätrenaissance-Philosophie in Deutschland 1570–1650: Entwürfe zwischen Humanismus und Konfessionalisierung, okkulten Traditionen und Schulmetaphysik*, ed. M. Mulsow, 57–83. Tübingen: Niemeyer.

Arthur, R.T.W. 2004. The Enigma of Leibniz's Atomism. In *Oxford Studies in Early Modern Philosophy*, ed. D. Garber, vol. I, 183–227. Oxford: Oxford University Press.

Burkhardt, H., and W. Degen. 1990. Mereology in Leibniz's Logic and Philosophy. *Topoi* 9 (1): 3–13.

Cook, R.T. 2000. The Logic of Leibniz's Mereology. *Studia Leibnitiana* Bd. 32 (1): 1–20.

Garber, D. 2009. *Leibniz: Body, Substance, Monad*. Oxford: Harvard University.

Hartz, G.A. 2006. *Leibniz's final system: Monads, matter, and animals*. New York: Routledge.

Kauppi, R. 1960. Über die Leibnizsche Logik. *Acta Philosophica Fennica, Helsinki* Fasc. XII.

Koslicki, K. 2008. *The Structure of Objects*. Oxford: Oxford University Press.

Lando, G. 2017. *Mereology. A Philosophical Introduction*. London: Bloomsbury Academic.

Leibniz, G.W. 1992. *Philosophische Schriften*. Band IV, ed. and trans. Herbert Herring, Frankfurt/Leipzig.

---

[92]See Koslicki (2008), Sattig (2015).

[93]A, followed by number of Series and volume = G. W. Leibniz. 1923. *Sämtliche Schriften und Briefe*. Darmstadt/Berlin. AG = G. W. Leibniz. 1989. *Philosophical Essays*, ed. and trans. Roger Ariew and Daniel Garber. Indianapolis/Cambridge: Hackett Publishing Company. C = L. Couturat (ed.). 1903. *Opuscules et fragments inédits de Leibniz*. Paris: Alcan. GM = C. I. Gerhardt (ed.). 1849–63. *G. W. Leibniz: Mathematische Schriften*, 7 vols., Berlin/Halle: A. Asher/W. H. Schmidt. LP = G. W. Leibniz. 1966. *Logical Papers. A Selection*, trans. and ed. with an introd. G. H. R. Parkinson. Oxford. *NE* = G. W. Leibniz. 1981. *New Essais on Human Understanding*, trans. and ed. P. Remnant and J. Bennett. Cambridge: Cambridge University Press.

———. 1998a. *Philosophical Texts,* trans. R. Francs and R. S. Whoolhouse. Oxford: Oxford University Press.

———. 1998b. *Recherches générales sur l'analyse des notions et des vérités. 24 thèses métaphysiques et autres textes logiques et métaphysiques*, introductions et notes par Jean Baptiste Rauzy, Paris.

———. 2000. *Die Grundlagen des logischen Kalküls,* herausgegeben und mit einem Kommentar versehen von Franz Schupp, unter der Mitarbeit von Stephanie Weber, Lateinisch-Deutsch, Hamburg, F. Meiner Verlag.

Lenzen, W. 1989. Arithmetical vs 'Real' Addition—A Case Study of the Relation Between Logic, Mathematics and Metaphysics in Leibniz. In *Proceedings of the 5th Annual Conference in Philosophy of Science*, ed. N. Rescher, 149–157. Lanham: University of America Press.

——— 2000. Guilielmi Pacidii Non Plus Ultra, oder: Eine Rekonstruktion des Leibnizschen Plus-Minus Kalküls, in Uwe Meixner, Alber Newen (eds.), *Philosophie der Neuzeit: From Descartes to Kant,* Paderborn *(Philosophiegeschichte und logische Analyse* 3), pp. 71–118.

———. 2004. *Calculus Universalis*. ed. Studien zur Logik von G. W. Leibniz. Paderborn: Mentis Verlag.

Lewis, C.I. 1960. *A Survey of Symbolic Logic*. New York: Dover. [first ed. 1918].

Lodge, P. 2001. Leibniz's notion of an aggregate. *British Journal for the History of Philosophy* 9 (3): 467–486.

Mercer, C. 2001. *Leibniz's Metaphysics: Its Origins and Development*. Cambridge: Cambridge University Press.

Mugnai, M. 1973. Der Begriff der Harmonie als metaphysische Grundlage der Logik und Kombinatorik bei Johann Heinrich Bisterfeld und Leibniz. *Studia Leibnitiana* V (1): 50–58.

Rescher, N. 1991. *Leibniz's Monadology: An Edition for Students*. Pittsburg: University of Pittsburgh Press.

Sattig, T. 2015. *The Double Lives of Objects*. Oxford: Oxford University Press.

Varzi, A. 2016. Mereology. In: *The Stanford Encyclopedia of Philosophy*. (Winter 2016 Edition), ed. Edward N. Zalta.

Chapter 3
Leibniz in Cantor's Paradise: A Dialogue on the Actual Infinite

Richard T. W. Arthur

Dialogus de anima brutorum inter Pythagoram et Cartesium in Elysiis campis sibi obviam factos. (Leibniz, 12 December 1676: A VI, 3, 582).

Abstract In this paper I present a fictional dialogue between Gottfried Leibniz and Georg Cantor on the actual infinite. The dialogue is set in the afterlife, and I use the authors' own words to the extent I can. Leibniz, enlisting a distinction due to the Scholastics, denied the actual infinite in the sense of a collection or set of terms (the categorematic infinite) in favour of a *syncategorematic* understanding: an actually infinite multiplicity of terms, syncategorematically understood, is one such that however many one supposes there to be, there are more; but there is no infinite number of them. When referring to "all" the terms in an infinite multiplicity, the "all" must be understood distributively, not collectively; Leibniz consistently rejects infinite collections or totalities as being incompatible with the Part-Whole axiom. One consequence of this is to show that Cantor's diagonal argument does not constitute a conclusive proof that there are more reals than natural numbers, since it depends on the premise that the real numbers form a totality, a collection such that none are left out; and this is established by the Power Set Axiom only on the hypothesis that the natural numbers form such a totality. Similarly, 1–1

I wrote this dialogue in the year 2000, since when it has had a certain life as an unpublished manuscript in exchanges with other scholars and in presentations to audiences, such as in Pisa in 2012, and on the internet. In the meantime, I have held off publishing it until now, because I had plans to write a second day on Russell and the continuum, a project which I have not yet been able to get to. I am indebted to several colleagues for their encouragement and generous critical feedback on earlier drafts of this dialogue, most especially Massimo Mugnai, Wayne Myrvold, Dean Buckner, Antonio Leon, Bill Harper and Jim Brown. For my understanding of Cantor's philosophy I have drunk deeply from Joseph Dauben's (1979) and Michael Hallett's (1984).

R. T. W. Arthur (✉)
McMaster University, Hamilton, ON, Canada

© Springer Nature Switzerland AG 2019
V. De Risi (eds.), *Leibniz and the Structure of Sciences*, Boston Studies in the Philosophy and History of Science 337,
https://doi.org/10.1007/978-3-030-25572-5_3

correspondence between the elements of two multiplicities does not establish that they constitute equal sets without the assumption that those infinite multiplicities are indeed consistent totalities. Consistency, on the Leibnizian model, would have to be shown by the provision of "real definitions", thus committing him to a kind of constructivism that would rule out Cantorian transfinite recursion. In contrast, it is shown that Leibniz's syncategorematic understanding of the actual infinite is not only consistent with this constructivism, but also with his own conception of the actually infinite division of matter, whereas Cantor's transfinite is not. Lastly, it is shown that Leibniz's claim that the universe, or any other collection of all unities, cannot itself be a unity, can be proved in an entirely analogous way to Cantor's proof that there is no ordinal number of all ordinal numbers. In sum, I use this dialogue to argue that the Leibnizian actual infinite constitutes a perfectly clear and consistent third alternative in the foundations of mathematics to the usual dichotomy between the potential infinite (Aristotelianism, intuitionism) and the transfinite (Cantor, set theory), and one that avoids the paradoxes of the infinite.

3.1 Introduction

In circumstances too bizarre to relate here, I recently chanced upon the following conversation between Georg Cantor and Gottfried Leibniz, apparently on the occasion of their first encounter in the Afterlife. It seemed safe to assume that it would be worth recording for posterity and, looking at the transcript now, that assumption seems to be borne out. Rather surprisingly, Leibniz was by no means the naive understudy one might have supposed he would be: he sets out a consistent conception of the actual infinite and poses some pointed objections to Cantor's interpretation of it as transfinite.

Although I was unable to begin my recording immediately, missing the many 'sehr geehrter Herr Professor's and other extravagant greetings and praises that went back and forth, my transcription (and translation from the mix of Latin, German and French in which they spoke) begins immediately afterwards.[1]

Cantor: I always remember with joy what you said of the *actual infinite*, that you were so much in favour of it that instead of admitting that Nature abhors it, as is commonly said, you hold that Nature affects it everywhere, the

---

[1] Throughout this piece I have set the actual words of the participants in bold with different font. Some inessentials, such as tense, have been altered to ease the fit. I have mixed up quotations from different periods of Leibniz's work, as I believe his views on the infinite underwent no substantive change during the last forty of so years of his life; the consistency of the resulting pastiche must serve as my only warrant for this here. Much the same applies to Cantor. All translations from Leibniz's Latin and French and from Cantor's German are my own, except where noted. I did not have access to Cantor's *Nachlass*; all quotations from it are from Dauben or Hallett as noted.

better to indicate the perfections of its Author.[2] This remark gave me great inspiration!

Leibniz: I am very glad. Indeed, I believe that there is no part of matter which is not, I do not say divisible, but actually divided; and that consequently the least particle ought to be considered as a world full of an infinity of different creatures.[3]

Cantor: Here again I took inspiration from your ideas,[4] for I believe that you were correct in your opposition to the Newtonian philosophy, and in pursuing a far-reaching *organic* explanation of nature.[5] I fully agree with you that there is an actually infinite number of created individual essences, not only in the universe but also here on our Earth, and, in all likelihood, even in every extended part of space however small.[6] I too hold that in order to obtain a satisfactory *explanation of nature*, one must posit the ultimate or properly *simple* elements of matter to be *actually infinite in number*.[7] In agreement with you, I call these simple elements of nature *monads* or *unities*. But since there are *two specific, different types of matter interacting with one another*, namely *corporeal matter* and *aetherial matter*, one must also posit two different classes of *monads* as foundations, *corporeal monads* and *aetherial monads*. From this standpoint the question is raised (a question that occurred neither to you nor to later thinkers): what *power* is appropriate to these types of matter with respect to their elements, insofar as they are considered *sets* of *corporeal* and *aetherial monads*. In this connection, I posit the corporeal monads, as discrete unities, to be as many as the natural numbers; and the aether, as continuous, to be composed of aetherial monads equinumerous with the points on a line. That is, I frame the hypothesis that the *power* of the corporeal monads is (what I call in my researches) the *first* power, i.e. $\aleph_0$, the first of the transfinite cardinal numbers; whilst the power of aetherial matter is the *second*, i.e. I posit the number of

---

[2]Cantor is here quoting from Leibniz's Letter to Foucher (January 1692), *Journal des Sçavans*, March 16, 1693, (GP I 416). He quotes this whole passage from p. 118 of Erdmann's collection of Leibniz's works in his "Grundlagen einer allgemeinenen Mannigfaltigkeitslehre" (Leipzig 1883), p. 179 in Cantor's *Gesammelte Abhandlungen* (hereafter cited as CGA). I shall adopt the convention of giving the original date of publication, but with page numbers referenced to CGA. Thus this would be (Cantor 1883, CGA 179).

[3]This continues the passage from Leibniz's Letter to Foucher quoted by Cantor (GP I 416).

[4]Cf. Dauben (1979, 292): "Leibniz was the source of greatest inspiration, as Cantor made clear by his choice of terminology: he called the ultimate components of matter which he hypothesized 'monads'".

[5](Cantor 1883, 177); quoted in (Dauben 1979, 293). Cf. Cantor's remarks about the organic world in the unpublished paper found by Ivor Grattan-Guinness (Grattan-Guinness 1970, 85): "Of special interest to me, however, is the application of mathematical type theory to study and research in the field of the organic."

[6]Cantor, Letter to Cardinal Franzelin, 22 January 1886 (CGA 399).

[7]This and all remaining quotations in this paragraph are culled from (Cantor 1885a, CGA 275–276).

aetherial monads to be equal to $\aleph_1$, the second cardinal number, which I believe (but have never been able to prove) to be the power of the continuum.[8]

Leibniz: I am most flattered that you honour me as the inspiration for your views. I also deeply sympathize with your opposition to the materialism implicit in the Newtonian philosophy, and your attempts to show the superiority of the organic philosophy I have always favoured. So it is with all the greater regret that I have to say that I must withhold my assent from these hypotheses of yours. For, setting aside my dissent from the notions that a continuum can be an actual existent, and that monads can compose a material continuum, it seems our deepest disagreement lies in the nature of the actual infinite itself. For where you believe in the existence of actually infinite numbers of monads I, on the contrary, believe that there is no such thing as infinite number, nor an infinite line nor any other infinite quantity, if one takes them as true wholes.[9]

Cantor: I am aware of your refusal to countenance infinite number. But I must confess I have never understood your position. For even though I found many places in **your** works where you come out against infinite numbers, I was still in the happy position of being able to find other pronouncements of yours where, seemingly in contradiction to this, you declare yourself to be unequivocally *for* the actual infinite (as distinct from the Absolute),[10] as in the case of the infinitude of substances we have already noted. But surely if there are actually infinitely many monads then there must be an infinite number of them! And what about your Infinitesimal Calculus? You must agree that this is a testament to the necessity of infinite number for mathematics!

Leibniz: On the contrary, in spite of my *Infinitesimal Calculus*, I admit no genuine (*véritable*) infinite number, even though I confess that the multiplicity (*multitude*) of things surpasses every finite number, or rather every number.[11] When it is said that there is an infinity of terms, it is not being said that there is some specific number of them, but that there are more than any specific number.[12] So, I understand a quantity to be *infinite* provided it is greater than any that can be assigned by us or designated by numbers.[13] Thus, properly speaking, it is true that there is an infinity of things, if this is understood in the sense that there are always more of them than can be assigned.[14]

Cantor: Here, it seems to me, you are following Aristotle. For he claimed that this was how mathematicians understood the infinite, as just a kind of potential for

---

[8]Cf. (Dauben 1979, 292).

[9]Leibniz, *New Essays on Human Understanding*, 1704, (A VI 6, 157/RB 157); quoted by Cantor (1883, CGA 179).

[10](Cantor 1883, CGA 179).

[11]Leibniz, Letter to Samuel Masson [1716], (GP VI 629).

[12]Leibniz, Letter to Johann Bernoulli, January 23rd 1699, (GM III 566).

[13]Leibniz, *De Quadratura Arithmetica*, (Leibniz 1993, 133).

[14]Leibniz, *New Essays*, chap. xvii: 'Of infinity'; (A VI 6157).

indefinite extensibility: "In point of fact, they neither need the infinite nor use it, but need only posit that a finite line may be produced as far as they wish" (*Physics*, 207b, 31–32). Likewise he allowed that one can have an infinite by division, since this amounts to saying only that no matter how many divisions one makes it is always possible to make more. But according to him this means that "this infinite is potential, never actual" (*Physics*, 207b, 12–13). However, you wish to claim that there is an actual infinity of divisions, not merely a potential one, or an indefinite one, as Descartes would say.[15]

Leibniz: Regarding your last remark, I have always held that Descartes' "indefinite" is not in the thing, but the thinker.[16] In any case, contrary to his own recommendation in Part I of his *Principles* that instead of the term 'infinite' we use the term 'indefinite', or that whose limits cannot be found by us, in Part II of the same work (§36) Descartes himself admits matter to be really divided by motion into parts that are smaller than any assignable, and therefore actually infinite.[17] Aristotle's views are more subtle. I think he is right to hold both that there is no infinite number, nor any infinite line or other infinite quantity, yet that there is an infinite by division, provided this is taken in the sense that there are so many divisions that there are always more of them than can be assigned. The Scholastics were taking that view, or should have been, when they allowed a *syncategorematic infinite*, as they called it, but not a *categorematic* one.[18] That is, to say that matter is actually infinitely divided is to say that there are infinitely many actual divisions, and this "infinitely many" is understood in the syncategorematic sense that "there are not so many divisions that there are not more" (*non sunt tot quin sint plura*).[19]

---

[15]Cantor equates the potential infinite with the indefinite at (CGA 373). See note 20 below.

[16]Leibniz, *Theory of Abstract Motion* (1671), §1, (LLC 339). O. Bradley Bassler argues that this is an early view that Leibniz comes to reject (Bassler 1998, n. 35). I demur: the parts of the continuum, it is true, Leibniz does come to regard as indeterminate. But the parts of matter are always for him determinate, and not at all indefinite; they are the result of an actually infinite division. This is a view I believe he never relinquishes from 1670 till his death.

[17]Leibniz, "Notes on Descartes' *Principles*", LLC, p. 25 (A VI 3, 214).

[18]Leibniz, *New Essays*, A VI 6, 157. Likewise, in his letter to Des Bosses of September 1, 1706 (GP II 314–15) Leibniz says: "*There is a syncategorematic infinite* ... But *there is no categorematic infinite*, or one actually having infinite parts formally". But one could certainly be forgiven for identifying this syncategorematic infinite with the potential infinite on the basis of this letter, for Leibniz defines it here as a "passive potential for having parts, namely the possibility of dividing, multiplying, subtracting, adding". See Bassler's erudite footnote on the syncategorematic and categorematic (Bassler 1998, 855, n. 150), the references cited therein, and (Uckelman 2015).

[19]For this quotation and an illuminating discussion see Philip Beeley (1996, esp. pp. 59-60). According to Beeley, Gregory of Rimini also held that "every continuum has a plurality of parts, and not so many finite in number that there are not more (*non tot finitas numero quin plures*), and has all its parts actually and at the same time" (p. 59); this is the syncategorematic infinite (although Gregory also upheld a categorematic infinity). Peter Geach, in his comments on an essay of Abraham Robinson's (Geach 1967, 41–42), points out that the high Scholastic meaning of 'syncategorematic' pertained to a certain way of using words, and sharply distinguishes the infinite in the syncategorematic sense as "there are infinitely many" from its assimilation to the potential

Indeed, accurately speaking, instead of an infinite number, we ought to say that there are more than any number can express, and instead of an infinite straight line that it is a straight line continued beyond any magnitude that can be assigned, so that a larger and larger straight line is always available.[20]

Cantor: With all due respect to the distinctions of the medieval schoolmen, this "syncategorematic infinite" appears to be nothing different from what I call the potential infinite.[21] The potential infinite means nothing other than an undetermined, variable quantity, always remaining finite, which has to assume values that either become smaller than any finite limit no matter how small, or greater than any finite limit no matter how great.[22] As an example of the latter, where one has an undetermined, variable finite quantity increasing beyond all limits, we can think of the so-called time counted from a determinate beginning moment, whereas an example of a variable finite quantity which decreases beneath every finite limit of smallness would be, for example, the correct presentation of your so-named differential.[23] There is no doubt that we cannot do without *variable* quantities in the sense of the potential infinite. But from this very fact the necessity of the actual infinite can be demonstrated.[24]

Leibniz: I would like to see the demonstration.

Cantor: Well, in order for there to be a variable quantity in some mathematical study, the domain of its variability must strictly speaking be known beforehand through a definition. However, this domain cannot itself be something variable, since otherwise each fixed support for the study would collapse. Thus this domain is a definite, actually infinite set of values. Hence each potential infinite, if it is rigorously applicable mathematically, presupposes an actual infinite.[25]

Leibniz: This claim that the potential infinite presupposes an actual infinite was, as I am sure you know, already made by that eminent French mathematician, Blaise Pascal. He claimed that the world could not be potentially infinite in spatial extent without this presupposing an actual infinite. But since our finite minds cannot entertain such an actual infinite without falling into contradiction, and yet we have this demonstration of its existence, he took this to show that

---

infinite by late scholastics like Suàrez (and, following them, Leibniz: see his letter to Des Bosses quoted in the previous footnote). But Leibniz's position is, I argue here, consistent with the more precise use of the term, as by Ockham. For a recent discussion, see (Uckelman 2015).

[20]Leibniz to Des Bosses, 11 March 1706, (GP II 304–05).

[21]Cantor equates the potential infinite with the syncategorematic in his 1885b: "That the so-called *potential* or *syncategorematic infinite* (Indefinitum) gives basis for no such division ..." [gibt zu keiner derartigen Einteilung Veranlassung ..."] (Cantor 1885b, CGA 373).

[22](Cantor 1887–8, CGA 409).

[23](Cantor 1887–8, CGA 401); freely adapted. Compare Hallett's translation (Hallett 1984, 12).

[24](Cantor 1886, 9); adapted from Hallett's translation (1984, 25).

[25](Cantor 1886, 9); adapted from Hallett's translation (1984, 25).

there are truths whose comprehension lies beyond the reach of finite minds. As I have argued, however, there is a way of understanding the actual infinite that is free of such contradiction, although I do not accept that this requires us to embrace actually infinite number, or a determinate infinite collection of values. This syncategorematic infinite, however should not be confused with the true or hypercategorematic infinite. The true [*veritable*] *infinite* is not found in a whole composed of parts. However, it is found elsewhere, namely in the *Absolute*, which is without parts, and which has influence on compound things since they result from the limitation of the absolute.[26]

Cantor: You will not find an opponent in me concerning the Absolute, which I agree cannot in any way be added to or diminished. It is therefore to be looked upon quantitatively as an absolute maximum. In a certain sense it transcends the human power of comprehension, and in particular is beyond mathematical determination.[27] But in this it is distinguished from that infinite which I call the Transfinite. This is in itself constant, and larger than any finite, but nevertheless unrestricted, increasable, and in this respect thus bounded. Such an infinite is in its way just as capable of being grasped by our restricted understanding as is the finite in its way.[28]

Leibniz: I too have always distinguished the Immensum from the Unbounded, and that to which nothing can be added from that which exceeds any assignable number.[29]

Cantor: This corresponds to my own view, if by *assignable* you mean *finite*. For by the actual infinite I understand a quantum that surpasses in magnitude every finite quantity of the same kind.[30]

Leibniz: This puts me in mind of a distinction I used to make among three different degrees of infinity. In descending order, they are (1) the *absolutely infinite*, (2) the *maximum*, or greatest of its kind, and (3) the *mere infinite*. The first or absolutely infinite is that which contains everything.[31] This kind of infinite is in God, since he is all one and contains the requisites for existing of all else.[32] I believe we in are agreement on this kind of infinite.

Cantor: We are indeed. What surpasses all that is finite and transfinite is … the single completely individual unity in which everything is included,

---

[26]"Quelques remarques sur le livre de Mons. Lock intitulé *Essay of Understanding*," 1695?-7, (A VI 6, 7). In the *New Essays*, Leibniz writes: "The true infinite, strictly speaking, is only in the *absolute*, which precedes all composition and is not formed by the addition of parts" (A VI 6, 157); and to Johann Bernoulli he writes on June 17th, 1698: "The real infinite is perhaps the absolute itself, which is not made up of parts, but which comprises beings having parts eminently and in proportion to their degree of perfection." (GM III 500).

[27](Cantor 1887–8, CGA 405); quoted in Hallett's translation, (1984, 13).

[28](Cantor, *Nachlass* VI, p. 99); quoted from (Hallett 1984, 14).

[29]Leibniz, "Annotated Excerpts from Spinoza", (A VI 3, 281/LLC 115).

[30](Cantor 1887–8, 401). This passage is quoted more fully below.

[31]Leibniz, "Annotated Excerpts from Spinoza", (A VI 3, 282/LLC 115).

[32]Leibniz, "On Spinoza and On the Infinite", (A VI 3, 385/LLC 43).

which includes the "Absolute", incomprehensible to the human understanding. This is the *"Actus Purissimus"* which by many is called God.[33]
But I am not sure we agree about the second, since I believe the only maximum
is God himself. But do finish your explanation.

Leibniz: Well, by the *maximum* I mean that to which nothing can be added: for
instance, a line unbounded on both sides, which is obviously infinite, since
it contains every length[34]; or the whole of space, which is the greatest of
all extended things.[35] But perhaps our views do not differ as much as you think,
for I agree with you that there is no maximum in things. For I hold that two
notions that are [equally] excluded from the realm of intelligibles, are that
of a minimum, and that of a maximum: what lacks parts, and what cannot
be part of another.[36]

Lastly, there is the *infinite in lowest degree*,[37] which I usually call
the *mere infinite*,[38] which is that whose magnitude is greater than we
can expound by an assignable ratio to sensible things. An example
would be the infinite space comprised between Apollonius's Hyperbola
and its asymptote, which is one of the most moderate of infinities, to
which there somehow corresponds in numbers the sum of this space:
$^1/_1 + ^1/_2 + ^1/_3 + ^1/_4 + \ldots$, which is $^1/_0$. Only instead of this 0, or nought,
let us rather understand instead a quantity infinitely or unassignably
small, which is greater or smaller according as we have assumed the
last denominator of this infinite series of fractions (which is itself also
infinite) smaller or greater. For a maximum does not apply in the case
of numbers.[39]

Cantor: I take exception to the idea that the common infinite is greater than
expressible by a ratio to anything sensible: one must distinguish between
numbers as they are *in and for themselves, and in and for the Divine
Intelligence,* and how these same numbers appear in our restricted,
discursive comprehension and are differently defined by us for systematic
or pedagogical purposes.[40] Also, I am glad we agree that a maximum does
not apply to numbers, and I even agree with you that there is no number of all
numbers. But regarding natural numbers, I do not believe that their infinity is
"unassignable". I submit that one can indeed assign a number to all the natural
numbers, just as one can assign a limit to a converging infinite sequence even

[33]Cantor, Letter to G. C. Young, June 20, 1908; quoted from (Dauben 1979, 290).

[34]Leibniz, "Annotated Excerpts from Spinoza", (A VI 3, 282/LLC 115).

[35]Leibniz, "On Spinoza and On the Infinite", (A VI 3, 385/LLC 43).

[36]Leibniz, "On Minimum and Maximum", (A VI 3, 98/LLC 13).

[37]Leibniz, "Annotated Excerpts from Spinoza", (A VI 3, 282/LLC 115).

[38]Leibniz, "On Spinoza and On the Infinite", (A VI 3, 385/LLC 43).

[39]Leibniz, "Annotated Excerpts from Spinoza", (A VI 3, 282/LLC 115).

[40]Cantor, *Nachlass* VII, p. 3; quoted from (Hallett 1984, p. 28).

when there is no last number in that sequence. In each case it is the *first* whole number which follows all the numbers.[41]

Leibniz: I remain to be convinced that one can assign such numbers. I concede the infinite multiplicity of terms, but this multiplicity does not constitute a number or one whole; it signifies nothing but that there are more terms than can be designated by a number. Just so, there is a multiplicity or complex of all numbers, but this multiplicity is not a number, nor is it one whole.[42]

Cantor: You are hardly alone, although lately the weight of opinion seems to have come around to my way of thinking. But in my view all so-called proofs against the possibility of infinite number are faulty. As can be shown in every particular case, and also concluded on general grounds, their chief failing, and where their πρῶτον πςεῦδος [initial mistake] lies, is that from the outset they expect or even impose all the properties of finite numbers upon the numbers in question.[43] As an example, consider the following argument by Tongiorgi:

> Let us suppose that an infinite multiplicity is constructed by the accumulation of ones. This will be an infinite number, and will be equal to *the number which immediately precedes it* with one added to it. Now was *the preceding number* infinite, or was it not? You cannot say it is infinite, for it could be further increased, and was in fact increased by the addition of one. Therefore it was finite, and with one added, it became infinite. Yet from two finites an infinite has emerged; which is absurd. (§350; 2^0 Pesch §412, 3^0, 4^0)

Here it is falsely assumed (because one is accustomed to this with finite numbers) that an actually infinite number must necessarily have a whole number that is the first to precede it, from which it results by the addition of one.[44] But according to my theory of the transfinite, the smallest transfinite ordinal number w is preceded by all the finite ordinal numbers, which have no maximum. Thus this actually infinite number is not a natural number (or "inductive number", as Russell called it), since it succeeds all natural numbers and is regarded as the *limit* of those numbers, *i.e.* is defined as the next greater number than all of them.[45]

---

[41](Cantor 1883, 195).

[42]Leibniz, Letter to Bernoulli, 21st February 1699 (GM III 575). On Leibniz's philosophy of mathematics, the reader is referred to the excellent discussion by Samuel Levey (1999). He quotes this passage on p. 139.

[43]Cantor, "Über die verschniedenen Standpunkte . . .", (CGA 370–77,371–2); cf. (Dauben 1979, 125); Cantor 1887–8, CGA 396).

[44](Cantor 1887–8, CGA 394). I have translated the quotation from Tongiorgi out of the Latin in which Cantor had left it.

[45](Cantor 1883, 196).

Leibniz: I can appreciate why you would want to exclude actually infinite number from the sequence of all natural numbers. For I once thought I could prove that the number of finite numbers cannot be infinite[46] by a similar proof as follows: If numbers can be assumed as continually exceeding each other by one, the number of such finite numbers cannot be infinite, since in that case the number of numbers is equal to the greatest number, which is supposed to be finite. But then I realized that to this argument it must be responded that there is no such thing as the greatest number[47]; this kind of demonstration proves only that such a series is unbounded.[48]

Cantor: But this is music to my ears! So many of the so-called finitists have failed to appreciate this unboundedness, and have assumed that if the natural numbers are actually infinite, there must be a greatest such number.[49]

Leibniz: Yet it seems to me that a similar argument could be applied to the idea that there could be a greatest number even if the parts are *actually infinite in number*. For suppose I accept that numbers do go to infinity when applied to a certain space or unbounded line divided into parts. Now here there is a new difficulty. Is the last number of a series of this kind the last one that would be ascribed to the divisions of the unbounded line? It is not, otherwise there would also be a last number in the unbounded series. Thus if you say that in an unbounded series there exists no last finite number that can be written in, although there can exist an infinite one: I reply, not even this can exist if there is no last number.[50]

Cantor: I agree that in a series such as $^1/_2$, $^1/_4$, $^1/_8$, $^1/_{16}$, $^1/_{32}$, etc., there is no last number, not even an infinite one. If there were, it would be an actual infinitesimal. But there are no such things.

Leibniz: I am glad we agree! I had occasion to argue this with one of the most eminent mathematicians of my time, Johann Bernoulli. He claimed that it was

---

[46]Leibniz, "On the Secrets of the Sublime", (A VI 3, 477/LLC 51).

[47]The importance of this proof has been noted by Bassler, who observes that here "Leibniz distinguishes between there being no greatest number and the number of finite numbers being infinite" (Bassler 1998, 850).

[48]Leibniz, "On the Secrets of the Sublime", (A VI 3, 477/LLC 53). Cf. also *De Quadratura Arithmetica*, p. 133: "Thus I call *unbounded* that in which no last point can be assumed, at least on one side; whereas we understand a quantity, whether bounded or unbounded, to be *infinite*, provided it is greater than any that can be assigned by us or designated by numbers." This is part of an addition to the Scholium that Leibniz subsequently cancelled.

[49]An example is given by Bassler in fn. 15 of his article, quoting the following sophism of Henry de Gand: "Infinities are finite. Proof: two are finite, three are finite, and so on to infinity; therefore infinities are finite." This assumes that there is an infinitieth finite number.

[50]Leibniz, "Infinite Numbers", (A VI 3, 503/LLC 99). On Leibniz's proof in "On the Secrets of the Sublime" that the number of finite numbers cannot be infinite, Bassler comments: "what he does not consider is the possibility of nonetheless admitting that there is an (indefinite) infinity of finite numbers" (1998, 853, n. 12). But this passage from "Infinite Numbers", written only a few weeks later, shows that Leibniz did consider the possibility of an infinity of finite numbers, in the sense of there being more than can be numbered.

inconsistent of me to claim that any finite portion of matter is already actually divided up into an infinity of parts, and yet deny that any of these parts is infinitely small.[51] To this I replied that, even if we suppose a line to be divided into fractions of its length so that its $^1/_2$, $^1/_4$, $^1/_8$, $^1/_{16}$, $^1/_{32}$, etc., are actually assigned, and all the terms of the series actually exist, it still does not follow that there also exists an infinitieth term.

Cantor: Still, I hold that there can be infinite sequences such that there is an infinite number *after* the whole sequence. You don't know how many attempted proofs I have seen of the impossibility of infinite number that begin by presupposing that an actual infinite *must be unincreasable* in magnitude; an erroneous assumption that one finds propagated not only in the *old* philosophy following on from that of the *Scholastics*, but even in the *more recent* and, one can almost say, *most recent* philosophy.[52] On this basis it is easy to prove that actually infinite number leads to contradiction. I find articulated in several places in Gutberlet's *Works*, for example, the wholly untenable thesis that "in the concept of supposed infinite magnitudes lies the exclusion of all possibility of increase". This can be admitted only in the case of the Absolute Infinite. The transfinite, by contrast, although conceived as definite and greater than every finite, shares with the finite the character of unbounded increasability.[53]

Leibniz: I do not understand how this increasability is compatible with the unvarying nature of the infinite you spoke of above when you claimed the potential infinite presupposed a definite, actually infinite set of values.

Cantor: By an Actual Infinite I understand a quantum that is *not variable*, but rather fixed and determined in all its parts, a genuine constant, but which at the same time exceeds in magnitude *every finite quantity* of the same kind.[54]

Leibniz: But in what sense can the same infinite number be "fixed and determined in all its parts", a *genuine constant*, and at the same time be increasable?

Cantor: In exactly that sense that 3 can be increased to larger numbers by the addition of new unities.[55]

Leibniz: But when a unity is added to 3, it is no longer 3 but 4. So it is not 3 that is increasable, but the series of which it was supposed to be the last number. This I grant. Thus concerning the infinite series of numbers we were discussing above, the only other thing I would consider replying to that line of reasoning is that the number of terms is not always the last number of

---

[51] This and the following quotations are from Leibniz's letter to Johann Bernoulli of August (?) 1698, (GM III 536).

[52] (Cantor 1887–8, CGA 404–5); cf. (Hallett 1984, 41).

[53] (Cantor 1887–8, CGA 394); quoted from (Hallett 1984, 41). The reference is to Gutberlet's *Das Unendliche metaphys. Und mathem. Betrachtet* (Mainz 1878).

[54] (Cantor 1887–8, CGA 401); cf. (Hallett 1984, 12).

[55] (Cantor, *Nachlass* VI, 47–48); quoted from (Hallett 1984, 41).

the series. That is, it is clear that even if finite numbers are increased to infinity, they never—unless eternity is finite, i.e. never—reach infinity. This consideration is extremely subtle.[56]

Cantor: But here we are *in* eternity, my dear Leibniz, so we should be able to find out! Seriously, though, I am amazed at the progress you had already made in this kind of reasoning in 1676. It is such a shame that you did not publish more of your reflections. For instance, in what you say here about the series (I) of positive real whole numbers 1, 2, 3, ..., ν, ..., you anticipate me by recognizing that the number [*Anzahl*] of so constituted numbers ν of the class (I) is infinite even though there is no greatest among them. For however contradictory it would be to speak of a greatest number of class (I), there is yet nothing objectionable in conceiving of a *new* number, which we will call ω, which is to be the expression of the fact that the whole domain (I) [of the positive real whole numbers] is given in its natural succession according to law.[57] One might also say that in distinguishing the number of terms of (I) from the last term of the series, you come very close to anticipating my distinction between *cardinal* and *ordinal* numbers[58]—with the exception that what I call the first infinite ordinal number ω is not the last term in the series, but the first whole number which follows all the numbers ν of the series, that is to say, greater than each of the numbers ν.[59]

Leibniz: I admit that lack of time is no longer an obstacle for us. In fact I have taken advantage of our situation to study your "Characteristic of the Transfinite", if I may call it that. Cardinality, if I understand correctly, measures *how many* are in a given collection, without regard to their order; whereas when order is taken into account, many different infinite collections of the same cardinality have different ordinal numbers. But I regret to report that I am in no way persuaded of the existence of infinite numbers of either kind.

Cantor: Given time—or its complete absence—I am sure I can persuade you! The reason I am so confident is that you, like me, believe in the Principle of Plenitude.

Leibniz: You mean the principle that God would create as much as possible. This principle I also call the harmony of things: that is, that there exists the greatest quantity of essence possible. From which it follows that there is

---

[56]Leibniz, "Infinite Numbers", (A VI 3, 504/LLC 101).

[57](Cantor 1883, 195).

[58]Commenting on the same passage, Samuel Levey has noted Leibniz's near anticipation of the cardinal/ordinal distinction: "In seeing his way clear to the fact that the number of terms in the series of natural numbers—the cardinality of the naturals—is not itself in the series, but rather lies outside it, Leibniz places himself well ahead of the majority of his peers on the topic. Further, taken at face value, his claim that 'the number of terms is not always the last number of the series' touches quite directly on the concept of cardinality, and conceives of a series' cardinality as a *number.* In the crucial case of the infinite series, the number of terms is the *cardinal number* "infinity" (waiving Cantor's distinctions between higher and lower transfinite cardinals), despite the fact there is no corresponding infinitieth element in the series." (Levey 1998, 84).

[59](Cantor 1883, 195).

more reason for existing than not existing, and that all things would exist if that were possible.[60]

Cantor: Precisely! Now, in the transfinite there is a vastly greater abundance of forms and of *species numerorum* available, and in a certain sense stored up, than there is in the correspondingly small field of the unbounded finite. Consequently, these transfinite species were just as available for the bidding of the Creator and his absolutely inestimable will power as were the finite numbers.[61] Otherwise put: since God is of the highest perfection one can conclude that it is possible for Him to create not just a *Finitum ordinatum*, but a *transfinitum ordinatum*. Therefore, in virtue of His pure goodness and majesty, we can infer the necessity of the actually resulting creation of a *transfinitum*.[62]

Leibniz: A necessity in virtue of God's goodness is what I call a hypothetical necessity. But you have not yet established even the possibility of his creating your *transfinitum*.

Cantor: I urge you to appreciate that these transfinite cardinal and ordinal numbers possess a definite mathematical uniformity, just as much as the finite ones do, a uniformity that is discoverable by men. *And all these particular modes of the transfinite have existed from eternity as ideas in the Divine Intellect.*[63] Such a transfinite imagined both *in concreto* and *in abstracto* is free from contradiction, and thus possible, and therefore just as much creatable by God as a finite form.[64]

Leibniz: One can imagine that it is possible that $1 = 2$ if division by zero is allowed. This does not mean that it is something that God could instantiate in the world.

Cantor: Then what sense can be made of your claim that matter is actually infinitely divided if there is not an actually infinite collection of the parts that have been divided?

Leibniz: Exactly the sense that I was trying to explain before, when I spoke of the Scholastic distinction between the categorematic and the syncategorematic. You made light of this distinction, assimilating the latter to the potential infinite. But it seems to me the distinctions of the Scholastics are not entirely to be scorned. For I believe that there is an actual infinity of parts into which any piece of matter is actually (not merely potentially) divided, but that there is no totality or collection of all these parts. And this kind of actual infinite can be distinguished from the actual infinite understood categorematically using what is now called predicate calculus.

Cantor: How? I would like to see how this could be done.

---

[60] Leibniz, "On the Secrets of the Sublime", (A VI 3, 472/LLC 45).

[61] (Cantor 1887–8, CGA 404, fn.); (Hallett 1984, 23).

[62] (Cantor 1887–8, CGA 400); (Hallett 1984, 23).

[63] Cantor, Letter to Father Ignatius Jeiler, 13 October 1895: *Nachlass* VII, 195; quoted from (Hallett 1984, 21).

[64] Cantor, Letter to Jeiler, 1895: *Nachlass* VII, p. 195; (Hallett 1984, 20). Cf. (CGA 396).

Leibniz: I was on the point of showing you. Here, I'll write on this slate so you can see. To say of the prime numbers, for instance, that there is an actual infinity of them understood *syncategorematically*, is to say that no matter how large a number N one takes, there is a number of primes M, where $M > N$. Thus if x and y are numbers that can be assigned to count the primes, with x finite, then [*sound of writing on a chalkboard*] $\forall x \exists y \, y > x$. But to assert their infinity *categorematically* would be to assert that there exists some one number of primes y which is greater than any finite number x, i.e. that [*sound of writing*] $\exists y \forall x \, y > x$. Thus the first way of expressing it does not commit you to infinite number, since x and y may both be finite. But the categorematic expression commits you to the existence of a number greater than all finite numbers.

Cantor: But if x and y may both be finite, in what sense does the first expression commit you to an actual infinite? This seems to me to be the potential infinite, beloved of the constructivists.

Leibniz: The primes are actually infinitely many in precisely the sense that if one assumes that their multiplicity is finite one may derive a contradiction, as did Euclid in his proof.

Cantor: But if the assumption that the number of primes is finite leads to a contradiction, why should one not conclude that their number is infinite, i.e. is an infinite number?

Leibniz: I grant you that there are actually infinitely many. But from Euclid's proof one cannot validly conclude that there is a categorematic infinity. For he begins his proof by supposing that there is a greatest prime, i.e. that there is some prime such that all primes are less than or equal to it. That is, if x and y are numbers that can be assigned to count the primes, and $Fx = x$ is finite, then there is an x such that any y different from x must be less than or equal to it: in symbols, $\exists x \forall y [Fx$ & $(y \neq x \rightarrow y \leq x)]$.[65] Then he shows how to construct a prime greater than this, thus contradicting the initial supposition. (One simply forms the product of this supposed greatest prime with all the preceding primes and adds one; this new number is not divisible by any prime without remainder, and is therefore a prime number greater than the supposed greatest.)

Cantor: Granted.

Leibniz: But the negation of the original supposition is [*more sounds of writing*] $\neg \exists x \forall y [Fx$ & $(y \neq x \rightarrow y \leq x)]$. This is equivalent to $\forall x \exists y \neg [Fx$ & $(y \neq x \rightarrow y \leq x)]$, or $\forall x \exists y [Fx \rightarrow \neg (y \neq x \rightarrow y \leq x)]$, or $\forall x \exists y [Fx \rightarrow (y \neq x$ & $y > x)]$: for any finite number of primes, there is a number of primes different from and greater than this. But this is the syncategorematic infinite, and from this one cannot infer the categorematic infinite, i.e. that there is a number of primes different from and greater than any finite number: "$\forall x \exists y \, y > x$" *ergo* "$\exists y \forall x \, y > x$", *non valet*.[66]

---

[65]I am guessing from the context that these (or something similar) are the formulas Leibniz wrote.

[66]I am indebted to my colleagues Nick Griffin, Gregory Moore and David Hitchcock for observing that the finiteness of x and its distinctness from y needed to be made explicit in this proof.

Cantor: I think this is now usually called the *quantifier shift fallacy*.

Leibniz: I believe so. I once suspected the great Locke to be guilty of it when he reasoned: "Bare nothing cannot produce any real being. Whence it follows with mathematical evidence that something has existed from all eternity."[67] I find an ambiguity here. If by 'something has existed from all eternity' he means that *there has never been a time when nothing existed*, then I agree with it, and it does indeed follow from the previous propositions with mathematical evidence. But if there has always been something it does not follow that one particular thing has always been, i.e. that there is an eternal being.[68] Still, if Locke did commit this error, it was not because he was deficient in logic: we can all reason badly, no matter how eminent we may be as logicians. Even Bertrand Russell committed this fallacy, and it would be hard to find a more distinguished logician than him.

Cantor: I am surprised to hear that! Lord Russell was, next to Ernst Zermelo, perhaps the ablest champion of my theory of the transfinite, and his and Whitehead's *Principia Mathematica* was a pinnacle in the history of logic.

Leibniz: Yet in his *Principles of Mathematics* (1903) he wrote: "Of some kinds of magnitude (for example ratios, or distances in space and time), it appears true that [i] there is a magnitude greater than any given magnitude. That is, [ii] any magnitude being mentioned, another may be found greater than it."[69] Now 'That is' is a particle denoting logical equivalence; but although [ii] follows from [i], to assert that the converse is valid is to commit the quantifier shift fallacy. Russell conflates the syncategorematic with the categorematic infinite.[70]

Cantor: I am grateful here for your indirection or tact in not accusing me of this fallacy. Actually, I do not say that the transfinitude of the primes (for example) directly follows from the falseness of the assumption of their finitude. Rather I prefer to say that, just as the actual infinity of monads is the foundation of extended matter, so the categorematic infinite is the foundation of the syncategorematic. There would not be an infinity of divisions of the continuum without an actually infinite collection of monads, each one underpinning a given part.

Leibniz: I know you put much stock in the fact that your transfinite numbers would be instantiated in reality ...

---

[67]Leibniz, *New Essays*, (A VI 6, 435); quoted from Locke, Essay, Bk. IV, ch. x, §3.

[68]Leibniz, *New Essays*, (A VI 6, 436). In terms of predicate logic, if $xTy = x$ exists at time y, then *there has never been a time when nothing existed* is $\neg\exists y \forall x \neg xTy$, which, as Leibniz says is equivalent to $\forall y \exists x\, xTy$, *there has always been something*. But to infer from this that $\exists x \forall y\, xTy$, *one particular thing has always been*, is to commit the quantifier shift fallacy.

[69](Russell 1903, 188).

[70]As has been pointed out to me by Wayne Myrvold and Dean Buckner, in the light of his subsequent discussion (188–89) my 'Leibniz' should probably have given Russell the benefit of the doubt here: Russell likely intended (ii) as a more exact expression of what was expressed somewhat stiltedly in (i), rather than as an equivalent form.

Cantor: Yes, that is so! I believe the suppression of the legitimate actual infinite can be viewed as a kind of shortsightedness that robs us of the possibility of seeing that, just as the actual infinite in its highest, absolute manifestation has created and sustains us, so in its secondary transfinite form it occurs all around us and even inhabits our minds themselves.[71] So, for instance, the various number-classes (I), (II), (III) etc. are representatives of powers which are actually found in corporeal and intellectual nature.[72] For apart from pure mathematics (which in my view is nothing other than pure set theory),[73] there is also *applied set theory*, by which I understand what one cares to call *theory of nature*, or *cosmology*, to which belong all so-called natural sciences, those relating to both the inorganic and the organic world … [74]

Leibniz: Yes, yes, but what if I were to demonstrate to you that the actually infinite division of matter, as I conceive it, while consistent with my actual infinite that is syncategorematically understood, is incompatible with there being a transfinite number of parts?

Cantor: Please give the demonstration, and I will comment when I see it.

Leibniz: According to my conception, the parts of matter are instituted and individuated by their differing motions. Each body or part of matter would be one moving with a motion in common. But this does not rule out various parts internal to that body having their own common motions, which effectively divide the body within. Accordingly I ask that I be allowed the following three premises:

(1) Any part of matter (or *body*) is actually divided into further parts.
(2) Each such body is the *aggregate* of the parts into which it is divided.
(3) Each part of a given body is the result of a division of that body or of a part of that body.

Cantor: Go on.

Leibniz: From (1) and (2) it follows that every body is an aggregate of parts. Incidentally, if one accepts the further premise that

(4) What is aggregated cannot be a substantial unity.

this immediately yields one of my chief doctrines, that matter, being an aggregate, cannot be substantial. But I will not pursue that further here.

To resume: we have the result that every body is an aggregate of parts, each of which is an aggregate of further parts, and so on to infinity. That is, by recursion,

---

[71](Cantor 1885b, 374–5); cf. Rudy Rucker's translation (Rucker 1983, 46).

[72](Cantor 1883, 181); cf. (Hallett 1984, 18), and the distinction between the two kinds of reality of numbers: *intrasubjective* or *immanent reality* and *transsubjective* or *transient reality*.

[73]Cantor, "Principien einer Theorie der Ordnungstypen" [1885]; in (Grattan-Guinness 1970, 84).

[74]Cantor, "Principien", in (Grattan-Guinness 1970, 85).

(5) Everybody is actually infinitely divided.

Cantor: —in your syncategorematic sense.

Leibniz: Exactly. And from (1) and (3) it follows that

(6) Any part of a given body must be reachable by repeated division of the original body.
This is the crux of the matter—each part of a body must be reachable by repeated division of the original body. It is either a part, a part of a part, or a part of a part of a part, and so on. This makes for a recursive connection between a body and any of its parts: Call P_r any part reachable by r divisions of the original body; then by (6)

(7) For all $r > 1$, the part P_r must have resulted from the division of a part P_{r-1}.

Cantor: Here it seems to me that in introducing this concept of "reachable by division" you are illicitly appealing to time. I must declare that in my opinion reliance on the concept of time or the intuition of time in the much more basic and general concept of the continuum is quite wrong.[75]

Leibniz: I am sorry if I gave the impression that this "reaching" a part by a division is a process occurring *in time*. Indeed, it cannot be, since each motion causing these divisions within matter is an instantaneous motion. It is just that in order for a piece of matter to be a part of the original body, it must be a result of a division of the body.

Cantor: Very well. But it still seems to me that if the body is actually infinitely divided at an instant, then all the infinite divisions simply are there. And if this is so, then there is a number of divisions ω, greater than any finite number r of divisions. Here ω is the first ordinal number $> r$ for all finite numbers r, i.e. the first transfinite ordinal. But you were going to prove that this is incompatible with actually infinite division?

Leibniz: Yes. Here is the demonstration, which I have adapted from one communicated to me by a scholar from Spain.[76] Suppose that a given body has an actual infinity of parts in your categorematic sense, i.e. suppose that there is a number of divisions ω, greater than any finite number r of divisions. Let P_ω be a part resulting from the ωth division ...

Cantor: Excuse me for interrupting, but here I must correct you. If there is a number of divisions ω, this means that there will be a well-ordered sequence of divisions of order type ω. But in that case there will be a number of divisions greater than any finite number, but no infinitieth division, or part P_ω. More formally, the ω-sequence of parts $\{P_1, P_2, P_3, \ldots\}$ does not issue in an ωth part P_ω: that would require a sequence of divisions of order type $\omega + 1$, namely $\{P_1, P_2, P_3, \ldots,$

---

[75](Cantor 1883, CGA 191), quoted from the translation in (Hallett 1984, 15).

[76]I owe this form of argument to Antonio Leon, who used it in his article of his published on the internet which he generously brought to my attention: "On the infiniteness of the set of natural numbers", http://www.terra.es/personal3/eubulides/infinit.html (now expired) He interpreted it to show that the natural numbers cannot be infinite in Cantor's sense.

P_ω}. This relates to what we were discussing earlier, when I agreed with your answer to Bernoulli.

Leibniz: I take your point. Still, if I understand you correctly, the transfinite really *begins* with ω: the first transfinite division would be the first after all the finite ones, assuming there is such a thing.

Cantor: This is true in the sense that ω is the first ordinal number after all the finite ordinal numbers. That it exists follows from the *second principle of generation* of whole real numbers, which I define more exactly as follows: if there is some determinate succession of defined whole real numbers, among which there exists no greatest, on the basis of this second principle of generation a new number is obtained which is regarded as the *limit* of those numbers, i.e. is defined as the next greater number than all of them.[77] Thus ω is the first ordinal greater than all the natural numbers in their natural sequence.

Leibniz: This is what I had in mind by referring to a "transfinite division". So long as we are considering a part P_r with r finite, we have not yet attained your realm of the transfinite. But if we suppose that there is a part P_t occurring as a result of a transfinite division, so that t is the first ordinal greater than every finite r, then $t = \omega$.

Now I adapt my Spanish correspondent's proof as follows: suppose there is a division issuing in an ωth part P_ω, where ω is the least ordinal number greater than any finite number r. The point is that by 7) above, P_ω must have resulted from the division of a part $P_{\omega-1}$. But ω - 1 cannot be infinite, otherwise ω would not be the first ordinal number $> r$ for all r, contrary to your supposition. But neither can it be finite, since one cannot obtain an actually infinite division by one more division of a finite aggregate of parts. Therefore there cannot be a transfinite division, i.e. one issuing in an ωth part P_ω.

Cantor: It seems to me that all your friend has proved by this kind of argument is that one cannot obtain the transfinite through ordinary recursion from the finite.

Leibniz: I think he has shown much more. For recursive connectibility is a crucial property of numbers for many mathematical proofs, and the demonstration shows that it is lacking in your transfinite ordinals.

Cantor: Here it seems you are unaware of my generalization of ordinary, Fermatian recursion. In standard proofs by mathematical induction, one standardly shows first that F holds for some first natural number, and then that if F holds for n, it also holds for $n + 1$, and concludes that it must hold for any natural number. But there is a generalization of this which we may call by the happy name "transfinite recursion". Here one first shows that F holds for 0, and then that, for any ordinal number k, if F holds for all ordinals less than k, it also holds for k, and concludes that it must hold for all ordinals. It is readily seen that this is applicable to all

---

[77](Cantor 1883, 196).

ordinals, transfinite as well as finite, so that Fermatian recursion is but a special case.[78]

Leibniz: That is very ingenious, and I thank you for enlightening me. But note that this generalized type of recursion only applies to the transfinite on the hypothesis that transfinite numbers exist. In this it differs from standard recursion, which in a sense generates the applicability of F to the successor numbers. Similarly, as I understand the notion of a 'part', it must be generated by a division, and this requires standard recursion. But a "transfinite part", if I may call it that, is separated from the whole of which it is putatively a part by a kind of rift. This rift between the finite and the transfinite is, in my view, a mark of the inapplicability of the transfinite to reality, at least in this case of infinite division.

Cantor: I cannot agree, for it remains the case that infinite division as described by you can be represented in my theory by an actually infinite ω-sequence of parts $\{P_1, P_2, P_3, \ldots\}$ whose power or cardinality is $\aleph_0$. Indeed it must be so represented if the division is completed. If, however, the division is incomplete, then the infinity in question is merely potential. It is one whose parts are variable quantities which are decreasing towards any arbitrary smallness, but which always remain *finite*. I call this infinite the *improper infinite*.[79] Thus I still am not convinced that your syncategorematic but actual infinite is not a contradiction in terms. Are you not presupposing the categorematic actual infinite when you use the universal quantifier?

Leibniz: I allow that this is how it is usually understood. But one may interpret it, as I have here, to mean "For each x ..." rather than "For all x ..." In this way one sees the link with another useful distinction from the Schools, namely that between the distributive and the collective use of a term like "every". For when we say "every man is an animal" or "all men are animals", the acceptation is distributive: if you take that man (Titius) or this man (Caius), you will discover him to be an animal.[80] According to this distinction, then, there is an actual infinite in the mode of a distributive whole, not in that of a collective

---

[78]This is a somewhat simplified version of transfinite induction (not Cantor's own term), which I owe to Wayne Myrvold. I am also in his debt for alerting me to the theory, and to its significance for this discussion. For Cantor's rigorous formulation, see his "Beiträge (1895–97)", (CGA 282–356).

[79](Cantor 1883, 165).

[80]Leibniz, Preface to an Edition of Nizolius (1670); quoted from *Philosophical Papers and Letters*, ed. and trans. Leroy Loemker (Dordrecht: D. Reidel, 1969), p. 129. Leibniz goes on to say that "If ... 'every man', or 'all men', is taken as a collective whole, and the same as the whole genus man, then an absurd expression will result ...: 'the whole genus man is an animal'." Of course, with the replacement of the logic of class inclusion by predicate logic, the criticism no longer applies in this form. On the one hand, Leibniz's analysis of an A-statement as a (possibly infinite) conjunction of conditionals is vindicated; on the other, one may interpret an infinite domain of the universal quantifier as an actually infinite set (as do Platonist mathematicians), or as a potentially infinite set (as do constructivists), or as an actual infinite understood distributively (as does Leibniz).

whole. Thus something can be enunciated concerning all numbers, but not collectively.[81]

Cantor: I disagree. If I say "The apostles are 12", the number 12 does not hold of the apostles individually.[82] Any number applied to a multiplicity must be understood collectively: there is a 1–1 correspondence between this multiplicity *M* (collectively) and any other multiplicity of 12 elements. The cardinality of a set is what you get by abstracting away from all other properties. If with a given set *M*, which consists of determinate, well-differentiated concrete things or abstract concepts which are called the elements of the set, we abstract away not only from the particular character of the elements but also from the order in which they are given, there arises in us a determinate general concept that I call the *power* of *M* or the *cardinal number* of the set *M* in question.[83] Thus, for example, the set of colours of the rainbow (red, orange, yellow, green, blue, indigo, violet) and the set of pitches in the octave (C, D, E, F, G, A, B) are equivalent sets and both stand under the same cardinal number *seven*.[84]

Leibniz: I allow that *numbers* must be understood collectively when applied to finite multiplicities. But the universal quantifier is not always used collectively, even when the collection is finite. For instance, when one says "All the children in my class weigh less than 40 kg," one most likely means that each child does, not that all the children put together do. Nothing is being said about the collection of children in the room: the quantifier is being used distributively. It is similar when the universal quantifier is used with an infinite multiplicity, like the natural numbers. Here something can be enunciated concerning all numbers, but not collectively. Thus it can be said that to every even number there is a corresponding odd number, and vice versa; but it cannot accurately be said that the multiplicity of even numbers is therefore equal to that of odd ones.[85]

Cantor: I grant you that we cannot infer the existence of infinite collections from a proof that they are not finite, since that would involve the fallacy we noted above. But the existence and equality of infinite collections can still be posited. Indeed, I believe that this is how we must proceed. Mathematics is completely free in its development, and bound only by the self-evident consideration that its concepts must not only be free from contradiction but also stand in ordered relations, fixed through definitions, to previously formed concepts that are already present and tested.[86]

---

[81](GP II 315)

[82]I am indebted to Massimo Mugnai for this objection (private correspondence).

[83](Cantor 1887–8, CGA 411).

[84](Cantor 1887–8, CGA 412).

[85]Leibniz, (GP II 315)

[86](Cantor 1883, CGA 182); cf. (Hallett 1984, 16).

Leibniz: I agree with this. But I do not believe these very concepts, the number of all even numbers, and the number of all numbers, are consistent with "concepts already present and tested". For the number of all numbers implies a contradiction, which I show thus: To any number there is a corresponding number equal to its double. Therefore the number of all numbers is not greater than the number of even numbers, i.e. the whole is not greater than its part.[87] Galileo showed the same thing using the number of all squares, and avoided the contradiction by denying that "equal to" and "lesser than" can be applied to the infinite. Here the axiom on which the proof depends, *Totus est majus sua parte*, is not only "present and tested", but something I believe I was the first to prove.

Cantor: The old and oft-repeated proposition "*Totum est majus sua parte*" may be applied without proof only in the case of *entities* that are based upon whole and part; *then* and *only then* is it an undeniable consequence of the concepts "totum" and "pars". Unfortunately, however, this "axiom" is used innumerably often without any basis and in neglect of the necessary distinction between "reality" and "quantity", on the one hand, and "number" and "set", on the other, precisely in the sense in which it is *generally false*.[88] Your example may help to explain. Let *M* be the totality (v) of all finite numbers v, and *M'* the totality (2v) of all even numbers 2v. Here it is undeniably correct that *M* is *richer* in its entity, than *M'*; *M* contains not only the even numbers, of which *M'* consists, but also the odd numbers *M''*. On the other hand it is just as unconditionally correct that *the same* cardinal number belongs to both the sets *M* and *M'*. Both of these are certain, and neither stands in the way of the other if one heeds the distinction between *reality* and *number*.[89]

Leibniz: This puts me in mind of an objection raised by a distinguished metaphysician of more recent times, who accused me of equivocation in deriving a contradiction from cases such as this.[90] For, he claimed, the sense in which the number of even numbers can be said to be less than the number of numbers is according to a criterion of equality that "if B is a proper subset of A, then there is a greater number of elements in A than in B"; whereas the sense in which they are equal is that their elements can be put into one-one correspondence. Once this difference in the meaning of equality is taken into account, he alleged, "there is no contradiction, as Leibniz [had] supposed [there to be]".[91]

---

[87]Leibniz, (GP I 338).

[88](Cantor 1887–8, CGA 416–417).

[89](Cantor 1887–8, CGA 412).

[90](Benardete 1964, 44–48). I am indebted to Mark van Atten for drawing this passage to my attention (although I was unable to convince him that Leibniz is not guilty of equivocation), and to Adam Harmer for his insights into how Benardete's criticism is misplaced (both private communications).

[91](Benardete 1964, 47). Here I have switched A and B for convenience. Benardete also distinguished a third criterion of equality, "If all the elements of [B] can be put into 1–1 correspondence

Cantor: That is a little different from my objection. But I take it you have a reply?

Leibniz: Yes, and I think it also answers your criticism. I deny that I am taking equality in more than one sense. I certainly agree with you that any two aggregates that form wholes and whose elements can be put into 1–1 correspondence, are equal. But I do not believe one can determine which of two aggregates is greater or equal unless they are both wholes. This is the basis of a proof I once proposed of the Euclidean part-whole axiom by means of the following syllogism, which depends on defining "less than" in terms of "[proper] part" of:

Whatever is equal to a part of *A* is less than *A*, by definition.
B is equal to a part of *A* (namely, to *B*), by hypothesis.
Therefore *B* is less than *A*.[92]

This is the sense in which we can say that the aggregate of even numbers is less than the aggregate of natural numbers: B, the aggregate of all even numbers, is equal to itself (by any criterion equality must be reflexive), and is a proper part of A, in that A is the union of B with the odd numbers. One cannot generate any contradiction unless one assumes that A and B are wholes, one of which can be part of the other. For it is only on this basis that one can infer that one has fewer elements than the other.

Cantor: I completely concur! Only I deny that infinite sets (or aggregates, as you call them) *are* entities that are based upon whole and part. In this way I avoid the contradiction . . .

Leibniz: . . . whereas I believe it to be no less true in the infinite than in the finite that the part is less than the whole.[93] If one accepts (as you do not) the supposition that the infinite aggregate of natural numbers forms a whole to which this axiom or the definition from which I prove it applies, then this entails a contradiction. I therefore reject the supposition: that is, I concede the infinite multiplicity of terms, but this multiplicity does not constitute a number or a single whole. It signifies only that there are more terms than can be designated by a number.[94] Consequently I do not concede your claim that the multiplicity of even numbers is equal to the multiplicity of natural numbers or of odd ones if one interprets this equality as sameness of number.[95]

Cantor: Still, I think you misunderstood my remark about relations to concepts already present and tested. If every new concept had to agree with all the old definitions, there would be no new mathematics. What I mean is that if one

---

with a proper subset of [A], then [A] contains a greater number of elements than [B]." If 1–1 correspondence is taken to define equality when applied to sets of numbers taken as wholes, then the latter constitutes an application of Leibniz's definition of "less than" in his proof of the part-whole axiom. But it is not a separate criterion of equality.

[92]Leibniz, *Initia Rerum*, GM VII 20

[93]Leibniz, *Pacidius Philalethes*, (A VI 3, 551/LLC 179).

[94]Leibniz, Letter to Bernoulli, (GM III 575).

[95]Leibniz (GP II 315).

can define a new concept—such as, here, equality of two infinite sets—in such a way that it is internally consistent, and stands in a determinate relation to old ones, then it should be allowed; and, because of the principle of plenitude, every consistent concept will be instantiated. In particular, in introducing new numbers [mathematics] is only obliged to give them definitions through which [*Cantor coughs*] such a relation is conferred on them to the old numbers that they are determinately distinguishable from one another in given cases.[96] So, not only do I believe that the idea of an infinite number is consistent; I believe also that there are different "sizes" of what I call the transfinite. For instance, although the actually infinite set (v) of all positive finite whole numbers v is equivalent to the set (μ/v) of all positive rational numbers μ/v, where μ and v are prime relative to one another, the set of *all real* number magnitudes is *not* equivalent to the set (v).[97] Thus, whereas the number of all rational numbers is the same as the number of natural numbers, the number of all real numbers is greater than either of these numbers, as I think I have demonstrated.

Leibniz: Yes, I have heard of these celebrated proofs of yours, and am eager to discuss them with you. I think it will pay us to go through them most carefully. I believe that you establish the first claim by pairing off the rational numbers with the natural numbers in such a way that one can be sure that every rational is counted?

Cantor: Precisely! One simply makes an infinite array as follows:

$$
\begin{array}{cccccc}
1/1 & 2/1 & 3/1 & 4/1 & 5/1 & \ldots \\
1/2 & 2/2 & 3/2 & 4/2 & 5/2 & \ldots \\
1/3 & 2/3 & 3/3 & 4/3 & 5/3 & \ldots \\
1/4 & 2/4 & 3/4 & 4/4 & 5/4 & \ldots \\
1/5 & 2/5 & 3/5 & 4/5 & 5/5 & \ldots \\
\ldots & \ldots & \ldots & \ldots & \ldots & \ldots
\end{array}
$$

Now if one goes through this list by beginning at the top left, moving to the right to 2/1 and then cutting diagonally down to 1/2, then down to 1/3, and then cutting diagonally back to through 2/2 to 3/1, across to 4/1, and then cutting diagonally down to 1/4, proceeding down to 1/5 and then back up through the next diagonal 3/3 to 5/1, and so forth, one can proceed through the whole list without missing out a single rational number. If it be objected that 2/2, 3/3 etc., or 1/2, 2/4, 3/6 etc., are different numerals for the same number, we can simply agree to cross out all recurrences of the same number before counting them.

Leibniz: I agree that this shows that the rational numbers are countable, or *denumerable*, as you say, in the sense that they can be paired off with the natural

[96](Cantor 1883, CGA 182); an inessential elision where Cantor coughs.

[97](Cantor 1887–8, CGA 412).

numbers that we use for counting. But this does not prove that they form a *totality* or that there is a *collection* of them.[98]

Cantor: But you will grant that I may define terms as I choose, provided no contradiction follows?

Leibniz: Of course. If one denies this, science becomes impossible.

Cantor: I am glad to see that you are not to be counted among those who reject the law of excluded middle, *tertium non datur*. Suppose, then, that I define the power of a collection or set, whether finite or not, as follows: We call two sets M and N "equivalent", and denote this by $M \approx N$ or $N \approx M$, if it is possible to set them in such a relation to one another that every element of one of them corresponds to one and only one element of the other. Two sets M and N have the same cardinal number if and only if they are equivalent.[99] Thus the previous proofs establish that the natural numbers, the even numbers and the rational numbers all have the same power or cardinal number. Now by abstraction we can say that all sets with the same cardinality have the same number of elements. That is, abstracting from the nature or specific character of the elements, as well as from the order in which they are given, all sets of the same cardinality have the same *cardinal number*. The cardinality of the set of natural numbers I define to be $\aleph_0$.

Leibniz: You may define things so. But you have not proved the possibility of your definitions, nor that $\aleph_0$ can properly be regarded as a number.

Cantor: I think you are working with too restricted a concept of number. No one has established the possibility of irrational numbers, but we all think these exist; and I believe that one can say unconditionally: the transfinite numbers *stand or fall* with the finite irrational numbers; they are like each other in their innermost being: for the former, like the latter, are determinate, delimited forms or modifications (ἀφωρισμένα) of the actual infinite.[100] Indeed, the provision of a foundation for the theory of irrational quantities cannot be effected without the use of the actual infinite in some form.[101] I assume from your silence that I may proceed. Suppose now that I take all the real numbers between 0 and 1. Now suppose we have some rule that enables us to compile an exhaustive list of these numbers ...

---

[98]Dean Buckner (in private correspondence) has drawn my attention to a passage from Wittgenstein's *Philosophical Remarks* in which he makes a similar point: "§ 141. Does the relation $m = 2n$ correlate the class of all numbers with one of its subclasses? No. It correlates any arbitrary number with another, and in that way we arrive at infinitely many pairs of classes, of which one is correlated with the other, but which are *never* related as class and subclass. Neither is this infinite process itself in some sense or other such a pair of classes. ... $m = 2n$ contains the *possibility of correlating any number* with another, *but doesn't correlate all numbers with others*." (Wittgenstein 1980, 161).

[99](Cantor 1895–7, CGA 283). Cf. (Cantor 1883, 167): "To every well-defined set there belongs a determinate power, whereby two sets are ascribed the same power if they can be mutually and univocally ordered one to another element for element."

[100](Cantor 1887–8, CGA 395–396); cf. (Lavine 1994, 93); cf. (Hallett 1984, 80).

[101](Cantor, *Nachlass* VI, 64); quoted from (Hallett 1984, 26).

Leibniz: But already I cannot accept this, for it is precisely the character of the infinite that it cannot be exhausted.

Cantor: I grant this without reservation. It is indeed precisely the character of the transfinite that it can always be increased. But what I meant is illustrated by the example of the rational numbers. There we had a rule that allowed us to list all the rational numbers—in principle, or given all eternity (which, you will agree, we have at our disposal). The rule allows us to say that no rational will be missing from the list. I call such an enumerable collection of elements a *denumerable set*: each element has a corresponding natural number, and for each natural number there is a corresponding element. There is, for example, an nth prime number for any natural number n.

Leibniz: I have already agreed that the rational numbers can be enumerated in this way, but not that they form a collection.

Cantor: Assume that it is possible to form such a collection—for how else would you prove the notion contradictory? Now suppose that the real numbers form such a denumerable set. Suppose, for example, that we can enumerate the totality of all the real numbers within any interval $(\alpha \ldots \beta)$ as follows. Let there be any two symbols m and w that are distinct from one another. Now consider a domain [*Inbegriff*] M of elements $E = (x_1, x_2, \ldots, x_\nu, \ldots)$, which depend on infinitely many co-ordinates $x_1, x_2, \ldots, x_\nu, \ldots$, such that each of these co-ordinates is either m or w.[102] (Indeed, if we take m and w as 0 and 1, each of these elements E can be regarded as a numeral expressing the real number as a binary sequence. I am sure this would be acceptable to you in the light of your pioneering work on binary numbers.) Now M is the totality of all elements E. I claim that such a set M is not of the power of the sequence $1, 2, 3, \ldots, \nu, \ldots$ That is shown by the following theorem:

If $S = E_1, E_2, \ldots, E_\nu \ldots$ is any simply infinite sequence of elements of the set M, then there is always an element E_0 that corresponds to no E_ν.

To prove this, let

$$E_1 = (a_{11}, a_{12}, \ldots a_{1\nu}, \ldots),$$
$$E_2 = (a_{21}, a_{22}, \ldots a_{2\nu}, \ldots),$$
$$\ldots \ldots \ldots \ldots \ldots \ldots \ldots \ldots \ldots$$
$$E_\mu = (a_{\mu 1}, a_{\mu 2}, \ldots a_{\mu \nu}, \ldots),$$
$$\ldots \ldots \ldots \ldots \ldots \ldots \ldots \ldots \ldots$$

Here the $a_{\mu\nu}$ are m or w in a definite manner. Produce now a sequence.

$$b_1, b_2, \ldots b_\nu, \ldots,$$

---

[102](Cantor 1890–91, CGA 278); cf. (Lavine 1994, 99).

so defined that b_ν is different from $a_{\nu\nu}$ but is also either m or w (i.e., 0 or 1). Thus if $a_{\nu\nu} = 0$, then $b_\nu = 1$, and if $a_{\nu\nu} = 1$, then $b_\nu = 0$. If we now consider the element.

$$E_0 = (b_1, b_2, b_3, \ldots),$$

of M (which element we may call the *antidiagonal* of S), we see at once that the equality.

$$E_0 = E_\mu$$

can be satisfied for no whole number value for μ. Otherwise, for the μ in question and for all whole number values of ν, $b_\nu = a_{\mu\nu}$, and therefore in particular $b_\mu = a_{\mu\mu}$ would hold, which is ruled out by the definition of b_ν. It follows immediately from this theorem that the totality of elements of M cannot be brought into the form of a sequence $E_1, E_2, \ldots, E_\nu \ldots$; we would otherwise be faced with the contradiction that a thing E_0 both and is not an element of M.[103] Thus we have the result that M, the totality of all the elements E, has a power (or cardinality) greater than the power of N, the number sequence 1, 2, 3, $\ldots$, ν $\ldots$. That is, the number of real numbers $\mathfrak{R}$ between, say, 0 and 1 is greater than can be enumerated, even using the infinity of natural numbers: $C(\mathfrak{R}) > C(N)$, or, $C(\mathfrak{R}) > \aleph_0$.

Leibniz: This excellent demonstration does indeed prove that there is a contradiction in assuming that the number of all real numbers between 0 and 1 (or any other interval) is a denumerable totality. But to say that the number of real numbers is greater than the number of natural numbers is to say that both sets can be treated as consistent wholes such that one is greater than the other. However, as I already argued, Galileo showed long ago that if one treats infinite sets as wholes, then they cannot be compared as to greater and lesser.

Cantor: Yes, yes, I know the demonstration in his *Due Nuove Scienze*. Since to each (natural) number there is a corresponding square, and vice versa, they are equal. But there are natural numbers that are not squares, e.g. 2, 3, 5, 7, etc. So the number of natural numbers is greater than the number of squares. But you are forgetting that I have abandoned the definitions characteristic of finite numbers, and have chosen (after Dedekind's example) to call sets equal if they can be set in 1–1 correspondence, notwithstanding Bolzano's observation that an infinite set is one that can be put into such correspondence with its proper subset.

Leibniz: I had not forgotten. But by what right can you say that $C(\mathfrak{R}) > C(N)$ just because there are real numbers that are left out of the enumeration corresponding to natural numbers, when you have denied that one can infer that $C(N) > C(S)$ on the grounds that there are natural numbers that are not squares?

Cantor: The difference is that if we drop the Euclidean axiom of wholes for infinite sets, namely that the whole (here, the set) is greater than the part (here,

---

[103] (Cantor 1890–91, CGA 278–279); cf. (Lavine 1994, 99–102).

the proper subset), then equality of infinite sets is decided solely on the basis of 1–1 correspondence. In the case of the squares, the assumption of a 1–1 correspondence with the natural numbers does not lead to a contradiction. But in the case of real numbers, it does.

Leibniz: But what can one conclude from the contradiction if one has no assumption about wholes and parts? Surely you have to assume that the set $S = \{E_1, E_2, \ldots, E_\nu \ldots\}$ lists *all the real numbers*?

Cantor: Yes, I believe you are right. The proof assumes that this set numbers all the elements of $\mathfrak{R}$ between 0 and 1, that they form a denumerable totality. Otherwise, if it were not denumerably complete, E_0 might be one of the elements that is missing from it.

Leibniz: Allow me to make this assumption explicit, in a manner suggested to me by my correspondent from Spain. (I shall use your terminology, and speak of infinite collections as sets [*Menge*].) We may call a subset T of a given infinite set M a *complete enumeration* of M iff (i) T is a denumerable subset of M, i.e. its elements can be put into 1–1 correspondence with those of N (*denumerabilty*); and (ii) no element of M is left out of T (*completeness*). Then we can agree to call M a *denumerable totality* if and only if there exists a complete enumeration of M.

Cantor: Fine. In my argument I assumed that T was such a complete enumeration of all the elements of $\mathfrak{R}$ between 0 and 1, which would be your set M. Now if one forms the antidiagonal t_0, then by construction $t_0 \notin T$, but $t_0 \in M$. Therefore there are more elements in M than are contained in the complete enumeration T, i.e. non-denumerably many.

Leibniz: But it seems to me that T is not a complete enumeration by this definition, since t_0 is an element of M left out of T, so that condition (ii) above fails. Nor may we avoid this by including the antidiagonal in a new enumeration, forming the subset T' of M which contains all the elements t_i of T and also its antidiagonal t_0, so that $T' = T \cup \{t_0\}$. For T' is a denumerable subset of M, since its elements can be put into 1–1 correspondence with N: $t'_1 = t_0$, $t'_n = t_{n-1}$, $n > 1$. (I believe Professor Hilbert used an argument like this concerning a certain infinite hotel).

Cantor: Obviously I can see why this still does not give us a complete enumeration: T' itself may be diagonalized, so that a new element t'_0 of M exists that is not included in the enumeration. Therefore T' is not itself a complete enumeration of M.

Leibniz: Exactly! And this argument can be reiterated to infinity. So we have proved that *there are no complete enumerations of M*. Yet, on the other hand, *each such subset T is a denumerable one*. Thus M is enumerable by infinitely many denumerable subsets, although none of these is a complete enumeration. So your conclusion should not have been that M is nondenumerable, but that it has no complete enumeration, and is therefore *not a denumerable totality*. This seems to me to indicate what I have been urging all along: you cannot treat an infinite multiplicity as a complete collection of elements.

Cantor: I am not persuaded. For I do not see why, once I have made the assumption that T lists all the real numbers, any further assumption is needed about completeness: if *all* of them are included, this *means* that none are left out. So I do not believe that you have shown any contradiction in the notion of an infinite collection. Moreover, I do not have to make a separate assumption that the real numbers form a set, provided only that you allow me that the natural numbers constitute a totality or set N, together with the assumption I made above that if L is a set, then so is the domain of all functions from L into an arbitrary fixed pair, such as 0 and 1.[104] Indeed, the above proof seems remarkable not only because of its great simplicity, but especially also because the principle employed in it is extendible without further ado to the general theorem that the powers of well-defined multiplicities have no maximum, or what is the same, that for any multiplicity L another M can be placed beside it that is of greater power than L.[105]

Leibniz: But, to take up my previous thread again, how can you claim that one set or whole is greater than another without some principle to the effect that the whole is greater than the part?

Cantor: The basis for the claim that any transfinite set M is greater than another L is that L is equal to a subset of M, and is not equal to M. Here equality of two sets is to be understood in terms of their being of the *same power*, that is, (as I explained before), L is equal to M if a one-one correspondence between their respective elements is possible.

Leibniz: If I may be so bold, I suggest this amounts to a replacement of the classical definition of *part*, but *not* a rejection of the Euclidean axiom that the whole is greater than the part. Going back to my proof of the Euclidean axiom,

Whatever is equal to a [proper] part of *A* is less than *A*, by definition.
B is equal to a part of *A* (namely, to *B*), by hypothesis.
Therefore *B* is less than *A*.

Here 'whole', as applied to collections or sets, is to be understood as the *set A*, and 'part' as one of its *proper subsets B*. As I understand your proposal, you wish to replace the major proposition of the syllogism with the following definition:

Whatever is equal to a subset of A and is not equal to A, is less than A,
where two sets are considered equal whenever a one-one correspondence between their respective elements is possible. But this is simply to define a *(proper) part* of a transfinite set A as *a subset of A that cannot be put into 1–1 correspondence with A*. Indeed, this definition will also be valid for finite sets,

---

[104]Lavine notes that although this proof of Cantor's is superior to his previous ones in that it "is independent of any notion of limit or of transfinite number", it "does, however, require a new set-existence principle: if L is a set, then so is the domain of all functions from L into an arbitrarily fixed pair" (Lavine 1994, 94). As he observes, this is "equivalent to the Power Set Axiom in common use today: the subsets of a set form a set" (95), since any one function from a set L to a fixed pair is fully determined by the subset of L that is taken by the function to a fixed member of the pair. So the set of all such subsets is canonically identifiable with the domain of all such functions from L into some fixed pair.

[105](Cantor 1890–91, CGA 279); cf. (Lavine 1994, 101).

since a proper subset of a finite set A is a subset of A that cannot be put into a one-one correspondence with it. I submit, therefore, that you do after all accept the Euclidean axiom that the whole is greater than the part, with *(proper) part* redefined to agree with your generalized definition of equality of sets.

Cantor: That is very ingenious. But it seems to me much more natural to regard any subset as a part, but to abandon the Euclidean axiom. Once I have defined 'less than' as above, I really have no need for any notion of 'whole' and 'part' of infinite sets or collections.[106]

Leibniz: But you will grant, I think, that some infinite multiplicities cannot consistently be regarded as wholes in the sense of completed collections?

Cantor: You are correct. If we start from the concept of a determinate multiplicity (a system, a domain [*Inbegriff*]) of things, it is necessary, as I discovered, to distinguish between two kinds of multiplicities (I always mean *determinate* multiplicities). For a multiplicity can be so constituted that the assumption of a "being together" of *all* its elements leads to a contradiction, so that it is impossible to construe the multiplicity as a unity, as "one finished thing".[107]

Leibniz: This is what I was urging you to concede just now about the denumerable totalities on which your diagonal argument can be seen as based.

Cantor: I do not accept that those were examples of such *absolutely infinite* or *inconsistent multiplicities*, as I call them. The latter have to do with the absolute, and are not susceptible to number. As one can easily convince oneself, "the domain of everything thinkable", for example, is such a multiplicity.[108]

Leibniz: I think I see the reason for your calling this inconsistent. The idea that this domain includes the collection of *all* thinkable things is immediately inconsistent with the fact that we can think of the domain itself, and this is necessarily something different from its elements.[109]

Cantor: Yes, this relates to an argument given in 1872 by my friend Richard Dedekind for the actual infinity of possible rational thoughts. It depends on the idea that a rational thought cannot represent itself. Then, if s is some such possible rational (i.e. non-self-representative) thought, then so is "s is a possible

---

[106]It should be noted that Cantor himself defines *part* as follows: "'We call 'Part' or 'Subset' [*Teilmenge*] of a set *M* every *other* set M_1 whose elements are alike elements of *M*." (Cantor 1895–7, CGA 282). For 'Cantor''s response here, I am indebted to Brad Bassler.

[107]Cantor, Letter to Dedekind, 28 July 1899; (CGA 443). Hallett notes (1984, 166) that Grattan-Guinness has shown that this "letter" is really an amalgam by Zermelo of several letters written at different times. See (Grattan-Guinness 1974).

[108](CGA 443).

[109]Cf. (Rucker 1983, 51): "Again, the reason that it would be a contradiction if the collection of all rational thoughts were a rational thought T is that then T would be a member of itself, violating the rationality of T (where "rational" means non-self-representative)."

thought", and so on to infinity.[110] Bernard Bolzano had given a similar argument regarding truths.[111]

Leibniz: But this seems an unfortunate example, since it seems it would also make Bolzano's "set of all absolute propositions and truths" an inconsistent totality: yet that was precisely the example he gave to show the existence of an absolutely infinite set.[112]

Cantor: This may be unfortunate, but it must be conceded. Similarly, it would seem to rule out the young Wittgenstein's view that "the world is the totality of facts"; for if that were a non-self-representative fact, it should have been included in itself, which is a contradiction. Allow me to spell out the argument more formally:

It depends on what we may call the *principle of the genetic formation of sets*, according to which a set is formed in two distinct stages: (i) some multiplicity of elements is given, and (ii) if some of these can be consistently combined into a unity, then this unity is a set.[113] Now let us call T the totality of thinkable things. If T is 'one finished thing', it must contain *all* thinkable things: that is, we must assume that all thinkable things being given, they may be collected together in thought to form T. But T itself is a thinkable thing not contained in the original totality. And if now we include T in the totality, another totality T' could be formed that would include T; but clearly the same argument could be applied to T'. Thus there is no escape: either all thinkable things do not form a totality, or T itself is not a thinkable thing.

Leibniz: And if I am not mistaken, Russell found in a paper of Burali-Forti a similar argument against the notion that there could be a set of all ordinals—one that you had already anticipated.

Cantor: That is true. I mentioned it in a letter to Dedekind a couple of years before Burali-Forti published his argument. It depends on the simple technical idea of a well-ordering: a multiplicity is called *well-ordered* if it fulfills the condition that every sub-multiplicity has a first element; such a multiplicity I call for short a *sequence*. Now I envisage the system *of all numbers* and denote it Ω. The system Ω in its natural ordering according to magnitude is a "sequence". Now let us adjoin 0 as an additional element to this sequence, and certainly if we set this 0 in the first position then Ω' is still a sequence:

---

[110](Dedekind 1963, p. 64); cf. (Rucker 1983, 50, 335).

[111]"The class of all true propositions is easily seen to be infinite. For if we fix our attention upon any truth taken at random . . . , and label it A, we find that the proposition conveyed by the words 'A is true' is distinct from the proposition A itself." (Bolzano 1950, §13, 84–85); cf. (Rucker 1983, 50).

[112](Bolzano 1950, 84).

[113]For the delineation of this principle as implicit in Cantor's view of sets I am indebted to Rudy Rucker, who calls it "the *genetic formation of sets principle*" (Rucker 1983, 208).

$$0, 1, 2, 3, \ldots \omega_0, \omega_0 + 1, \ldots, \gamma, \ldots,$$

of which one can readily convince oneself that *every* number occurring in it is the [*ordinal number*][114] of the *sequence of all its preceding elements*. Now Ω' (and therefore also Ω) cannot be a consistent multiplicity. For if Ω' were consistent, then as a well-ordered set, a number δ would belong to it which would be greater than all numbers of the system Ω; the number δ, however, also belongs to the system Ω, because it comprises *all* numbers. Thus δ would be greater than δ, which is a contradiction. Thus *the system Ω of all ordinal numbers is an inconsistent, absolutely infinite multiplicity.*[115]

Leibniz: So you rejected my proof that there is no number of all numbers, which I showed to follow from the axiom that the whole is greater than the part, since this number would be both equal to and greater than itself. Yet now you wish to persuade me that there is no ordinal number Ω of all ordinals, on the grounds that Ω, like Ω', is a multiplicity necessarily greater than any of its parts, of which it is itself one—that is, that the ordinal number belonging to it would be both equal to and greater than itself!

Cantor: This shows that not every multiplicity of elements can be regarded as a set.

Leibniz: How then do you define a set?

Cantor: By a 'manifold' or 'set' I understand in general any 'many' [*Viele*] that can be thought of as one [*Eines*], that is, any domain of determinate elements which can be united into a whole through a law.[116]

Leibniz: This agrees well enough with my definition of a whole or *aggregate*: If, when several things are posited, by that very fact some unity is immediately understood to be posited, then the former are called *parts*, the latter a *whole*. Nor is it even necessary that they exist at the same time, or at the same place; it suffices that they be considered at the same time. Thus from all the Roman emperors together, we construct one aggregate.[117]

Cantor: But you countenance only finite aggregates?

Leibniz: No. The above definition is valid also for infinite aggregates, provided only they are united into a whole by the mind. Take, for instance, an infinite series.

---

[114]Cantor had "*type* [*typus*]", defined as follows: "if the simply ordered multiplicity is a *set*, then I understand by *type* the general concept under which it stands, and also under which all and only *similar* ordered sets stand… If a sequence F has the character of a set, then I call the type of F its '*ordinal number*', or '*number*' for short." (Letter to Dedekind, 28th July, 1899, CGA 444)

[115]Cantor to Dedekind (CGA 444–5).

[116](Cantor 1883, CGA 204, fn. 1). Cantor gives a similar definition in 1895: "By a 'set' we understand every collection into a whole M of determinate, well-differentiated objects m of our intuition or our thought (which are known as the 'elements' of M)." (Cantor 1895–7, CGA 282). Cf. (Hallett 1984, 33).

[117](A VI 4, 627/ LLC 271).

Even though the sum of this series cannot be expressed by one number, and yet the series is produced to infinity, nevertheless, since it consists in one law of progression, the whole is sufficiently perceived by the mind.[118] Note that this only involves the infinite in its distributive mode: given the first term and the law of the series, every subsequent term is determined. But this does not require that the mind be able to view all the terms as a collection.

Cantor: Your talk of a law of progression accords well with my own intentions in defining sets in terms of a law. For when I say a set is any domain that can be united into a whole through a law, I am thinking primarily in terms of the concept of well-ordering I just explained.[119] Thus I hold that in a similar way to how [the whole real number] v is an expression for the fact that a certain finite number [*Anzahl*] of unities is unified into a whole, one may conceive ω to be the expression of the fact that the whole domain (I) [of the positive real whole numbers] is given in its natural succession according to law.[120]

Leibniz: I once held a similar view, as when I criticized Spinoza's claim that "there are many things which cannot be equated with any number". For if the multiplicity of things is so great that they exceed any number, that is, any number assignable by us, this multiplicity itself could be called a number, to wit, one that is greater than any assignable number whatever.[121] But shortly afterwards I came to the view I articulated above, that there is no number of all numbers at all, and that such a notion implies a contradiction.[122] That is, an infinity of things is not one whole, or at any rate not a true whole.[123] For when it is said that there is an infinity of terms, it is not being said that there is some specific number of them, but that there are more than any specific number.

Cantor: I agree that if one allows every well-defined infinite collection to count as a totality, then a contradiction can be obtained. That is why I do not

---

[118]Leibniz, "On the True Proportion of the Circle to its Circumscribed Square, Expressed in Rational Numbers" (1682), (GM V 120).

[119]This is argued persuasively by Shaughan Lavine in his *Understanding the Infinite*, pp. 84–86. He observes that although the above passage has seemed to many like a classic statement of the Comprehension Principle, it continues: "I believe that in this I am defining something which is related to the Platonic εἶδος or ἰδέα as well as to that which Plato in his dialogue *Philebus or the Highest Good* calls the μικτόν. He counterposes this to the ἄπειρον, i.e. the unlimited, indeterminate, which I call the improper infinite, as well as to the πέρας, i.e. the limited, and explains it as an ordered "mixture" of the latter two." Quoting relevant passages from the *Philebus* in support, Lavine argues that "Cantor's typical use of the word *law* in the *Grundlagen* is 'natural succession according to law', which suggests quite a different picture: a 'law' is for Cantor a well-ordering or 'counting' . . .".

[120](Cantor 1883, 195).

[121]Leibniz, "Annotated Excerpts from Spinoza" (Feb-April 1676), (LLC 111).

[122]Leibniz, *Pacidius Philalethi*, (1676), (LLC 179).

[123]Leibniz, "Infinite Numbers," (April 1676), (LLC 101). Cf. Leibniz, *New Essays*, p. 159: The idea of an infinite number is absurd, "not because we cannot have an idea of the infinite, but because an infinite cannot be a true whole."

allow such inconsistent multiplicities as the set of all the ordinals Ω or the totality of all cardinals to enter into set theory. If, on the other hand, the totality (*Gesamtheit*) of elements of a multiplicity can be thought without contradiction as "being together", so that their collection into "*one* thing" is possible, I call it a *consistent multiplicity* or a "set".[124]

Leibniz: I agree, one cannot simply assume the existence of a set or any other arbitrarily defined entity without first establishing its possibility. One needs what I call a real definition.

Cantor: What is that, may I ask?

Leibniz: A real definition is one according to which it is established that the defined thing is possible, and does not imply a contradiction. For if this is not established for a given thing, then no reasoning can safely be undertaken about it, since, if it involves a contradiction, the opposite can perhaps be concluded about the same thing with equal right. And this was the defect in Anselm's demonstration, revived by Descartes, that a most perfect or greatest being must exist, since it involves existence. For it is assumed without proof that a most perfect being does not imply a contradiction; and this gave me the occasion to recognize what the nature of a real definition is. So causal definitions, which contain the generation of the thing, are real definitions as well,[125] as are recursive definitions in mathematics. This is why the example I gave above of infinite division is apposite. The transfinite, however, is not susceptible to recursive definition.

Cantor: Consistency may be assumed if we know of no reason to deny it.

Leibniz: Then allow me to give another example. We are agreed, I believe, that monads are the true unities underlying all perceived phenomena, and that there are actually infinitely many of them.

Cantor: Yes. As I was saying earlier, I maintain that the *power* of *corporeal matter* is what I have called in my researches the *first* power,[126] i.e. that the number of corporeal monads is $\aleph_0$, the first transfinite cardinal.

Leibniz: Then you must hold that their collection into one thing or whole is possible.

Cantor: True: we must think of *corporeal matter* at each *moment of time* under the *representation* of a point-set P of the first power, and the aetherial matter in the same space under the *representation* of the next occurring point-set Q of the second power.[127] Each of the sets P and Q must be a consistent *totality*, otherwise the whole notion of a transfinite cardinal would fail.

Leibniz: Call it a totality you will. I say that the collection of all substantial unities or the accumulation of an infinite number of substances, is, properly

[124]Cantor, Letter to Dedekind, 28 July 1899, (CGA 443).

[125]Leibniz, (A VI 4, 1617/LLC 305–07).

[126](Cantor, 1885a, 276).

[127](Cantor, 1885a, 276).

speaking not a whole any more than infinite number itself.[128] And I will prove it to you. For we both agree that there is an actual infinity of true unities, and these can be aggregated, at least in thought. Now suppose the aggregate of these unities is itself a unity; let us call this unity U. Now according to the principle of the genetic formation of sets that you mentioned above, this unity is not included in U itself. Yet if it is the aggregate of *all* unities, it must be included in itself. Therefore U is not a true unity. No entity that is truly a unity is composed of a plurality of parts.[129]

Cantor: Here I believe a distinction must be made. Your argument shows that the collection of all corporeal monads cannot itself be a corporeal monad. But it does not prove that matter cannot be a consistent collection of a different order. The first transfinite ordinal, remember, is not a natural (or inductive) number, and the same applies to the first transfinite cardinal. What I assert and believe to have demonstrated in my various works and earlier researches is that after the finite there is a *transfinite*, (which one can also call *suprafinite*), that is, an unlimited ascending scale of determinate modes, which by their nature are not finite but infinite, but which, just like the finite, can be determined by definite, well-defined *numbers* distinguishable from one another.[130]

Leibniz: I, on the other hand, believe that all the paradoxes of the infinite have their source in the same error, namely, that of treating an infinite collection of elements as a whole. Allow me to explain, by giving a brief synopsis of our argument, drawing together more tightly our main points of contention. You assume that all the natural numbers can be regarded as a collection or a completed whole, or "one finished thing". Without this assumption it would not be possible for there to be a first number after all of them, as you define ω_0, the first transfinite ordinal number. I have objected that the number of all numbers implies a contradiction, since the number of numbers in the part, or proper subset, would be equal to the number of numbers in the whole collection. You have evaded this objection by denying the applicability of the part-whole axiom to infinite collections—or, as I would rather explain it, by redefining a *proper part* of a transfinite set S as a subset P of S that cannot be put into element by element correspondence with S. Yet in your proof of the actual infinitude of the multiplicity of all ordinal numbers Ω, you acknowledge that if Ω were a set or consistent totality, then a set Ω' could be formed, of which Ω would be a part by this definition. But the number corresponding to Ω' would not only be greater than that corresponding to Ω, but also equal to it: the whole would be equal to the part, a contradiction. You therefore declare Ω to be an *absolutely infinite* or *inconsistent multiplicity* or one that it is impossible to construe as a unity, as "one finished thing". I would suggest that consistency demands either that

---

[128]Leibniz, *Theodicy* 249, (GP VI 232).

[129]Leibniz, (A VI 4, 627/LLC 271).

[130](Cantor 1883, 176); cf. (Hallett 1984, 39).

every well-ordered multiplicity of ordinals is a consistent totality (which, as we have seen, leads to a contradiction) or that none of them is.

Again, your diagonal argument proves that if the infinite multiplicity of natural numbers forms a consistent totality with cardinality $\aleph_0$, then (assuming the power set axiom) there is necessarily a totality which is greater, to which there corresponds the greater cardinal number $2^{\aleph_0}$ (which, if I understand correctly, cannot be proved equal to $\aleph_1$, the first cardinal greater than $\aleph_0$.) But for each such totality there will be such a totality greater than it, including one greater than the totality of all thinkable totalities. But this is self-inconsistent. So, therefore, is the original totality, $\aleph_0$, by *modus tollens*.

Cantor: If I understand you correctly, you are asking how I know that the well ordered multiplicities or sequences to which I ascribe the cardinal numbers $\aleph_0, \aleph_1, \ldots \aleph_\omega, \ldots \aleph_\aleph, \ldots$ are really "sets" in the sense of the word I have explained, i.e. "consistent multiplicities". Is it not thinkable that *these* multiplicities are already "inconsistent", and that the contradiction arising from the assumption of a "being together of all their elements" has *simply not yet been made noticeable*? My answer to this is that the same question can just as well be raised about finite multiplicities, and that a careful consideration will lead one to the conclusion that even for finite multiplicities *no* "proof" of their consistency is to be had. In other words, the fact of the "consistency" of finite multiplicities is a simple, unprovable truth; it is (in the old sense of these words) "the axiom of finite arithmetic". And in just the same way, the "consistency" of those multiplicities to which I attribute the alefs as cardinal numbers is "the axiom of extended transfinite arithmetic."[131]

Leibniz: I am not so sure that it is impossible to demonstrate the consistency of finite multiplicities. But in any case, we know of no inconsistent finite multiplicities, whereas you admit there are inconsistent infinite ones.

Cantor: That may be so, but, notwithstanding your objections, I entertain no doubts about the truth of the Transfinite, which with God's help I have recognized and studied in its diversity, multiformity and unity for so long.[132] My theory stands as firm as a rock; every arrow directed against it will quickly return to its archer.[133]

Leibniz: How can you be so sure?

Cantor: Because I have studied it from all sides for many years; because I have examined all objections which have ever been made against infinite numbers; and above all, because I have followed its roots, so to speak, to the first infallible cause of all created things.

---

[131]Cantor, Letter to Dedekind, 28th August, 1899; (CGA 447–8); cf. (Rucker 1983, 254).

[132]Cantor, Letter to Father Ignatius Jeiler, Whitsun 1888; *Nachlass* VI, 169; quoted from (Hallett 1984, p. 11).

[133]This and the following quotation are taken from Cantor's letter to Heman, June 21, 1888, *Nachlass* (I), p. 179; quoted from (Dauben 1979, 298).

Leibniz: Suffice to say that God alone is infallible. I have enjoyed our conversation, which I am sure we will continue on another occasion. Good day, to you, most eminent Sir!

+ + +

3.2   Afterword

Since Leibniz was too gracious to press his point further against Cantor, perhaps I may summarize the argument on his behalf, and say a few words on what I take to be its significance. If the above is a fair statement of his position, then I think it must be granted that

1). Leibniz's construal of the infinite as an actual infinite, syncategorematically understood, and as a distributive whole but not a collection or set, constitutes a perfectly clear and consistent third alternative in the foundations of mathematics to the usual dichotomy between the potential infinite (intuitionism, constructivism) and the transfinite (Cantor, set theory).

   This should be contrasted with the following passage from one of the foremost modern proponents of intuitionism, Michael Dummett: "In intuitionist mathematics, all infinity is potential infinity: there is no completed infinite. ... [This] means, simply, that to grasp an infinite structure is to grasp the process that generates it, that to refer to that structure is to refer to that process, and that to recognise the structure as being infinite is to recognise that the process will not terminate."[134] To this it may be replied on Leibniz's behalf: there is also an actual infinite which is not a collection or set. All the terms of an infinite series are actually given once the law and the first term are given, provided this "all" is not understood collectively.

2). This conception of the infinite as not involving infinite number is appropriate to Leibniz's conception of the actual infinite division of matter; whereas Cantor's transfinite is not, since one cannot get to an wth part by recursively dividing.

   This can be compared with some criticisms of Leibniz's philosophy of the infinite in a recent article by one of his ablest and most intelligent commentators, Samuel Levey.[135] Levey charges that "the deep cause of the difficulty in Leibniz's theory of matter—what leads him into error and what prevents

---

[134](Dummett 1977, 55–56). Cf. (Lavine 1994, 176): "The intuitionist endorses the actual finite, but only the potentially infinite ... The constructivist of my first example endorses the actual denumerable, but only the potentially non-denumerable." But Hidé Ishiguro (private communication) dissents from Dummett's view, holding that Brouwer's conception of the infinite is more nearly syncategorematic.

[135](Levey 1999, esp. 152ff).

him from seeing it—is his constructivism" (Levey 1999, 156). In a nutshell, Levey sees Leibniz as endorsing an *actualism* in his account of matter that is undermined by his tendency toward a *constructivist* view of the infinity of numbers. "An infinity of parts comes to view only from the outside perspective, where the completed infinity can be seen all at once ... [A]s it falls outside the series of finite levels but encompasses them all, we might call it the 'omega level' ... (156) To the constructivist about the infinite, however, *there is no omega level.* Infinity is always potential and incomplete ... Leibniz lets his constructivism come between him and the parts of matter. By thinking in constructivist terms about the division of matter into parts, he loses sight of the omega level and the problems that should appear most vividly there." (157) But the above argument shows that the actual, distributive infinite is understood from the inside: there is no omega level of the division of matter, but this does not make it any less actual. Rather than "dividing into two isolated accounts, the actual infinity of nature and the constructive infinity of mathematics" (157), Leibniz gives an integrated account that avoids equally the paradoxes of the infinite and the shortcomings of constructivism.[136]

3). Through this example of the actually infinite division of matter one can see that recursive connection is closely linked to Leibniz's demand for real definitions, according to which the possibility of the thing defined must be established in the definition.[137] Thus in Leibniz's causal definition of an infinite aggregate, each element is generable by recursion from some initial element (body).

This aligns Leibniz's philosophy of mathematics closely with constructivism and intuitionism, as Levey has observed. But the same example of infinite division shows that the generation *need not be thought of as a process in time.* So long as there is a law according to which further elements are generated (as in the case of the Euclidean generation of primes), an infinity of elements can be understood without any commitment to either an infinite collection (Weierstraß, Cantor) or an unending temporal process (Aristotle, Kant, Brouwer, Weyl).

It was further argued that:

4). Cantor's diagonal argument does not constitute a conclusive proof that there are more reals than natural numbers, since it depends on the premise that the real numbers form a totality, a collection such that none are left out; and this is established by the Power Set Axiom only on the hypothesis that the

---

[136]Here I would not be understood as rejecting all of Levey's analysis. I believe he is right in seeing a tendency in Leibniz toward a constructivist view of numbers, but do not agree that this results in any vacillation about the actuality of infinity. Also, much of what Levey has to say about limits, the division of the continuum, and the origins of Leibniz's monadism is pure gold. But that is another matter, and there is not the space to go into it here.

[137]This was already noted long ago by Nicholas Rescher, in his *The Philosophy of Leibniz*: "Leibniz, on wholly logical grounds, rejects the notion of infinite number, holding that a definition must involve a proof of the possibility of the thing defined, as he had maintained earlier against Descartes' ontological proof" (Rescher 1967, 105).

natural numbers form such a totality. Similarly, 1–1 correspondence between the elements of two multiplicities does not establish that they constitute equal sets without the assumption that those infinite multiplicities are indeed consistent totalities.

5). Leibniz's claim that the universe, or any other collection of all unities, cannot itself be a unity, can be proved in an entirely analogous way to Cantor's proof that there is no ordinal number of all ordinal numbers; and this result, together with the inapplicability of the transfinite to Leibnizian aggregates (and, one might add, the unprovability of the continuum hypothesis), precludes the kind of application of the transfinite to physics that Cantor had envisaged.

6). Leibniz's insistence on the part-whole axiom, rejected by Bolzano and Cantor for infinite sets, can be seen to be justified even in Cantorian set theory, provided *whole* is identified with Cantor's *consistent totality* or *set*, and *part* is re-identified as a subset that cannot be put into 1–1 correspondence with the whole. Although this allows one to avoid Galileo's paradox, since the set of even numbers is not a "part" of the set of natural numbers in this sense, it cannot avoid the paradox of the number of all ordinals without some independent criterion for disqualifying the multiplicity of all ordinals from forming a set.

Whether this Leibnizian approach to the foundations of mathematics is adequate to the construction of any interesting mathematics is a topic for another project. Obviously, it will be crucial to see what kind of account of continuum can be sustained in opposition to the point-set foundation, since the Leibnizian approach eschews infinite sets.

Bibliography

Arthur, R.T.W. 2015. Chapter 7: Leibniz's Actual Infinite in Relation to his Analysis of Matter. In *Interrelations Between Mathematics and Philosophy*, ed. G. W. Leibniz, 137–156. Springer: Archimedes Series, Norma Goethe, Philip Beeley and David Rabouin.

Bassler, O. Bradley. 1998. Leibniz on the Indefinite as Infinite. *The Review of Metaphysics* 51 (June): 849–874.

Beeley, Philip. 1996. *Kontinuität und Mechanismus*. Stuttgart: Franz Steiner.

Benardete, José A. 1964. *Infinity: An Essay in Metaphysics*. Oxford: Clarendon Press.

Bolzano, Bernard. 1950. *Paradoxes of the Infinite*. London: Routledge/Kegan Paul.

Cantor, G. 1879–83. *Über unendliche lineare Punktmannigfaltigkeiten*, 1–4, CGA 139–164.

———. 1883. *Grundlagen einer allgemeinenen Mannigfaltigkeitslehre*, CGA 165–209.

———. 1884. *Die Grundlagen der Arithmetik*, CGA 440–451.

———. 1885a. *Über verschiedene Theoreme aus der Theorie der Punktmengen in einem n-fach ausgedehnten stetigen Raume Gₙ*, CGA 261–277.

———. 1885b. *Über die verschiedenen Standpunkte in bezug auf das actualle Unendliche*, CGA 370–77.

———. 1886. Über die verschiedenen Ansichten in Bezug auf die actualunendlichen Zahlen. *Bihand Till Koniglen Svenska Vetenskaps Akademiens Handigar* 11 (19): 1–10.

———. 1887–8. *Mitteilungen zur Lehre von Transfinitum*, CGA 378–439.

———. 1890–91. *Über eine elementare Frage der Mannigfaltigskeitslehre*, CGA 278–281.

———. 1895–7. *Beiträge zur Begründung der transfiniten Mengenlehre*, CGA 282–356.

———. 1962. Gesammelte Abhandlungen, ed. Ernst Zermelo. Berlin, 1932; repr. Hildesheim: Georg Olms. Cited as CGA.

Dauben, Joseph Warren. 1979. *Georg Cantor: His Mathematics and Philosophy of the Infinite*. Boston: Harvard University Press.

Dedekind, R. 1963. (German original, 1872). *Essays on the Theory of Numbers*. New York: Dover Publications.

Dummett, Michael. 1977. *Elements of Intuitionism*. Oxford: Clarendon Press.

Geach, Peter. 1967. Comments on an essay of Abraham Robinson's. In *Problems in the Philosophy of Mathematics*, ed. Imre Lakatos, 41–42. Amsterdam: North-Holland Publishing Co.

Grattan-Guinness, Ivor. 1970. An Unpublished Paper by Georg Cantor: Principien einer Theorie der Ordnungstypen; Erste Mittheilung. *Acta Mathematica* 124: 65–107.

———. 1974. The Rediscovery of the Cantor-Dedekind Correspondence. *Jahres-bericht der deutschen Mathematike-Vereinigung* 76: 104–139.

Hallett, Michael. 1984. *Cantorian Set Theory and the Limitation of Size*. Oxford: Clarendon Press.

Lavine, Shaughan. 1994. *Understanding the Infinite*. Cambridge, MA: Harvard University Press.

Leibniz, G.W. 1981. *New Essays on Human Understanding*. Trans. Peter Remnant and Jonathan Bennett. Cambridge: Cambridge University Press. Cited as NE.

———. 1849–63. *Leibnizens Mathematische Schriften*. C. I. Gerhardt. Berlin/Halle: Asher and Schmidt, reprint ed. Hildesheim: Georg Olms, 1971. 7 vols. Cited as GM.

———. 1875–90. *Die Philosophische Schriften von Gottfried Wilhelm Leibniz,* ed. C.I. Gerhardt. Berlin: Weidmann; reprint ed. Hildesheim/New York: Georg Olms, 1978. 7 vols. Cited as GP.

———. 1923. Sämtliche Schriften und Briefe, ed. Akademie der Wissenschaften der DDR. Darmstadt/Berlin: Akademie-Verlag; cited by series, volume and page, as A VI 2, 123, etc.

———. 1993. De Quadratura Arithmetica, [1676]. Ed. Eberhard Knobloch. Göttingen: Vandenhoek & Ruprecht.

Levey, Samuel. 1998. Leibniz on Mathematics and the Actually Infinite Division of Matter. *The Philosophical Review* 107: 49–96.

———. 1999. Leibniz's Constructivism and Infinitely Folded Matter 134 162, New Essays on the Rationalists, Rocco Gennaro Charles Huenemann New York, Oxford University Press

Rescher, Nicholas. 1967. *The Philosophy of Leibniz*. Englewood Cliffs: Prentice Hall.

Rucker, R. 1983. *Infinity and the Mind: The Science and Philosophy of the Infinite*. (originally published by Birkhäuser, 1982). New York: Bantam.

Russell, Bertrand. 1903. *Principles of Mathematics*. New York: W. W. Norton.

Uckelman, Sara L. 2015. The Logic of Categorematic and Syncategorematic Infinity. *Synthese* 192: 2361–2377.

Wittgenstein, Ludwig. 1980. *Philosophical Remarks*. Chicago: University of Chicago Press.

Chapter 4
Leibniz on the Continuity of Space

Vincenzo De Risi

Abstract The present essay describes Leibniz's foundational studies on continuity in geometry. In particular, the paper addresses the long-debated problem of grounding a theory of intersections in elementary geometry. In the early modern age, in fact, several mathematicians had claimed that Euclid's *Elements* needed to be complemented with additional axioms in order to ground the existence of the intersection points between straight lines and circles. Leibniz was sensible to similar foundational issues in the Euclidean tradition, and dedicated several studies to investigate a good definition of the continuity of space, in order to ground a general theory of intersections of curves and surfaces. While Leibniz's researches on continuity in relation with the Calculus have been extensively studied in the past, the present essay deals with the less-known Leibnizian notion of a continuous space, as it is to be found in several unpublished writings preserved in Hannover. The subject widens as to encompass the relation between mereology and *analysis situs*, Leibniz's studies on a geometrical characteristics, and Leibniz's theory of space at large.

4.1 Introduction

The notion of *continuity* spreads throughout Leibniz's thought to such an extent that it may be impossible to find a single topic of debate, among the hundreds that the great man tackled in his life, in which he did not use it to foster his ideas or defend his positions. Continuity represented for him a formidable heuristic tool in mathematics and in logic, in physics and the sciences of life, in metaphysics and epistemology, and on this notion he was able to build some of his most daring intellectual constructions, as well as several important arguments that he employed

V. De Risi (✉)
Laboratoire SPHère, Paris, France

Max Planck Institute for the History of Science, Berlin, Germany

© Springer Nature Switzerland AG 2019

V. De Risi (eds.), *Leibniz and the Structure of Sciences*, Boston Studies in the Philosophy and History of Science 337,
https://doi.org/10.1007/978-3-030-25572-5_4

111

in defending his views. Given the breadth of the use of continuity in Leibniz's works, it does not come as a surprise that it may be difficult to pinpoint a common meaning underlying all the applications of the word, and in fact it is easy to conclude, when browsing Leibniz's papers, that he was working rather with a family of connected and yet different (and sometimes pretty vague) notions of continuity which do not admit of reduction to any precise and common core.

If, however, we restrict our investigations to the *mathematical* meaning of continuity, we may be able to mark out a certain number of exact definitions of this concept that Leibniz was able to fashion and formalize in the hope, and indeed in the need, to ground his new analytical discoveries. Such mathematical definitions of continuity may be broadly arranged into two groups, dealing respectively with the concepts of function and of space. Leibniz was able to provide a very good definition of a *continuous function* (even though he never arrived at any formal definition of a function itself) or of continuous transformation, which may be easily compared with Weierstrass's celebrated definition of this latter, through the ε-δ formalism, which is still in use today.[1] Such definition was perfectly natural within the context of the new infinitesimal Calculus that Leibniz was inventing and developing and was largely unprecedented in the mathematical literature, which had had no need, in previous centuries, of dealing with this kind of transformations in such an exact way.

On the other hand, Leibniz also needed a completely different concept of continuity: one that might apply not to transformations or functions but rather to space and figures. He needed, in short, a good definition of what it means to say that a line, a surface, or space itself are (or may be) continuous objects. This notion of the continuity of things (rather than functions), far from being invented only in the seventeenth century, had been a topic of discussion since antiquity and had long since represented a lively source of debate and controversy. In the course of the centuries many different meanings had accumulated and accreted around this notion, which remained, on into modernity, more the object of a vague, intuitive understanding than of any exact formulation. As a matter of fact, when it was fully formalized in the nineteenth and twentieth centuries, this notion of continuity simply exploded in a huge array of different meanings, each of them capturing a few features of the intuitive (historical) understanding of a continuous whole. The effect of this formal analysis of the concept of continuity was that the expression "continuous space" (or surface, line, etc.), which had been so common at the time of Leibniz and during many centuries, disappeared altogether and has, today, no meaning at all in contemporary geometry. We rather speak, for instance, of metrically complete spaces, or connected spaces, arcwise-connected spaces, geodesically-connected spaces, locally compact spaces, dense spaces, separable spaces, regular or semi-regular spaces, normal spaces, and so

---

[1] These very famous Leibnizian definitions may be found in his *Lettre sur un principe général*, published in 1687 (GP III, pp. 51–55); and the related essay *Principium quoddam generale* (A VI, 4C, n. 371, pp. 2031–2039; GM VI, pp. 129–35).

forth. Some of these notions are connected to one another through complex relations and difficult topological theorems, and it proved impossible to preserve any unitary concept of spatial continuity. The space and figures employed in classical geometry (from antiquity up to the early modern age) are "continuous", indeed, in several of the above-mentioned senses. The notion of "connectedness", for instance, the naïvely-grasped meaning of which is just that a figure comes in one piece, was certainly easy to conceive of, and thematize, as an aspect of continuity; but it is also a notion insufficient to justify many unspoken geometrical assumptions of classical mathematics. The notion of "completeness" is considerably less intuitive and considerably more difficult to formalize; it is also much more important; yet it fails to capture the intuitive notion of continuity, as a discrete set may well be a complete one (e.g. $\mathbb{Z}$, the set of integer numbers). "Density" is another notion that had been easily intuitively grasped ever since antiquity (at least in its less formal meaning of infinite divisibility); and, while it is not strong enough to justify many continuity assumptions in Euclidean geometry, a figure (or space) which is *both* dense and complete may well be a reasonable candidate for being "continuous" in the intuitive sense. The latter notion might perhaps also be expressed through "compactness", since a totally limited subspace within a metric space is complete if, and only if, it is compact (and figures were limited in Greek geometry); nonetheless, the notion of compactness is quite an abstract one, and proved especially difficult to define in general terms. The topological "separation" properties also play a role in this conceptual framework, as they may help, for instance, in formalizing the notion of "contiguity" as distinguished from continuity; but they are not immediately useful in defining the latter. Even the famous Axiom of Archimedes (which rules out the possibility of infinitesimal magnitudes) has several connections with the notion of continuity; once again, though, it cannot, taken alone, be considered to be a viable expression of this latter notion ($\mathbb{Z}$ trivially fulfills such an axiom). In short, an entire galaxy of different notions is needed in order to formalize and axiomatize the "continuity" of space. It is not surprising, therefore, that, as long as early modern mathematicians were committed to expressing a *single* concept which would capture and draw all these meanings together, their efforts were doomed to failure.

Leibniz himself, then, who had, on the one hand, been able to produce *a single* definition of a continuous function and apply it throughout his works, was obliged, on the other hand, to experiment with *several* definitions of the continuity of space and spatial objects, each of which may be seen as an attempt to grasp one of the above-mentioned concepts. It is remarkable, however, that Leibniz's last and most fully articulated definition of the continuity of space has several points in common with the modern understanding of metrical *completeness*. The latter was first defined by Richard Dedekind in 1858 and presented in print in his celebrated work on *Stetigkeit und Irrationale Zahlen* (1872), in order to define the real number field (which is the smallest complete extension of the field of rational numbers). Such notion, and its variants due to Cantor and Weierstrass, still plays the most important role in the domain of the foundations of mathematics, and was accepted by Hilbert in his axiomatization of continuity in his *Grudlagen der Geometrie* (1899).[2]

It is true, indeed, that Dedekind stressed that his own definition of the real numbers was un-geometrical in the sense that it did not have recourse to any consideration of geometrical magnitudes (such as segments of incommensurable length). He also strongly protested whenever anyone compared his construction of the real field to those of others (such as Bertrand's) which grounded arithmetical properties on the continuity of segments. On the other hand, however, Dedekind's case was that the continuity of segments had not itself been established beforehand, and that therefore any geometrical construction of real numbers could not help but miss the target of a rigorous foundation. He even claimed that the only way he saw to define the continuity of geometrical magnitudes was to reshape those considerations on ordering that he had applied to numbers into analogous considerations on the disposition of the points of a segment. So that the only way to *directly* ground the continuity of geometry would have been to mimic Dedekind's own definition for number theory.[3] Such a (very simple and yet extremely powerful) definition of geometrical continuity is formulated as follows:

> If all points of the straight line fall into two classes such that every point of the first class lies to the left of every point of the second class, then there exists one and only one point which produces this division of all points into two classes, this severing of the straight line into two portions.[4]

---

[2] See HILBERT (1968, first ed.1899), §8, pp. 30–33; the Axiom of Completeness was added in the French first edition (1900) and the second German edition (1903), and that on Linear Completeness (a variation of the previous one) in the seventh German edition (1930). On the history of Hilbert's completeness axiom, see GIOVANNINI (2013).

[3] This is hinted in the first ed. *Vorwort* to Dedekind's *Was sind und was sollen die Zahlen?*, where Dedekind discusses Jules Tannery's construction of continuity and denies Tannery's claim that a similar definition of real numbers had been given by Bertrand—since Bertrand had employed geometrical considerations, and therefore had to appeal to the continuity of space. The continuity of space, however, was simply assumed by Bertrand (and others) as "intuitively" given—something which Dedekind argued to have been a serious mistake, since a dense space, for instance, would still be intuited in the same way. On the contrary, notwithstanding the fact that we may subjectively begin, when examining what continuity is, from a notion formed with the help of spatial intuition (as Dedekind himself did in his work), we may only claim to have an *exact definition* of the continuity *of space* when we have already moved to logical considerations and a proper definition of the continuity of *numbers*: "All the more beautiful it appears to me that without any notion of measurable quantities [i.e. spatial magnitudes] and simply by a finite system of simple thought-steps man can advance to the creation of the pure continuous number-domain; and only by this means, in my view, is it possible for him to render the notion of continuous space clear and definite" (DEDEKIND 1932, vol. 3, p. 340; Engl. transl. Beman). Dedekind's explicit reference was to his previous work on *Stetigkeit und irrationale Zahlen*, and in particular to its §3, in which he had given (see the next *footnote*) a geometrical reformulation of his definition of the completeness of real numbers through Dedekind's cuts.

[4] The passage comes from §3 of *Stetigkeit und irrationale Zahlen*, and the original runs: "Zerfallen alle Punkte der Geraden in zwei Klassen von der Art, daß jeder Punkt der ersten Klasse links von jedem Punkte der zweiten Klasse liegt, so existiert ein und nur ein Punkt, welcher diese Einteilung

Such an axiom is in fact sufficient to rigorously prove all propositions of elementary geometry—and much more.[5] Dedekind's idea of the completeness of a straight line has a remarkable historical relevance for our study, as most of the attempts to define continuity in the early modern age were purely geometrical, and aimed at establishing the continuity of segments rather than the completeness of real numbers. Dedekind's idea that a formalization of continuity in geometry should be formulated in a similar way to the one employed in building the real field permits us to draw insightful comparisons between these modern endeavors and Leibniz's (and others') earlier attempts. We should note that Dedekind's geometrical formulation of his notion of completeness assumes that a line is actually composed by points (in some set-theoretical or topological sense of "composition"), and that this latter fact was one that was generally denied in antiquity and in the early modern era. We see, therefore, that any formulation of completeness attempted before the nineteenth century had necessarily to find a different way of approaching such a definition.

It should be noted, in fact, that, while Leibniz was naturally led to define the continuity of a function by his studies on infinitesimal analysis, his researches on the definition of the continuity of space were only partially elicited by these new discoveries and were rather discussed in Leibniz's logical and epistemological writings on the foundations of geometry. These papers, which Leibniz often denominated, quite generally, as his new enterprise of an *analysis of situation* (*analysis situs*), dealt with many foundational issues in geometry, extending from the proof of Euclid's axioms to the definitions of the basic relations of similarity and congruence, from a discussion on tridimensionality to new demonstrations of ancient theorems—and on, indeed, to many other topics besides.[6] Most of these Leibnizian writings referred to Euclid's *Elements*, which was the text that had always to be commented upon when dealing with the foundations of mathematics. The issue of the continuity of space, therefore, found its natural place in this kind of investigation. It is important to understand, in any case, that Leibniz's foundational studies on continuity were quite new, and that, even if the tradition of commentaries on Euclid could already boast a history of two millennia, the issue of continuity had only seldom been referred to therein, so that, in the seventeenth century, this issue was looked upon as a remarkably new problem in the foundational domain.

The *locus classicus* for the debate, at the time, was Euclid's demonstration of *Elements* I, 1.

---

aller Punkte in zwei Klassen, diese Zerschneidung der Geraden in zwei Stücke hervorbringt" (DEDEKIND 1932, vol. 3, p. 322; Engl. transl. Beman).

[5]An explicit proof that Dedekind's completeness allows one to prove the existence of the intersection points needed in elementary geometry is in HEATH (1925), vol. 1, pp. 237–40; see also VITALI (1923, pp. 210–12).

[6]For a treatment of Leibniz's investigations on *analysis situs* and geometry in general, see my DE RISI (2007, 2015). The main Leibnizian essays on *analysis situs* are still to be found in the fifth volume of the nineteenth-century edition of Leibniz's mathematical writings, edited by Gerhardt (GM VII, pp. 141–211). A useful collection of Leibniz's early essays on the same topic (in the original Latin and French translation) may be found in J. ECHEVERRÍA, M. PARMENTIER (1995).

4.2   Euclid's Gaps and Aristotle's Continuity

The first proposition of Euclid's *Elements* teaches how to draw an equilateral triangle on a segment assumed as its side. The problem is solved by tracing two equal circles having as their centers the ends of the given segment and as their radius the length of the segment itself. The two circles intersect at a point above the segment, and the triangle is drawn by connecting, with straight lines, the newly-found point with the two endpoints of the given segment. The possibility of the construction is grounded on Euclid's Postulate 3, that allows the drawing of circles with any center and radius, and on Postulate 1, that allow the connection, by straight lines, of any pair of points. The equality of all three sides of the triangle is grounded, in turn, in the definition of a circle, all the radii of which are equal to one another (thus $AB = AC$ and $AB = BC$), and in Euclid's Common Notion 1, which states that things equal to the same thing are also equal to one another (thus $AC = BC$).

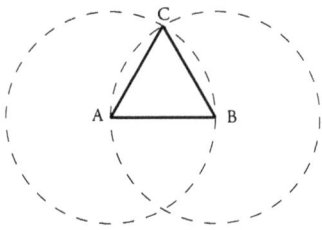

Despite the number of principles employed by Euclid to ground his first and simplest proof, modern interpreters object that one extremely important passage was overlooked, namely the assumption of the existence of the point of intersection C of the two circles. This requirement, the existence of the point of intersection, is read, in turn, as a requirement of continuity: the plane underlying Euclid's constructions should be sufficiently continuous to guarantee the existence of every point of intersection between circles, and of this one in particular. Euclid, however, provided no justification, either in his system of principles (postulates and common notions), or in any of his proofs, for the existence of the intersection points. In particular, the notion of continuity was never thematized in any ancient mathematical text, and neither Euclid, Archimedes, nor Apollonius ever mentioned this notion in their works (or defined geometrical objects as "continuous"). If they had any sort of implicit theory about intersections, it was surely based on other grounds than a notion of continuity.[7]

This notwithstanding, the notion of continuity (ϲυνέχεια) was explicitly thematized in a few *philosophical* discussions during antiquity, and Aristotle played here the most prominent role. Even though Euclid and the other Greek mathematicians were probably unaware of (or uninterested in) Aristotle's contribution on the subject, the importance of his conception grew with the passing of the centuries, and already at the time of the Roman Empire several commentators on ancient mathematical texts were blending the thought of Euclid with certain Aristotelian elements. In this connection, continuity made its appearance in a few mathematical

---

[7] As a matter of fact, I do think that Euclid and other Greek mathematicians may have had a solution to the problem of intersections, that was grounded on a subtle combination of diagrammatic inferences and the definition of a point. I have presented some evidence for this claim in my DE RISI (forthcoming a).

texts or paratexts of late antiquity, which were in turn transmitted to the Middle Ages and the Renaissance. In the early modern age, when a fuller discussion on space and continuity in geometry was gaining ground, the influence of Aristotle and his definition of this notion became paramount: so that it is impossible to ignore Aristotle's contribution to the topic even if it was not originally intended to implement the foundations of mathematics. Leibniz himself, in his attempt to build a new mathematical definition of the *continuum* could not help but begin by reinterpreting the old Aristotelian definition.

The first definition, or perhaps a characterization, of continuity that can be found among Aristotle's works states that a continuous quantity (a magnitude) is such that all its parts are themselves magnitudes, and are thus composed of further parts that are also magnitudes (and so on). This directly entails the infinite divisibility of a continuous quantity, and seems to amount (in modern terms) to equating continuity with *density*.[8] Aristotle's aim in proposing such a definition seems to have been that of critiquing several forms of atomism extant at the time in natural philosophy. In these passages, he always treats continuity as a monadic property.[9]

Aristotle's argument for infinite divisibility also aimed to prove that a continuous whole cannot be composed of points. This latter consequence of Aristotle's definition of continuity is very important and marks a milestone in the history of thought. It represented the main counter-thesis advanced by innumerable generations of philosophers and mathematicians against the Zenonian theses on the composition of the *continuum*. Aristotle, by rejecting the possibility of composing a magnitude from indivisibles, became committed to accepting extended quantities as the basic elements of geometry and thus to regarding points (or lines with respect to surfaces, and surfaces with respect to bodies) as mere boundaries of more primitive continuous wholes. In this respect, the modern set-theoretical approach to geometry, which starts from a set of points and constitutes a "continuous" manifold (whatever "continuity" may mean in its different contexts) by means of a further set of topological (or ordering) relations among points, is completely extraneous to the foundational perspective of antiquity. Aristotle's understanding of a magnitude is in fact more closely related to those "post-modern" mathematical theories, such

---

[8]We might also make a distinction between infinite divisibility and density, as the latter mathematical property seems to imply that a magnitude is composed by points—which is not the Aristotelian assumption. I will nonetheless conflate the two expressions from time to time, as they seem to be equivalent under many foundational respects, and the modern notion of density helps to shed light on the actual strength of certain axioms and assumptions. I thank Richard Arthur and Viktor Blasjo for stressing this point with me.

[9]This meaning of continuity as density is to be found in *De cael.* A 1, 268a7–8, where it seems to be presented as a definition. The idea that a continuous whole cannot be resolved into elements that have no parts (ἀλλ'οὐθὲν ἦν τῶν ϲυνεχῶν εἰς ἀμερῆ διαιρετόν) is also expressed in *Phys.* Z 1, 231b12, and is restated in *Phys.* Z 2, 232a23–25. In both passages of the *Physics*, the definition appears as a characterization of the definition given by the notion of a common boundary. The simpler idea that a continuous whole is indefinitely divisible (τὸ εἰς ἄπειρον διαιρετὸν ϲυνεχές) is to be found in *Phys.* A 2, 185b10–11; Γ 1, 200b20–21; and Z 8, 239a23. A *magnitude* (which is a continuous quantity) is said to be indefinitely divisible in *Phys.* Z 6, 237a33–34.

as pointless geometry, region-based topology, or the theory of *locales*, which (not unaware of the Aristotelian position on this issue) have developed tools to deal with continuity which are alternative to the set-theoretical approach.[10] All ancient geometers seem to share this point of view (even if they may have not shared, or even known of, Aristotle's arguments for it), and their mathematical constructions are grounded in a multi-sorted ontology, in which points, lines, surfaces, and solid bodies remain distinct from, and irreducible to, one another.[11] This has important consequences for the mathematical theory of intersections.

The Aristotelian notion of density is liable to a purely mathematical treatment. It comes as no surprise that already in late antiquity a few mathematicians were to define continuity, in geometrical treatises, as density (of course, there is no reason to think that they had in mind Aristotle's discussion in particular). Density, however, is not suitable to ground the Euclidean theory of intersections, since this latter requires some stronger continuity principle. In modern terms, a plane that should only contain points with rational coordinates would be dense (i.e. its parts would be infinitely divisible), but the intersection of the two circles in *Elements* I, 1 could still be missing.

In the Aristotelian *corpus*, however, we also find a more fundamental definition of continuity. The latter states that two things are continuous with one another if they touch one another and their boundaries are one and the same.[12] The subtlety of the definition mainly consists in its difference from the Aristotelian notion of contiguity: two bodies are said to be contiguous if they touch, but their boundaries are not one and the same. The definition of "touching", or being in contact, seems to be given by the overlapping of the boundaries of a thing (i.e. these are together, ἅμα, and therefore occupy the same place, τόπος). This definition would allow two contiguous things to be in contact while still allowing them to move one with regard to the other. This is the case inasmuch as their boundaries are not, here, unified in

---

[10]Among the numerous mathematical studies in these fields, I will mention here ROEPER (2006), which gives a theory of continuity without quantifying on points, but rather on extended regions. Roeper's system manages to recover Dedekind completeness, in this different setting, through a strong axiom on the completeness of a Boolean algebra employed in the formalization. A different axiomatization of linear regions which also entails completeness is given in the important HELLMAN, SHAPIRO (2013), which is more historically sensitive than Roeper's paper. The authors have further developed their views in LINNEBO, SHAPIRO, HELLMAN (2015), which limits the use of the actual infinite in order to better comply with Aristotelian conceptions of mathematics (but, in doing so, it loses completeness). The paper HELLMANN, SHAPIRO (2015), further shows how to extend to two dimensions the previous results for linear regions. The latter three articles have been recently collected in the volume HELLMAN, SHAPIRO (2018). For a survey of the most important modern contributions on pointless continuity (which date at least from Whitehead's seminal studies from the 1910s), see CALOSI, GRAZIANI (2014).

[11]I will not discuss here Archimedes' *Method* and his theory of indivisibles, which presents several interpretive difficulties. In any case, it seems not to have had any impact on the theory of intersections itself.

[12]The Aristotelian definition of continuity by the notion of a common boundary is especially expounded on in *Phys*. E 3, 227a10–13.

a stronger sense and may, therefore, "slide" one upon the other. So, for instance, a wheel is contiguous to (but not at the same time continuous with) the path in which it revolves, and the water is contiguous to the body submerged in it. A wooden table, on the other hand, is continuous inasmuch as two parts of it are not only in contact with one another but are held together by *one* common boundary.

It should be clear that such a definition was originally conceived as a relation between physical objects and was employed by Aristotle to explain his conception of motion.[13] Its use was later extended by commentators to encompass other domains, such as the continuity of colors and sounds or the continuity of geometrical objects; but its original aims seem not to have been so general. Outside of the physical domain, in which the motion of bodies immediately reveals the divide between contact and continuity (or concurrent motion), it is hard to get a firm and precise grasp on Aristotle's distinction between contiguity and continuity. That is to say, it is hard to say exactly what it means for two boundaries to be not only in the same place but also unified in some further way, and a large amount of exegetical discussion has been spent, during the centuries, in attempting to better spell out this latter distinction. While several metaphysical solutions to the problem might perhaps be found, a purely mathematical treatment of the difference between two coinciding points and one single point is not easy to envisage (using the tools of classical geometry alone).[14] It should also be noted that Aristotle generally excluded the possibility that two extensions (διαςτήματα) could share the same place or be together in any sense. Indeed, in the *Physics* he even argued against a conception of place (τόπος) as tridimensional extension on the grounds that, if place were so defined, then it and the located body would be two extensions overlapping together.[15] This directly entails that Aristotelian magnitudes cannot have a common *part*. It is not entirely clear how this general impossibility of overlapping can possibly be dropped in the case of boundaries. It is clear, however, that boundaries of parts should be allowed to overlap and be common to different magnitudes.[16]

---

[13]The distinction between contiguity and continuity in relation to motion is especially worked out in *Phys.* Δ 4, 211a24-b4, where Aristotle says that if a body is contained in another body and continuous to it, the former body is *a part* of the latter; while if it is contained and contiguous to it, the latter body is the *place* of the former. In the first case, the bodies move together; in the second case, the first body moves in the second as in its place. The Aristotelian idea that two bodies are continuous with one another when they move together (see also *Metaph.* Δ 6, 1016a5–6; I 1, 1052a20) had an important posterity in the modern era and in Leibniz in particular.

[14]Ibn Rushd, for instance, will later claim that the Aristotelian distinction between contiguity and continuity is only viable in physics, whereas in mathematics two figures are either continuous or completely apart from one another (i.e. their boundaries are either together and one, or not together at all). See some references in SYLLA (1982).

[15]The argument on the impenetrability of place as διάςτημα τοῦ μεγέθους is to be found in *Phys.* Δ 4, 211b19–25. Aristotle also simply states the idea that two *mathematical* bodies cannot interpenetrate, and therefore cannot have parts in common, in *Metaph.* B 2, 998a12–15; M 2, 1076b1–2.

[16]A less metaphysical and more mathematical explanation of the impossibility of the overlapping of parts may be derivable from the fact that quantity is for Aristotle just a set of parts, in the sense

Moreover, Aristotle elsewhere argued that boundaries cannot "be in a place" in any proper sense. Much less, then, can they share a common place. These (and other) conceptual difficulties brought several Aristotelian commentators to modify, or utterly reject, Aristotle's definition.

A further important feature of Aristotle's definition of continuity should also be highlighted: rather than defining a *continuous whole*, Aristotle always considers the continuous connection of the parts forming it. In other words, continuity is not a property of a thing but rather a *relation* between two (or more) things which are disposed (or *situated*) in a certain way with respect to one another (i.e. with touching and identified boundaries). The latter aspect of the notion implies that continuity cannot be attributed, as such, to certain magnitudes and figures, but only to the points where they join with one another. It is true, however, that Aristotle himself spoke quite often of "continuous wholes", employing the term as a monadic predicate; even more often he talked about the *continuum* (τὸ ςυνεχές). Such an expression would appear to signify, with some abuse of language, that the *parts* of such a whole are continuous with one another (in the relational sense). Aristotle normally has in mind *actual* parts, that is to say a whole which is divided into (continuous) parts by means of some internal criterion. A broken line, for instance, forming an angle, may be said to be "continuous" insofar as the two segments constituting it are continuous with one another in the common vertex. Thanks to this kind of continuity, the broken line constitutes a unity of some sort, notwithstanding the fact that its unity, according to Aristotle, is not a unity *per se* (καθαὐτόν). A straight, unbroken line, on the other hand, does indeed possess a unity in itself, but could hardly be conceived of as continuous in the Aristotelian sense. This is so because it is not made up of actual parts and therefore no continuity in the relational sense may be attributed to it.[17]

It is also true that, in some other passages, Aristotle seems to hold that it is possible to consider something to be continuous (intended here as a monadic predicate) even if its continuously connected parts are only potential. In this sense

---

that measuring it amounts to counting its parts. This would be ill-defined, of course, if one were to admit the possibility of overlapping parts. Boundaries, however, have, in any case, "zero measure" and therefore they may freely overlap.

[17]The passages on the broken line in Aristotle are to be found in *Metaph.* Δ 6, 1016a2–3, and 12–13. Bonitz noticed that the expression ἡ κεκαμμένη γραμμή (in 1016a2–3) may refer to both a broken line and a curved line (as in ARIST. *De inc. anim.* 9, 708b22), but the following passage (1016a12–13) clarifies that Aristotle had in mind the former (cf. BONITZ 1848, p. 235). This passage has been a source of puzzlement for many commentators, especially since in *Phys.* Θ 8, Aristotle states that a motion in a broken trajectory should be considered to be discontinuous. Yet in the latter passage he seems to have in mind that a moving object, in order to make a turn at an angle, would have to stop; and coming to a stop is "the contrary of motion" and would therefore imply a discontinuity in motion. The problem, therefore, would be the continuity of motion but not the continuity of the line traced by the motion. In fact, a motion along a broken line is said to be continuous in *Phys.* E 4, 229a2. I thank Marco Panza for a discussion on this point that helped me to clarify my views. Panza has dealt with Aristotelian continuity in relation to the geometrical theory of intersections in PANZA (1992); and in a longer, forthcoming paper, an early draft of which he was so kind to share with me. Aristotelian and Euclidean continuity have been compared also by CAVEING (1982).

a table may be said to be continuous inasmuch as the potential parts of it are all continuous with one another. It should be clear, however, that the potential parts of a table are still thought of in physical, rather than mathematical, terms: Aristotle is thinking of a few determinate parts of the whole that have not yet been spelled out (and sometimes he talks about nails or glue to enforce continuity). Nowhere does he make any allusion to the idea that, in order to ascribe continuity to a whole, one should have to consider *all possible parts* of it, in abstract and mathematical terms, as well as their reciprocal continuity relations.[18] In this respect, Aristotle at no point suggests a *structural* definition of continuity (i.e. one in which continuity would be a monadic property stemming from the system of relations among all possible parts of a whole), and generally intends this notion rather as a simple relational property.

Compared with the characterization of continuity as infinite divisibility, Aristotle's "physical" definition of continuity has a less mathematical outlook. Nonetheless, a few modern interpreters adopt the optimistic stance of recognizing in the Aristotelian definition a topological (or "proto-topological", as they put it) notion of *connectedness*.[19] While the intuitive idea of topological connectedness is clearly hinted at by Aristotle's definition, I think it is a bit of a stretch to see an exact characterization of such a property in this latter. The notion of connectedness relies, in an essential way, on a distinction between closed and open sets and, while Aristotle certainly has and employs a notion of a boundary, this seems insufficient to clearly establish the distinction at stake here. Most of all, however, Aristotle's failure to consider all possible parts of a whole (i.e. his lack of a structural approach to the notion) seems to preclude our attributing to him the distinction of having been the first to formulate the *mathematical* concept of connectedness, which requires the possibility of quantifying on all possible subsets of a given set.[20] In any case, whereas the relation between mereological parthood and topological connectedness

---

[18]The Aristotelian notion of a *potential* part should not be conflated with the modern notion of a *possible* part. The distinction seems to acquire some importance if we consider that parts, for Aristotle, are impenetrable (see above). A segment AD, in which we specify two intermediate points B and C, can be conceived of as constituted by two (continuous, potential) parts AB and BD, and it can also be conceived of as constituted by two different (continuous, potential) parts AC and CD. But it is impossible to conceive of these partitions together, for in this case parts AB and AC would overlap with one another. We may also recall that Aristotle endorsed a "statistical" notion of possibility, according to which something is possible if it is real at a given time. This notion is scarcely applicable to mathematics in any case.

[19]For the recent history of the notion of connectedness, see WILDER (1978).

[20]The main effort in this direction has been made by WHITE (1988). White offered some more pieces of evidence for his thesis in his later, and lengthier, work WHITE (1992), which deals, however, mostly with the physical notion of continuity. White's main point in establishing the modern notion of connectedness from Aristotle's definition is grounded in the statement that Aristotle never considered the existence of *open intervals*. While White discusses the issue at some length (especially in the book), a contrary instance might be *Phys.* Θ 8, 263b9–264a6. I will not, however, try to follow out this difficult question any further. I may also note that the above-mentioned LINNEBO, SHAPIRO, HELLMAN (2015) claims that "in the Aristotelian setting" some form of connectedness "is all the 'completeness' one can ask for" (p. 229).

is mathematically rich,[21] and Aristotle's work may offer some reflections in this direction, the concept of connectedness itself is too weak to be of any help in elementary geometry. Each of the two "circles" in *Elements* I, 1, may well be topologically connected but their intersection point still be missing (think of two open rings).

A last point that should be emphasized concerning the mathematical use of Aristotle's definitions of continuity is that the interplay between a conception of continuity as, on the one hand, a relation and, on the other, a monadic property also had an important impact on the geometrical theory of intersections. As a matter of fact, without considering continuity in terms of relations, it would be impossible to establish the existence of the intersection between two figures, since from monadic premises such as "Circle α is continuous" and "Circle β is continuous" there would be no way to conclude "Point C exists".[22] In late antiquity, the Middle Ages and the early modern era, it was quite common to see a geometrical magnitude defined as a continuous quantity, without further qualification or explanation and without any reference to a theory of relations. In such formulations the possibility of actually employing continuity in a geometrical proof was simply lost.

Aristotle's relational conception of continuity avoided the pitfalls involved in a monadic conception of this latter. Yet, Aristotle's own syllogistic does not allow for a treatment of relations; there was, therefore, no way, in antiquity, to implement continuity as a relation into a deductive system. This is, of course, a wider historical problem and it can be said that *since* a logic of relations was not developed in antiquity, *therefore* any deduction concerning continuity was left unexpressed in the geometrical proof (and possibly assumed through inspection of the diagram). The lack of such a logic of relations, in turn, may have been rooted in a general metaphysical world-picture that took individual substances and their properties to be the only existing beings. Such a world-picture would only alter with the advent of the early modern age, when relations began to be thematized as independent entities, when a specific logic of such independent entities was first conceived of, and when *space*, as a system of relations (a structure), was introduced into both metaphysics and mathematics.

4.3   The Problem of Continuity in the Early Modern Age

In the early modern age, geometers became aware that Euclid's proof of *Elements* I, 1 presented a demonstrative gap of some sort, and, as a consequence, the theory of

---

[21] The entire field of *mereotopology* is devoted to exploring the connections between the notions of part-whole and continuity. In this field of study the concept of topological connectedness is often used as the main bridge between the two disciplines. For an introduction to the subject, see VARZI (1994); COHN, VARZI (2003); and PRATT-HARTMANN (2007).

[22] This very neat observation is made by FRIEDMAN (1992, pp. 60–61).

intersections first became an object of mathematical investigation. Their approach to the matter, however, was more pragmatic than theoretical, and they did not embark in attempting to provide a general definition of continuity that could have been employed in geometry. They opted for the easier way to add a few *ad hoc* axioms to the principles of Euclid's *Elements*, that could bridge the gap in the proof of *Elements* I, 1 (and cognate propositions). These principles are still in use in today's axiomatizations of elementary geometry, where they are called the Line-Circle and the Circle-Circle intersection principles:

> If a straight line has one point inside a circle, the line will meet the circle.
> If one of two circles has one point inside the other and one point outside it, the two circles will meet.

Such principles are in fact sufficient to ground the whole theory of intersections in elementary Euclidean geometry, and avoid to make any reference to the notion of continuity.[23]

The first instance of such principles is probably to be found already in Oronce Fine's *Protomathesis* from 1532, who added similar axioms in order to give a gapless proof of *Elements* I, 1. In the following century, however, the use of this kind of intersection axioms became widespread, and we find similar principles thematized and discussed in the essays on the foundations of geometry by Claude Richard (1645), Blaise Pascal (1655), Giovanni Alfonso Borelli (1658), Caspar Schott (1661), Gilles Personne de Roberval (1673–1675) and others.[24]

Despite the fact that such authors did not attempt a general definition of continuity, some of them *justified* the newly-introduced intersection axioms by stating that they derive from the fact that the geometrical figures themselves (such as the circles in *Elements* I, 1) are continuous—in some unspecified sense. This was a common view at the time: we may not possess a good definition of all the elementary geometrical notions, but we can make up for them by adding axioms that operationally characterize such notions. In a similar vein, for instance, several geometers of the seventeenth century claimed that no good definition of a straight line had ever been given, but that a few axioms about straight lines (that there is a straight line between two points, that two straight lines do not enclose a space, etc.) were sufficient to safely employ this notion in geometry. Continuity, then, could have been supplanted by the Line-Circle and Circle-Circle intersection axioms, without engaging in the most difficult task of giving an exact definition of this notion.

Leibniz was well aware of these developments. He had studied and annotated Borelli's book, and employed Schott's principles in his geometrical researches. Most importantly of all, he had had the chance, during a visit to Paris, to read the

---

[23] See HARTSHORNE (2000, pp. 104–16).

[24] For a quick survey of these editions of Euclid and their intersection axioms, see DE RISI (2016). I am expanding on the subject, detailing the various early modern solutions, also in my DE RISI (forthcoming b).

manuscript notes by Pascal on the foundations of geometry and to study Roberval's papers on the same subject.[25]

Nonetheless, Leibniz's approach to continuity was radically different from anything that had been attempted before and his work on the subject, however tentative and imperfect it may be judged to be, opened a new era in the examination of this notion in mathematics. The main revolution, compared to those preceding theorizations, consisted in the fact that Leibniz now attempted a directly mathematical definition of continuity (in general) rather than being content to accept a number of local and *ad hoc* statements which merely fixed up the gaps in the Euclidean proofs.

There are a number of reasons for this radically new attitude. Leibniz's studies on the Calculus are clearly the main driving force behind his investigation of continuity, since they required a general understanding of this notion. General philosophical concerns were also at stake, because Leibniz was able to capitalize upon his mathematical discussions of continuity, which had arisen out of the Calculus, turning them into a whole range of metaphysical principles of continuity, which may have been a little vague in their formulation but still had, in his eyes, a strong heuristic value for philosophical investigations. Mathematical epistemology also played an important role here. Leibniz's foundational project in geometry, in fact, relied not so much on the introduction of new axioms as on their progressive elimination by proofs. His ideal geometry was a science without axioms: a science which would be founded upon the Principle of Contradiction alone and which would show that pure logic is powerful enough to establish all the truths of Euclidean geometry. In pursuit of this aim, he wanted to prove continuity and intersections rather than merely assuming these latter on the basis of special principles. In this sense, the continuity principles of Pascal, Borelli, or Roberval seemed to him to be an intermediate stage in the process of perfection of the science; they represented the realization that something was missing in the ancient proofs, as well as an *ad hoc* way of "fixing up" these latter that still needed, however, to be fully worked out by "proving" these *ad hoc* statements from the geometrical definitions.

It should be added that, in the few years that elapsed between the works of Richard and Borelli and the investigations of Leibniz, geometry had been extended to a far wider domain of objects than those dealt with in Euclid's *Elements*. Whereas a foundational solution provided for the geometry of the *Elements* might still have been conceived to be applicable to a few other cases extending beyond the ambit of this latter (such as Apollonius's conic sections, or perhaps Descartes's algebraic geometry), it was clear that a discussion on continuity and intersections conducted at the end of the seventeenth century which aspired to ground the possibility of the new geometry emerging in this period could not possibly be restricted to the narrow domain of circles and straight lines alone. Leibniz's foundational studies, in particular, aspired to provide a foundation for the whole of mathematics, not just elementary geometry. He could not rest content with the "Circle-Circle" or "Line-

---

[25] See DE RISI (2015).

Circle" continuity principles and needed the full Dedekind completeness. In fact, in the many studies that Leibniz undertook on the text of the *Elements* and the foundations of Euclidean geometry, I have not been able to find any reference at all to *ad hoc* axioms with which the theory of intersections might be grounded.

Nonetheless, before turning to his discussion on continuity, it should be remarked that something similar to a set of *ad hoc* axioms for intersections is to be found in Leibniz's symbolic studies on a new *characteristica geometrica*. These latter researches were 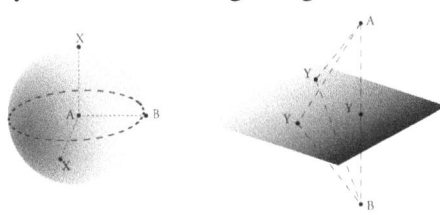 aimed at producing a formal system, similar to Cartesian algebra, which, rather than expressing quantities and magnitudes, would directly express points and geometrical relations. Leibniz began to work at developing such a logical system around 1679 and continued to do so for many years, without, however, ever achieving any real breakthrough. In this formal system, Leibniz could express geometrical points by means of letters, situational relations by means of a functional symbol, and other geometrical relations among situated elements, such as congruence, similarity or equality, by means of other operational symbols. A kind of universal quantifier, expressed by a mark over a letter, completed the set of symbols for most elementary applications. To give an idea of such a symbolic system, a sphere in space is here expressed as $\overline{X}$: A.B $\simeq$ A.X, meaning that the sphere is the set of all points X's such that the situational relation of A to B is congruent to the situational relation of A to X, the latter condition implying that the distance between A and B is equal to the distance between A and all X's. A plane in space, on the other hand, is expressed by $\overline{Y}$: A.Y $\simeq$ B.Y, meaning that it is the collection of points Y's that are at the same distance from the two points A and B. Leibniz thought that the whole of geometry should be expressed in such a symbolic system, which he called a *characteristica geometrica propria*, that is to say, a system directly representing geometrical objects and relations, without any admixture of numbers or of extraneous functions and relations such as those of ordinary Cartesian algebra. Leibniz's treatment of intersections in this system is given axiomatically and not by a general definition of continuity. In fact, in his first attempts at such a characteristics, he basically accepted a few symbolic axioms in order to guarantee the existence of the points of intersection required. Thus, given two planes $\overline{Y}$: A.Y $\simeq$ B.Y, and $\overline{Y}$: A.Y $\simeq$ C.Y, these will intersect in the straight line $\overline{\overline{Y}}$: A.Y $\simeq$ B.Y $\simeq$ C.Y, obtained by composing the two symbolic expressions; and the two spheres $\overline{X}$: A.C $\simeq$ A.X and $\overline{X}$: A.B $\simeq$ A.X would intersect in the circle $\overline{X}$: A.B.C $\simeq$ A.B.X, obtained from the previous expressions by a different composition rule. Leibniz gave similar combinatorial rules to show that two straight lines meet in a point, and two circles meet in two points.[26] It is clear that these very particular cases of intersections cannot be easily generalized to other figures, and that, while dispensing with a notion of continuity, they are to be considered largely insufficient to ground geometry. Leibniz never attempted to justify the combinatorial axioms themselves and it is not easy to fathom how they could have been proven from more elementary combinatorial assumptions (or from the Principle of Contradiction). In later years, as we will see, Leibniz also attempted to formalize his notion of continuity in the language of the *characteristica geometrica*.

4.4 Leibniz's Early Studies

We now turn to Leibniz's studies on the notion of continuity, in so far as they were aimed at defining a geometrical *continuum* and therefore grounding a general theory of intersections in elementary geometry. These Leibnizian researches often took their lead from the definitions of continuity offered by Aristotle, reworking them in just a few details, or simply reinterpreted the terms of the definition, giving to the old words a completely new meaning. By this progressive adjustment of the old definition, Leibniz finally achieved a new, mathematical definition of continuity (as completeness) which far surpassed any Aristotelian notion.

As a young man, Leibniz often repeated, even *ad literam*, the Aristotelian definition of continuity and contiguity.[27] Even in these early years he already attempted some reinterpretations, and sometimes addressed the classical problems of such a definition. In 1670, for instance, attempting to ground a theory of the composition of the *continuum* by indivisibles (something that Aristotle had denied), he asserted that such indivisibles had to be non-extended yet nonetheless composed of parts, so that they could touch each other in some parts and (following the Aristotelian definition) constitute something continuous.[28]

Later on, he rejected such a view, and attempted to define continuity as the togetherness of the boundaries of two bodies in the same place (as Aristotle had done in the case of contiguity), trying to explain the difference between continuity

---

[26]The rules may be easily extended to any number of dimensions and, even if Leibniz did not pursue this line of reasoning, he was applauded by Grassmann for having found such general formulas (see GRASSMANN 1847).

[27]See for instance his first essays on physics written in Paris, the *Propositiones quaedam physicae* from 1672: "Et c'est la difference entre les choses contiguës, qui se touchent seulement dont les extremitez sont ensemble, et entre les choses continues, qui se presentent, dont les extremitez sont devenuës un. Comme Aristotle meme l'observe" (A VI, 3, n. 2, p. 8). At the end of his Parisian stay, in the great dialogue *Pacidius Philaleti*, he discussed the whole matter once again (proposing his peculiar solution to the problem of the composition of the *continuum*), and mentioned once again Aristotle's definition and distinction: "Memini Aristotelem quoque Contiguum a Continuo ita discernere, ut *Continua* sint quorum extrema unum sunt, *Contigua* quorum extrema simul sunt" (A VI, 3, n. 78, p. 537).

[28]See the *Fundamenta praedemonstrabilia* of the *Theoria motus abstracti*, in A VI, 2, n. 41, p. 264.

and contiguity by the forces which impel such bodies. In this way, he attempted to replace Aristotle's metaphysical reference to unity by a physical explanation. He discarded, however, this theory as well in the following years.[29]

Leibniz also attempted to offer a less physical and more geometrical distinction between continuity and contiguity, by accepting a Scholastic development of the Aristotelian notion of contiguity, and defining contiguous bodies as bodies that are non-distant to one another:

> *Contiguous* things are those between which there is no distance.[30]

In other writings, in fact, he explicitly corrected the Aristotelian notion:

> From which one understands that the boundaries of contiguous things are not only "not one" but also "not together".[31]

In some mature writings Leibniz claimed that two contiguous bodies are separated by an "unassignable" space, that is to say, by a space smaller than any *given* space. The use of quantifiers in these last statements is, however, a little uncertain. On the one hand, if we accept the Archimedean Axiom, the distance in question here seems to have necessarily to remain finite, even though it is *vague* (inasmuch as it is not assignable) and Leibniz may in fact have endorsed this view

---

[29] See A VI, 3, n. 4, p. 95–96, from 1672. The theory was refuted in the *Pacidius Philaleti* from 1676. A fine account of Leibniz's early theory of the cohesion of matter and of this latter's relation to his later views on continuity is to be found in LEVEY (1999).

[30] The passage comes from a 1672 fragment on forces and physical bodies: "*Consistentia* corporum est quantitas virium necessariarum ad solutionem continuitatis. *Continua* sunt contigua cum aliqua consistentia. *Contigua* sunt inter quae distantia nulla est" (A VI, 3, n. 4, p. 94; translated in ARTHUR (2001, p. 19). It is not easy to make out whether, and if so how, Leibniz's conception of contiguous, undistant points was accepted by him in mathematics. We may mention that, in 1698, Leibniz stated very clearly to Johann Bernoulli the impossibility of giving two points that are infinitely close or next to one another: "...non tamen sequitur duo puncta dari sibi infinite vicina, et multo minus dari sibi proxima" (GM III, p. 536). In 1705, however, we find him writing to De Volder that "we cannot conceive of three continuous points in a straight line; but two may be conceived of: the extremities of two straight lines from which the continuous whole results" ("Tria puncta continua in eadem recta concipi non possunt. At duo concipiuntur: extremitas unius rectae et extremitas alterius rectae ex qua constat idem totum", in GP II, p. 279). The apparent oscillation in his views here may perhaps be accounted for by saying that the passage addressed to Bernoulli concerns ideal objects, while the passage to De Volder deals with the real world. In any case, the notion of contiguity was still there at the end of Leibniz's life. We may notice that the terminology of points that are next to one another is clearly derived from the indivisibilist scholastic tradition. We find similar passages, for instance, in J. WYCLIF (1893, vol. 3, p. 30): "nam est dare duo puncta immediate, ut patet de corporibus tangentibus, sic ubicaciones vel situaciones eorum sunt immediate", quoted in ROBERT (2009, p. 155 n. 116).

[31] The passage comes from Leibniz's 1693 remarks on the celebrated book by Libert Froidmont, the *Labyrinthus sive de compositione continui* (1631). These remarks of Leibniz's have been published as ALCOBA (1996), but they have been recognized as reading notes to Froidmont only by PALMERINO (2016). Our quote is from p. 194 of Alcoba's paper: "Unde etiam intelligitur contiguorum extrema non tantum non esse unum, sed ne quidem esse simul". Some remarks on the relations between Leibnizian and Aristotelian views on continuity may be found in CROCKETT (1999).

(he took it to be further evidence that bodies are not completely real, and therefore an argument in favour of phenomenalism). Nonetheless, it is not clear how such a notion could be formalized in geometry.[32] On the other hand, it is possible that these unassignable spaces are to be regarded as actually infinitesimal *vacua* between parts of matter. They would thus tend to push Leibniz toward a non-Archimedean geometrical system.[33] In fact, there seems to be no viable topological notion of an immediate contact of two boundaries in classical (Archimedean) mathematics. A space that would allow such contact would not be *normal* (in the technical topological sense, i.e. the sense of not satisfying axiom of separation T_4)[34] and such spaces cannot be metrized: should we accept that two boundaries might be at no distance at all from one another, the notion of distance would simply disappear. Indeed, even if we could manage (with some difficulty) to formalize such a notion of contact in mereology (as the Aristotelian and Leibnizian terminology of parts and wholes directly suggests we might), the resulting mereological structure could

---

[32]It should be noticed that already the above-mentioned book by Froidmont had claimed that the notion of contiguity could not be expressed by a contact or overlapping of boundaries, but rather only by a compenetration of *indeterminate* parts. Froidmont applied this solution to both physical and mathematical contact. It is possible that such a theory could be formalized making use of a formal notion of vagueness, but I will not further pursue this line of thought here, since Leibniz was to abandon such a view in his mature years. On Froidmont's theory of the *continuum*, see also PALMERINO (2015).

[33]The above-mentioned passage from Leibniz's reading notes on Froidmont clearly represents an opening toward non-Archimedean considerations (see pp. 194–96 of the quoted paper by ALCOBA 1996). After having defined contiguity as undistant contact, Leibniz gives the example of a set of smaller and smaller spheres, one inside the other, which are tangent at a single point. Take for instance three spheres α, β, γ, such that γ (the smaller one) is in β, which is in α (the larger one), while all three of them are tangent at point T. Leibniz says that the smaller, more internal sphere β can touch the larger external sphere α at point T, but the third even smaller sphere γ, inside the second β, *does not* touch the first larger sphere α (because the second one is *interposed* between them), even though it passes through point T as well. He concludes that if we take an infinity of such concentric tangent spheres, we will have a "full point" (*punctum plenum*), i.e. T, which contains an infinity of distinct and *ordered* points, and which therefore cannot be distinguished from an unassignable segment. In the following pages, Leibniz is naturally brought to consider sheaves of parallel, undistant straight lines, and he finds them more problematic. His notes conclude in a somehow inconclusive fashion, a few pages later, by stating that perhaps one has to deny the existence of actually infinitesimal lines.

[34]The T_4 separation axiom states that given any two closed sets, there are two *disjoint* open sets each containing one of the closed sets. While the topological subtleties involved in such a proposition could not possibly be grasped in the seventeenth century, the intuitive meaning of such a principle is that between the boundaries of two figures there must be "space enough", thus forbidding contiguity in the sense of "immediate contact". Roberval, in his *Eléments de géométrie* has a similar postulate, stating that "Quelque grand que soit un solide fini et terminé, l'esprit s'en peut encore représenter un plus grand et meme de sorte que le plus grand contiendra entièrement le moindre et le surpassera de toutes parts, le petit étant une portion du plus grand. On peut aussi entendre que ces solides inégaux soient entièrement séparés de lieu" (Book I, Post. 7; ed. JULLIEN 1996, p. 99).

not be strengthened into a metrical one, and would therefore be useless in classical Euclidean geometry.[35]

The notion of contiguity, however flawed from a mathematical point of view, allowed Leibniz to formulate in the Parisian years (1672–1676) a theory of matter that he endorsed (with small modifications) for the rest of his life. Leibniz denied the existence of atoms and accepted the infinite divisibility of matter. He also denied that bodies are actually divided, to infinity, into points (because he denied that points could possibly compose the *continuum*). Instead, he endorsed the (quite non-Aristotelian) view that bodies are infinitely divided *in actu*, but in a purely "syncategorematic" way. This latter notion aimed at expressing the idea that there is no final element in the division of matter (i.e. no point), even though there are more divisions of bodily parts than can possibly be expressed by any finite number. A useful comparison may be made here with an infinite series (a mathematical tool that Leibniz was gradually mastering during his Parisian years), in which there is no final element (i.e. no actual infinitesimal term), even though there are more terms belonging to the series than any finite number can possibly express, and these terms draw progressively closer and closer to zero.[36] Leibniz claimed, therefore, that matter is not divided, all the way down, into points, but is rather divided to infinity following a certain succession of smaller and smaller corpuscles, such that each of these corpuscles is always further divided in some (but not all) possible ways. In mathematical terms, we might say that the system of boundaries in matter is *dense* in this latter.[37]

A first consequence of Leibniz's thesis of the actually infinite division of bodies was that he began to regard continuous magnitudes not as faithful representations of real bodies (as Aristotle would have held them to be), but rather as abstract and ideal objects employed in mathematics by means of which we can attempt to model the

---

[35]For a few difficulties involved in capturing the notion of contiguity in a mereological setting with some topological elements (but no metrics), see, for example, CASATI, VARZI (1999); and BENNETT, DÜNTSCH (2007). See also the above-mentioned HELLMAN, SHAPIRO (2013), which offers an axiomatization for linear regions without points that allows for a kind of contiguity (i.e. "adjacency"). In this context, however, to be contiguous only means that no *region* is interposed between two other regions—which are therefore "adjacent" to one another. It should be noted that this notion of contiguity is much weaker than the Aristotelian one (not to mention the one recurring to a vanishing distance). Aristotle himself called "consecutive" two parts such that between them there is nothing of the same genus (two houses are consecutive to one another if no further house is between them, but they need not to be contiguous; the same applies to two lines separated by something which is not a line): see ARIST. *Phys.* E 3, 226b35–227a5.

[36]See for instance S. LEVEY (1998), ARTHUR (2001, 2013, 2018).

[37]The most important text from the Parisian period on this subject is the dialogue *Pacidius Philaleti* (A VI, 3, n. 78, pp. 529–71) but this cluster of ideas is restated several times in the mature writings. On Leibniz's notion of continuity in his early and Parisian years, see BEELEY (1996); ARTHUR (2001), which offers a rich anthology of Leibniz's writings on the topic, accompanied by a thoughtful and extensive commentary; and LEVEY (2003), which also discusses at some length the notion of contiguity and approaches Leibniz's later solutions to fractal geometry.

(actually discontinuous) world, but only up to a certain degree of approximation.[38] This prompted him, in the 1680s, to develop a more formal definition of continuity for mathematical objects, one which rested upon the notion of indeterminacy. While real bodies, he argued, are divided to infinity through a system of nested boundaries which separates them into smaller and smaller contiguous particles, an ideal, continuous extension is characterized by the absence of such actual divisions and is susceptible of being divided in whatever way the mathematician may wish to divide it.

This Leibnizian definition recurs extremely frequently in his mature works. Although it was slightly reformulated from time to time, the basic definitional system is the following: a *continuum* is a whole the parts of which are *indefinite* and have a reciprocal *position* with respect to one another.[39] The parts are indefinite because the continuum is not actually divided and no system of boundaries is given. In this respect, Leibniz is simply restating, as a definitional property, the old Aristotelian idea that a continuous whole is only potentially divisible: but he is applying this property of merely potential divisibility not to real things (which are, for him, not just potentially but actually divided) but rather to geometrical objects. He is thus mathematizing Aristotle's definition by limiting it to the domain of ideal entities. The further definitional requirement here—namely, that the parts of a continuous whole should have a position with respect to one another—is intended to exclude from the class of continuous wholes both numbers and intensive magnitudes (heat, force, etc.). Sometimes Leibniz simply states that a continuous whole has *partes extra partes*, where the preposition *extra* is intended to mark the positional nature of the relation.[40]

It should, lastly, be mentioned, that Leibniz, like many other mathematicians and philosophers before him, also happened to characterize, from time to time, continuity as density. He saw the possibility of finding new terms between any two terms of a set as a mark of continuity.[41] Nonetheless, it is in this connection that we find his first true breakthrough.

---

[38] See GARBER (2015).

[39] Cf. this passage from 1680–1684: "*Continuum* est totum cujus partes indefinite assumi possunt, et *habent positionem* inter se. In quo differt ab unitate, et a toto intenso, ut potentia, calore" (A VI, 4A, n. 97, p. 390). Several similar definitions, however, may be found among Leibniz's papers.

[40] See, for example, C 438: "*Continuum* est totum cujus partes sunt extra partes, et indeterminatae. Nempe *extra partes*, id est separatim perceptibiles, ut distinguatur à Graduali toto, cujus partes se penetrant; cum aestimatur intentio qualitatum; *Indeterminatae* vero sunt *partes* continui, quia nullae jam sunt assignatae, sed pro lubitu assignari possunt, ut distinguatur à contiguo".

[41] Already in 1671, Leibniz qualified the common definition of continuity as density by stating that a continuous whole is a whole such that between any two *parts* of it one can find a further *part*, which is, of course, a different property than simple density (because it quantifies over parts rather than points): "Continuum est Totum inter cuius partes quaslibet interjectae sunt aliae partes ejusdem" (*De natura rerum corporearum*, A VI, 2, n. 45₂, p. 307). We will see that this was to be an important feature of Leibniz's mature notion of continuity. In other texts, however, he was not so careful and simply referred to finding points between any two given points.

In fact, Leibniz's theory of the infinite division of matter is mathematically very relevant in so far as it may hint to a distinction between density and continuity. Leibniz, in fact, was obliged to attempt to characterize an infinite sequence of elements (i.e. the boundaries of bodies) within the *continuum* of space, which is dense in this latter but does not exhaust it (since, in addition to the boundaries, there is also matter in space).[42] This is a development of the utmost importance for a mathematical understanding of continuity and the first step toward the possibility of defining completeness in the modern sense. I was not able to find any trace of a distinction between continuity and density in any author preceding the work of Leibniz (and, truth be told, not even in essays written many decades after Leibniz's death). Leibniz's very peculiar conception of an actual, syncategorematic infinite, which found no sequel among his successors, seems to be the source of this important step forward.

The clearest passage stating a mathematical distinction between density and continuity, however, was written by Leibniz much later than his first introduction of a syncategorematic infinity. It is to be found in his mature treatise *Specimen geometriae luciferae*, which is the very essay in which Leibniz's new conception of continuity as completeness was (as we shall see) to make its first appearance. The *Specimen*, a text probably written in the mid-1690s,[43] is one of Leibniz's most daring explorations of the foundations of geometry and collects together a number of new and outstanding ideas in mathematics. In a section of it, Leibniz discusses continuous geometrical transformations and gives the example of the transformation of a circle into a family of ellipses having the same area. He says that the set of ellipses thus produced may be regarded as *continuous*, and that we may, in fact,

---

[42] We may note that Clarke (and possibly Newton himself, who was behind several of Clarke's observations in the famous correspondence with Leibniz) objected to this theory of matter by saying that an infinity of divisions in bodies would result in bodies disappearing: "If therefore carrying on the division *in infinitum* you never arrive at parts perfectly solid and without pores, it will follow that all bodies consist of pores only, without any matter at all—which is a manifest absurdity" (Clarke's *Fourth Reply*, postscript; in ROBINET 1957, p. 116). The argument seems to conflate density with continuity.

[43] The dating of the *Specimen* is extremely difficult, since there are no external (i.e. material) criteria for it, and one is forced to rely on internal content only. I accept the dating proposed by the *Leibniz-Archiv*, since this concords well with the idea that it is a development of Leibniz's thought from the very similar essay *Hic memorabilia...* (in DE RISI 2007, pp. 586–87) which we will discuss below and which is most likely also to be dated to around the mid-1690s (there might be taken as possible proof of this the use of the rare term "epharmostica", which appears in this text and then in Leibniz's fragment *De ordinatione cognitionum* in A IV, 6, n. 68, p. 500, a text dating from not earlier than July 1695; and in another couple of essays, such as the *Euclides in definitione Diametri Circuli* in LH XXXV, XII, 2, Bl. 107, and the *Analysis didactica* in LH IV, 7C, bl. 139–145, also surely dating from the 1690s). Moreover, the terminology employed by Leibniz in the *Specimen* is strictly connected with a *Scheda* on situation and extension (published in DE RISI 2007, pp. 588–89), also dating from a period within that timeframe (cf. the use of the notion of *congeneum*). The *Specimen* would seem also to have been written after 1693, because it appears to display a development with regard to the notion of continuity as compared to Leibniz's reading notes to Froidmont (see above, *note* 31).

give a geometrical structure to this set by constructing a solid body the sections of which are the ellipses of the set. Such a solid body, *composed* by the ellipses, would be itself continuous:

> We may conceive something as continuous not only among those things that exist at the same time, and in fact not only in space and time, but also in a transformation and in the aggregate of all states of a certain continuous transformation. For instance, if we assume that a circle is continuously transformed, and goes through all the species of ellipses preserving its magnitude, then the aggregate of all these states, that is to say of all the ellipses, may be regarded as continuous even if the ellipses are not in contact and do not exist together but one of them is produced by the other. We may, however, take a family of ellipses that are congruent to those obtained by the transformations, and compose a solid figure from all these ellipses, that is to say, a solid whose sections parallel to the base are all those ellipses taken in order.[44]

This is already a remarkable passage, for it shows that Leibniz was able to conceive of a solid as composed of "indivisible" surfaces. The continuity of such a solid seems to be guaranteed by the continuity of the transformation, which makes it possible to pick up all the ellipses (without any missing elements). In the immediately following passage Leibniz went on to state that, if we were to consider the transformations of a sphere into a family of spheroids, the outcome would be a four-dimensional continuous figure which, however, we would not be able to *exhibit*, "for in extension we do not have more than three dimensions". We might, however, express such a notion by employing other non-geometrical means (i.e. by exhibiting these extra dimensions using weight or speed, for example).

In any case, Leibniz says, the important point is that we may construct a (biunivocal, and continuous) correspondence between the family of ellipses and the points of a segment (the height of the solid thus composed). In the case of spheroids, their transformations into other spheroids of equal volume is not a one-parameter transformation, and therefore we should place them not in correspondence with a segment but rather with the points of a plane (for a two-parameter transformation), or a solid body, or a *n*-dimensional object. From a mathematical point of view, Leibniz seems to be remarkably close to the modern idea of a continuous fibration of a manifold.

He goes on to give some examples of these fibrations and then concludes by saying that it is easy to find a family of figures in which, between any two of them, we may always find another (given a rule for interpolation), these figures *not*, however, being *continuous* because they do not exhaust the points of the

---

[44]The text comes from GP VII, p. 285: "Possumus continuum aliquod intelligere non tantum in simul existentibus, imo non tantum in tempore et loco, sed et in mutatione aliqua et aggregato omnium statuum cujusdam continuae mutationis, v. g. si ponamus circulum continue transformari et per omnes Ellipsium species transire servata sua magnitudine, aggregatum omnium horum statuum seu omnium harum Ellipsium instar continui potest concipi, etsi omnes istae Ellipses non sibi apponantur, quandoquidem nec simul coexistunt, sed una fit ex alia. Possumus tamen pro ipsis assumere earum congruentes, seu componere aliquod solidum constans ex omnibus illis Ellipsibus, seu cujus sectiones basi parallelae sint omnes illae Ellipses ordine sumtae".

parameter (the segment, or plane, etc.). In this way, he actually manages to provide a mathematical discussion of the difference between density and continuity:

> Thus we understand the nature of a continuous transformation and that it is not enough, in order for such a continuous transformation to be said to obtain, that, between any two states, it be possible to find an intermediate one: it is possible, in fact, to conceive some progressions such that there are always further interpolations, and nonetheless they must not be conflated with a continuum.[45]

Even though the difference between continuity and density is possibly never stated again with this degree of clarity, it seems that Leibniz never fell back behind the point of this important breakthrough.[46] In fact, the *Specimen* is the first essay in which Leibniz is able to formulate his last, and most fully perfected, definition of continuity and it is entirely possible that such a novel definition arose together with, and as a result of, the clear conceptual distinction between continuity and density.

We should finally remark that the continuity of the figures composed by the fibrations seems to derive from the continuity of the parameter itself; that is to say, this continuity appears to be derived from the biunivocal correspondence with the points of a segment and the continuity of the segment itself. In this connection, Leibniz's distinction between density and continuity seems to offer some indication of the fact that the points of the segment are in some sense the "totality" of all possible points. In a still informal sense, Leibniz is treating the segment (i.e. the parameter, or the basis of the fibration) as a *complete* set of points.

This idea also resonates with some other considerations that Leibniz engaged in elsewhere with regard to transformations. We have already mentioned that Leibniz possessed a very good analytic definition of a continuous function, given by a kind of ε-δ formalism. While such a definition is far from sufficient to define any kind of continuity of space, Leibniz had also formulated a continuity principle based on his notion of a continuous function, stating that in continuous things the last

---

[45] Here the original passage: "Itaque ex his etiam mutationis continuae natura intelligitur, neque vero ad eam sufficit, ut inter status quoslibet possit reperiri intermedius; possunt enim progressiones aliquae excogitari in quibus perpetuo procedit talis interpolatio, ut tamen non possit inde conflari aliquod continuum..." (GM VII, p. 287).

[46] This point may be recognized in several quite indirect ways. One of them is that, in a few essays written after the *Specimen,* Leibniz states that the infinite divisibility (i.e. density) of a straight line may be inferred by its definition as a self-similar line (i.e. a line, each segment of which is similar—in mathematical terms—to the whole). This deduction of density is repeated in the *Nouveaux Essais*, II, XVII, § 3, from 1704 (A VI, 6, n. 2, p. 158); in a letter to Des Bosses dated February 14th, 1706 (GP II, p. 300); in another letter to Wolff from 1711 (BW, p. 141), and in the essay *In Euclidis πρῶτα* from 1712 (GM V, p. 206). In none of these passages, however, does Leibniz state that the straight line is *continuous* for this very reason. He seems to be aware that through his argument from similarity he may only derive a weaker property than continuity. Nevertheless, it must be noted that, in a very late letter to Des Bosses dated May 29th, 1716, Leibniz still claimed that if between any two points of an extension there is another point (i.e. if the set of points is dense), then the extension is continuous, thus relapsing into the common view: "Eo ipso, dum puncta ita sita ponuntur, ut nulla duo sint, inter quae non detur medium, datur extensio continua" (GP II, p. 515).

element of an infinite sequence may be treated just the same as the previous ones.[47]
This rather vague "postulate" (as Leibniz also called it) might be interpreted in
geometrical terms as stating that a continuous space should contain the limit of any
(converging) sequences in it, since otherwise this limit would not be an *extremum
inclusivum* but rather an *extremum exclusivum* (i.e. a limit external to the set)—
something which is ruled out by the hypothesis of continuity. With such a (rather
generous) interpretation of Leibniz's principle, one might attempt to claim that
continuity is here defined as completeness in Weierstrass-Cantor terms (existence
of the limits of all converging sequences).[48] Nonetheless, aside from the fact that
Leibniz's words would thereby be being stretched far beyond their immediately
evident meaning, it should be noted that Leibniz never applied such a principle to
space or to magnitudes. He clearly considered it a heuristic rather than a constitutive
principle, and as something that *may happen* (thanks to God's good will, which
should play no role in geometry) in *some* continuous things—in short, a contingent
characterization of continuity rather than a definition of it. In particular, Leibniz
never attempted to employ such a definition of continuity to prove the intersection
of lines. We will, therefore, abandon this line of research, since (however promising
one might wish to consider it to be) it seems not to have been intended by Leibniz
as a foundational approach to geometry. We rather turn now to the most mature
results of these researches, that is to say, to Leibniz's final definition of continuity
as completeness.

---

[47]Such a principle is stated, for instance, in a letter to Christian Wolff that Leibniz published
in the *Acta eruditorum* from 1713: "Atque hoc consentaneum est *Legi Continuitatis,* a me olim
in Novellis Literariis Baylianis primum propositae, et Legibus Motui applicatae: unde fit, *ut in
continuis extremum exclusivum tractari possit ut inclusivum,* et ita ultimus casus, licet tota natura
diversus, in generali lege caeterorum, simulque paradoxa quadam ratione, et ut sit dicam, *Figura
Philosophico-rhetorica* punctum in linea, quies in motu, specialis casus in generali contradistincto
comprehensus intelligi possit, tanquam punctum sit linea infinite parva seu evanescens, aut quies
motus evaneseens, aliaque id genus, quae *Joachim Jungius,* Vir profundissimus, *toleranter vera*
appellasset, et quae inserviunt plurimum ad inveniendi artem, etsi meo judicio aliquid fictionis et
imaginarii complectantur, quod tamen reductione ad expressiones ordinarias ita facile rectificatur,
ut error intervenire non possit: et alioqui Natura ordinatim semper, non per saltus procendes legem
continuitatis violare nequit" (now in GM V, p. 385). The passage has been recently commented
upon by JESSEPH (2015), and GLEZER (2017, pp. 50–55). A similar, very clear statement of
the same principle is also to be found in an unpublished essay written by Leibniz around the
Historia et origo calculi differentialis: "Assumo autem hoc postulatum: *Proposito quocunque
transitu continuo in aliquem terminum desinente, liceat ratiocinationem communem instituere,
qua ultimus terminus comprehendatur*" (now in the appendix of LEIBNIZ 1846, p. 40).

[48]This thesis has been advanced by ANAPOLITANOS (1990).

4.5   Leibniz's Definition of Continuity

The final stage of Leibniz's investigations into continuity began around 1695 and is to be found in the masterwork on geometry of his middle years, the *Specimen geometriae luciferae*.[49]

Leibniz's most mature definition states (loosely paraphrasing) that a continuous whole is a whole such that its parts overlap in a boundary, which is not itself a part. It is remarkable how close this Leibnizian definition is to the Aristotelian one in terms of contact and identification of the boundaries of two bodies; nonetheless, the understanding of the terms involved is quite different and by this time several centuries of philosophical work had left their sediments upon the old Aristotelian definition. For example, the reformulated definition now no longer refers to some given parts, but rather to all the possible parts of a whole. The parts may now overlap one another—something that was not possible according to the Aristotelian definition. We do not, however, need to insist on these (and many other) subtle philosophical variations, since Leibniz's massive advance in this field clearly consists above all in his having given a *mathematical* definition of the terms employed in this concept of continuity (parts, boundary, etc.). Such a definitional apparatus is to be found, indeed, in the *Specimen* itself. But the way is in fact prepared for it in innumerable essays and fragments from the late 1680s and early 1690s which set the stage for the new definition.[50]

The starting point is the reworking of the notion of a *part*. We need, then, to delve a little way into Leibniz's mereology.[51] It should be noted that Leibniz generally conceived of wholes and parts as collections (or sets, or aggregates) of elements—

---

[49]The only essays known to me that deal with these Leibnizian texts on continuity are GIUSTI (1990) and LEVEY (1999), which, however, discuss them only in relation to the standard (Aristotelian) notion of connectedness.

[50]In fact, Echeverría has found two Leibnizian fragments from 1679, in which Leibniz plainly states that a whole is continuous if its parts have a boundary in common: "*Continuum* est cuius partes habent terminum communem" (LH XXXV, I, 11, Bl. 54v., and similarly in LH XXXV, I, 12, Bl. 4r.; transcribed in ECHEVERRÍA 1980, vol. 2, pp. 293 and 329; also mentioned in ECHEVERRÍA 1990). In these fragments, however, the most important qualification is missing: namely, that stating that those parts must have in common *only* a boundary. As they stand, in fact, they seem to mean that, in order for there to be continuity, it is *sufficient* that the parts of the whole should have a boundary in common; nonetheless, they might have in common also an entire part. The notion is still naïve and seems to be driving at some idea of connectedness. No geometrical consequences, in any case, are drawn from this definition, and Leibniz did not attempt to employ it to prove *Elements* I, 1 or any other Euclidean proposition in need of a continuity assumption. In short, such fragments may just reflect the old Aristotelian definition of continuity.

[51]A fuller account of Leibniz's rather complex mereological thought is to be found in the essay by Massimo Mugnai in this volume, which deals especially with the mereological considerations in Leibniz's *Specimen geometriae luciferae*, which is also the most important text for his definition of continuity.

possibly, of points. Leibniz's mereology, therefore, at least as far as geometry is concerned, is a kind of mereology in a naïve set-theoretical environment.[52]

Leibniz defines a *part* as something that is *in* (relation of *in-esse*) the whole and is *homogeneous* with the whole. To *be in* the whole means simply that the part is a *proper* subset of the whole and Leibniz never considers the whole itself as one of its parts. In some other texts Leibniz employs the equivalent notion of *in-existere*.[53] The notion of *homogeneity* is, on the other hand, rather complex in Leibniz's writings. The usual geometrical meaning of the term is derived from Euclid's definition of ratio and is connected with the Archimedean Axiom: two magnitudes are homogeneous if they are capable, when multiplied, of exceeding one another.[54] Leibniz also proposed another, more general, meaning, saying that a thing is homogeneous to another thing if there is a third thing which is similar to the first and equal to the second.[55] *Similarity* is here the usual geometrical

---

[52]We will deal further with these complications (i.e. the interaction between set-theoretical and mereological elements) below. In a sense, the conceptual setting might be considered analogous to the one described in the celebrated D. LEWIS (1991). I employ here freely the word "set" or "collection", whereas Leibniz, of course, had no axiomatic approach to the notion of a set. It should be remarked, though, that he was able to provide some rather abstract definitions of membership. In one of his writings, he said that if, from the fact that *A,B,C* are posited, it immediately follows that *L* is also posited, then *L* should be considered to be the aggregate of *A,B,C* (*Inquirenda logico-metaphysica*, in A VI, 4A, n. 210, p. 998). In such a situation, each of the *A,B,C* is *in L* (relation of *in-esse*). We should notice that Leibniz also defines the *parthood* relation by means of similar statements and that he probably did not make any exact distinction between mereological and set-theoretical considerations. Nonetheless in the above-mentioned passage it seems clear that the aggregate *L* cannot be entirely reduced to the three elements that compose it, since it is explicitly stated that it is a fourth object, distinct from these latter.

[53]Cf. for instance the *Specimen geometriae luciferae*, in GM VII, p. 274. This definitional system is recurrent in several other writings from the 1690s onwards. It should be noted that Leibniz's definition of parthood in the early years was still very metaphysical, even when spelled out in geometrical writings, and could not be employed for the purpose of any mathematical construction. See for instance § 27 of the *Characteristica geometrica* from 1679, where Leibniz himself, after having defined a part as a requisite of the whole, different from it, immediate, and *in recto* with other co-requisites, plainly states that these are, however, just metaphysical determinations that are of no use in mathematics. Nor did he give any mathematical ones (GM V, pp. 151–52).

[54]This statement does not appear, as such, in the *Elements*. Rather, Euclid (or Eudoxus) states in the infamous Definition 3 of Book V (infamous inasmuch as it is very obscure, and has been the object of much speculation afterwards) that "a ratio is a sort of relation in respect of size between two homogeneous magnitudes", without however defining homogeneity. In the following Definition 4, however, Euclid states that "Magnitudes are said to have a ratio to one another which are capable, when multiplied, of exceeding one another", this latter condition being equivalent to the Archimedean Axiom. It was very easy, then, to put together the two definitions and to say that two magnitudes have a certain relation in respect of size (i.e. a ratio) to one another if they are homogeneous, that is to say, if they can be multiplied so as to exceed one another. In this way, "being homogeneous" came to mean, in some early modern texts, "being Archimedean".

[55]*Elements* VI, 25 shows how to construct a polygon which is similar to a given polygon and equal to another one. Leibniz's definition of homogeneity is, therefore, a kind of generalization of this Euclidean proposition, which however did not refer to "homogeneity" at all in the ancient tradition. In fact, the ancient tradition attributed *Elements* VI, 25 to Pythagoras, with a certain emphasis on

notion and *equality* means the identity of measure. Since Leibniz seems to have the idea that any continuous transformation splits into a similarity and an equality (i.e. is composed by them), this notion of homogeneity is a very broad one and encompasses any kind of continuous (or better, bi-continuous) transformation. No triangle of any area can possibly be homogeneous with a solid sphere, because no (finite) number of such triangles, taken together, can ever amount to, or surpass, the volume of a sphere and because there is no figure, similar to the given triangle, that has the same measure (i.e. volume) as a sphere.[56] In particular, a *boundary* is defined by Leibniz as something which is always necessarily non-homogeneous with the bounded figure (since it has a different number of dimensions), and therefore is not a *part* of the figure in the proper sense (even though it is *in* the figure). A *section* between two figures (e.g. the straight line in which two planes intersect) is a case of a common boundary, and it is, by definition, not a part of either figure; rather, it *is in* both of them.

A further, important condition of parthood that we may want to consider is that a part normally includes its own boundary. This latter condition is seldom (if ever) explicitly spelled out by Leibniz in the context of the definition of continuity but it represents an important feature of his philosophy of mathematics in general. Parts and wholes are often regarded by Leibniz as *closed* sets.[57] The only exceptions here are infinite sets: we have already mentioned, in passing, Leibniz's distinction between a syncategorematic and a categorematic infinite, and the latter implies that an infinite whole is to be considered a non-closed set. Such an infinite set should not even be a "whole" (or even a "set") in the proper sense and Leibniz, indeed, often spells out his conception of a syncategorematic infinite by saying that it does not constitute a whole. Nevertheless, from a mereological point of view, such infinite aggregates are still divisible into parts, and therefore "wholes" in a wider sense of the word. We should finally note that the mereological wholes and parts taken into consideration by Leibniz in his discussions on continuity are geometrical objects, not numerical sets. In this context, Leibniz seems to consider an object limited in its measure (a finite segment, for instance) as a finite, closed object, even though under certain considerations it might be regarded as an infinite (and therefore syncategorematic, non-closed) aggregate of points.

---

the importance of the discovery of the solution of such a problem (see the famous passage in PLUT. *Symp.* VIII, 2, 4).

[56] In the *Specimen geometriae luciferae*, Leibniz also claimed that the Archimedean notion of homogeneity can be derived from his own more comprehensive definition, given in terms of similarity (GM VII, p. 283).

[57] If this were not the case, of course, the whole array of problems with *contiguity* would not have arisen. I have offered in DE RISI (2007, p. 194 n. 65) an example of a non-closed set (a circle without a center) employed by Leibniz in A VI, 4A, n. 178, p. 847. See BREGER (1986), who also claims that Leibniz's parts are always closed sets. We should also note that the distinction between open and closed set was quite common in the seventeenth century, and in fact it permeated not only mathematical textbooks but, to a greater degree, philosophical treatises, wherein the Scholastics commonly distinguished between internal and external limits of bodies and figures (if the limit is "external", the body in question is missing its frontier: it is topologically open).

It is also true, however, that a definition of parts as closed sets would trivialize under many respects the definition of continuity as completeness, for any closed subset of a complete space is itself complete. The point of Leibniz's researches in this domain, in fact, seems to be that of making explicit this last requirement. Since he clearly conceived non-closed sets in his geometrical investigations, therefore, we will *not* add closure among the requirements for parthood.

Given the previous cluster of definitions, a first attempt to arrive at a new notion of continuity can be found in a short fragment which should also date from the mid-1690s. When writing it, Leibniz was clearly pleased at having discovered something quite new, as he wrote in the margins of the sheet of paper containing it:

> Here I stumbled upon something memorable: The notion of a *Continuum*, and parts, and homogeneity, without making use of similarity, motion or transformation.[58]

The definition of continuity is the following:

> A is *continuous* when any two things *B* and *C* however taken, such that they exhaust [*exhaurientia*] *A*, have some thing *D* in common which exist-in [inexistens] both *B* and *C*.[59]

The fragment goes on to define the notion of "exhausting", specifically by differentiating it from the notion of "co-equating" or "co-integrating": the former may be a covering of the original space with some overlapping, while the latter, if it overlaps at all, does not overlap in a part:

> Exhausting things are things, as *B* and *C*, that exist-in *A* and such that nothing is in *A* that is not in *B* or *C*, that is to say in the composition of *B* and *C*.
>
> (Note that it may happen that the exhausting things taken together are greater than the exhausted thing—if they have a part in common [. . .]).
>
> The latter are different from co-equating or co-integrating things [*coaequantia vel cointegrantia*], which together are equal to the whole, while the exhausting things may be greater in magnitude than the whole even if they do not encompass anything bigger. A section is a continuum in the co-equating things which is in each of the co-equating things. [. . .]
>
> If two exhausting things have nothing in common except what is necessary for continuity, what is common is called a section.[60]

---

[58] From the text *Hic memorabilia...*, to be found in LH XXXV, I, 14, Bl. 76, and published in DE RISI (2007, pp. 586–87): "Hic memorabilia nactus sum: Continui Notionem, et partis, adeoque et homogenei, non supponendo similitudinem, vel transformationem seu motum".

[59] "*Continuum* est *A* in quo utcunque sumta bina exhaurientia *B* et *C*, aliquid habent commune *D*, seu utrinque tam *B* quam *C* inexistens", in DE RISI (2007, p. 586). In the same essay, Leibniz had already defined the notion of *in-existere* by saying that *B* inexist to *A* in the case where, if the existence of *A* is posited, then the existence of *B* is also immediately posited as well. See above, *note* 52.

[60] "*Exhaurientia* sunt plura, ut *B* et *C*, existentia in uno *A*, talia ut nihil sit in *A*, quod non sit in *B* vel in *C*, seu si composito ex *B* et *C*. (Nota. Posse fieri, ut exhaurientia simul posita sint majora eo quod exhauriunt, si scil. habeant partem commune. [. . .] Ab his differunt *coaequantia* vel *cointegrantia*, quae simul aequantur toti, cum exhaurientia possint toti excedere magnitudine etsi nihil amplius contineant †. Itaque *sectio* est in coaequantibus continuum quod est in utroque coaequanti. [. . .] Si bina exhaurientia nihil habeant commune, quàm quod necesse est ad continuitatem, quod commune

In this convoluted passage, in which Leibniz is still looking for a good definition of several terms, continuity is defined by two parts having something in common, which may either be a part or not (since the requirement for continuity is that the parts exhaust the whole, not that they co-equate with the whole). In the last-quoted sentence, however, Leibniz insists that this common element may well be a section rather than a part, even though this is not taken, here, to be a necessary condition for continuity. As a matter of fact, this would be impossible in this context, since a few lines before the notion of "section" itself had been defined recurring to the *continuum*.

It is to be noted, moreover, that both exhausting and co-equating parts are coverings, in modern terms, and that a co-equating set of parts may be a partition (i.e. a covering with no common elements). A co-equating set of parts, however, may also be a covering (i.e. have some overlapping) provided that the common elements have *zero measure*, as is hinted at by the Leibnizian sentence on being "greater in magnitude". That is to say, the condition for having B and C co-equating with A is that $B \cup C = A$, and either (1) $B \cap C = \varnothing$ or (2) $B \cap C \neq \varnothing$ but $m(B \cap C) = 0$. Leibniz does not seem concerned with recurring to the notion of measure in defining continuity. His geometry is a metric theory, and he seems not to display any interest at the problem of giving a definition of continuity that might be applied in non-metric settings. On the contrary, the fragment goes to discuss many other issues and it is mainly concerned with an attempt to define a notion of part without recurring to homogeneity and *similarity*. Leibniz's preoccupation here might well be that of providing a notion of continuity that is applicable to non-spatial magnitudes, where the notion of similarity fails and this seems to him a much more important concern than the recourse to measure theory. It should also be noted that Leibniz clearly regarded it as an important logical step forward to distinguish between exhausting and co-equating parts in dealing with the theory of measure (as distinct, I mean, from the theory of continuity), where it is essential to partition a magnitude into a number of congruent, non-overlapping parts.[61]

It seems that such attempts to define parthood without similarity are given up in the following studies (at least, I am not aware of any other Leibnizian discussions on these themes). But the notion of continuity sketched here, which pleased Leibniz

---

est dicitur *sectio*", in DE RISI (2007, pp. 586–87). The same train of thought is to be found in another fragment dating from the same years: "*Exhaurientia* ipsius A sunt plura B, C, D si nihil sit in A quod non sit in B vel C vel D. Vicissimque nihil in ipsis quod non sit in A, et tamen sit in uno, quod non est in caeteris. *Continuum* est A, in quo utcunque sumta bina exhaurientia B et C, habent aliquid commune. … Exhaurientia quae non habent partem communem dicuntur *cointegrantia*" (*Euclides in definitione Diametri Circuli*, unpublished in LH XXXV, XII, 2, Bl. 107).

[61] This remark is, in fact, a recurrent one in Leibniz, and can be read in the *Nouveaux essais*, IV, VII, § 10: "Pour ce qui est de cet axiome, que le tout est égal à toutes ses parties prises ensemble, Euclide ne s'en sert point expressément. Aussi cet axiome a-t-il besoin de limitation, car il faut ajouter que ces parties ne doivent pas avoir elles-mêmes de partie commune; car 7 et 8 sont parties de 12, mais elles composent plus que 12" (A VI, 6, n. 2, p. 414). See also a passage in a text on *Mathesis generalis* in LH XXXV, I, 9, Bl. 9v, which has been recently published in RABOUIN (2018, pp. 166–67).

so much, is re-stated in more definite terms in the following months, this time explicitly ruling out the possibility that that intersection among parts which makes them continuous might be itself a part. This is explicitly stated prominently, if not for the first time, in the *Specimen geometriae luciferae*.[62] Leibniz's mature definition of continuity is formulated as follows:

> A continuum is a whole any two of the co-integrating [*cointegrantes*] parts of which (i.e. parts which, taken together, coincide with the whole) have something in common, and indeed in the case where the parts are not redundant and have no common part, i.e. when the sum of their magnitudes is equal to the total magnitude, they have at least a boundary in common.[63]

Here the final and most important condition—i.e. that the common element should be a section and not a part—is finally in place. Leibniz is very clear that such a definition must apply to a wider range of objects than just the geometrical magnitudes, and in fact he himself quickly moves on to discuss continuous transformations and fibrations, as we have seen above (the example of the family of ellipses). The reference to the measure of the parts is, however, still present. It is to be noted, moreover, that in this definition (and the following one) Leibniz refers

---

[62] A definition of continuity similar to the one in the *Specimen geometriae luciferae* is to be found in a paper by Leibniz whose dating is still uncertain but which may date back to the mid-1680s. This is a draft accompanying a longer essay called *Prima geometriae principia* and apparently somehow related to it. Here a continuous whole is defined by using the notion of co-integrating parts, and by clarifying that the parts should have a common boundary which is not itself a part. In the lack of a complete edition of Leibniz's geometrical writings, it is hard to make any definite chronological statement about the invention of this definition. It seems safe to claim, though, that the great *Specimen* from the mid-1690s is the first important paper that puts this notion at the very center of the foundational discussion. In any case, the above-mentioned fragment of the *Prima geometriae principia* reads: "Continuum est totum cujus partes duae cointegrantes habent aliquid commune, quod dicitur *Terminus*. ... Appello autem partes cointegrantes quae simul sunt aequales toti, et nullam habent partem communem. Terminus itaque non est pars nec est homogeneus terminatis. Alioqui foret pars" (unpublished in LH XXXV, I, 5, Bl. 5r; I thank Javier Echeverría for mentioning this text to me).

[63] "Continuum est totum, cujus duae quaevis partes cointegrantes (seu quae simul sumtae toti coincidunt) habent aliquid commune, et quidem si non sint redundantes seu nullam partem communem habeant, sive si aggregatum magnitudinis eorum aggregatio totius aequale est, tunc saltem habent communem aliquem terminum" (GM VII, p. 284). We may note that in the *Specimen* Leibniz did not define the term "*cointegrans*", which had been defined in the other (probably contemporary) fragment which we have just examined. Had the notion been defined in identical manner in the two texts, the Leibnizian addition in the *Specimen* "and such indeed that..." would have been pleonastic, since co-integrating parts have, by definition, no part in common. Since, however, Leibniz had not defined the notion, he possibly felt the need for this further specification. In the case, however, that "*cointegrans*" here should signify no more than just "exhausting" (i.e. a covering allowing parts in common), we may note that the clause introduced by "and such indeed that..." is actually a second condition for the definition. It is not introduced as a disjunction (two parts should have a part in common *or* at least a boundary), but as a further requirement. The latter interpretation is confirmed by an incidental passage in the *Specimen* in which Leibniz mentions a continuous space the parts of which cannot have any part in common (GM VII, p. 266).

to "wholes" which have parts. It is not immediately clear whether these must have a boundary or not, or be limited or not.

It seems that Leibniz repeated such a definition sev- 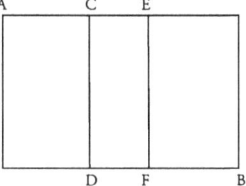 eral more times starting from these years onward.[64] The last and possibly most perfected occurrence of it is probably to be found in the mature Leibnizian masterpiece on geometry, the *In Euclidis* πρῶτα from 1712. This treatise summarizes three decades of Leibniz's work on the foundations of geometry in the form of a detailed commentary on the definitions and the axioms of the *Elements*. In it Leibniz offers a variation on his classical definition of a *section*, which recurs to the previous notions of co-integrating:

> The section of a magnitude is anything which is common to two parts of the magnitude which have no part in common. Let AB be a magnitude and its parts be AD and BC, which have the straight line CD in common. The latter is in AD and is in BC, even though these parts have no part in common. In fact, CD *is not* a part of AD, nor a part of BC. The evidence is that AD + BC + CD is not greater than AD + BC, that is to say, the former does not have to the latter any ratio of the greater to the lesser, since AD and BC are parts equating the whole. On the other hand, if the parts *should have a common part*, as for instance the parts AF and BC of the whole AB have the common part CF, then AF + CB would be greater than AB [. . .], and nonetheless AF ⊕ CB, that is to say AF and CD taken together, are not greater than AB. It is important to mark this difference between the addition of quantities, and the taking-together [*simul-sumtionem*] of things. As the mental addition of *quantities* [*quantitatum*] is designated by +, so the real addition of *magnitudes* [*magnitudinum*], or of their quanta [*quantorum*], is designated by ⊕.[65]

Needless to say, the discussion of the notion of a section is functional to the definition of continuity, which is still defined by a mereological structure such that the whole is decomposed into parts which have no part, but rather a section, in common:

---

[64] Among the unpublished papers from the 1690s, see for instance the short essay in LH XXXV, VII, 30, Bl. 128, stating: "*Continuum* est, cuius duae partes cointegrantes habent commune quod dicitur terminus. Partes cointegrantes voco quae simul sunt aequales toti, nec habent partem communem . . . ".

[65] "*Sectio* magnitudinis est quidquid est commune duabus Magnitudinis partibus partem communem non habentibus. Esto Magnitudo AB, ejusque partes AD, BC, quibus communis recta CD, quae est et in AD et in BC, etsi hae partes non habent partem communem. Nam CD *non est* pars ipsius AD nec ipsius BC. Hujus indicium est, quod AD + BC + CD non est majus quam AD + BC, seu non habet ad ipsum rationem majoris ad minus, cum AD et BC sint partes aequantes totum. Sed si partes *partem communem haberent*, uti ex. gr. AF et BC partes ipsius AB habent partem eommunem CF, tunc AF + CB majus foret quam AB (cum sint partes complentes quidem totum AB, ut faciunt AD et BC, sed non aequantes, ut etiam faciunt AD et BC), et tamen AF ⊕ CB seu AF et CB simul sumta, non sunt plus quam AB. Quod discrimen inter additionem quantitatum, et simul-sumtionem rerum probe notandum est. Et ut additio mentalis *quantitatum* designatur per +, ita additionem realem *magnitudinum*, seu ipsorum quantorum designo per ⊕, donec aliquid commodius occurrat." (GM V, p. 184).

> In order for quantity to be continuous, two requirements must be fulfilled: first, any two parts which, taken together, equal the whole must have something in common which, however, is not itself a part; second, the parts must be, as is commonly said, exterior to one another, that is, it must be possible to take any two parts (which together do not equal the whole) which have nothing in common, not even a minimum.[66]

The last requirement of the second definition is the *partes extra partes* clause, which was not included in the previous definitions, since in the essay which we are discussing now Leibniz is concerned only with geometry (Euclidean geometry, in fact), and gives a more restricted definition of continuity which applies only to geometrical magnitudes.[67] He also explicitly wanted to exclude angles from the definition of continuous things, since he denied that angles are magnitudes, that is to say parts of the plane (stating rather that they are relations between lines).[68] On the other hand, in the *In Euclidis* πρῶτα we have the advantage that any reference to measure has disappeared, and the definition is given in purely mereological and situational terms. In this respect, it is much neater and more mathematically worthy than the preceding one, even though the reference to parthood needed here still requires the notion of homogeneity, and this latter requires, in turn, the notions of similarity and equality (i.e. identity of measure) so that the definitional advantage might be not so great.[69]

Despite these differences, however, and despite the fact that Leibniz's formulations remain a bit vague as regards their exact meaning (as they stand they may be formalized in many different ways), Leibniz's actual *use* of them in geometrical demonstrations makes quite clear that the last definitions (the ones in the *Specimen* and in the *In Euclidis* πρῶτα, along with other fragments in between the two) all amount to the same idea: a geometrical whole is continuous if, and only if, any two arbitrarily-chosen parts of it are such that if the two parts together cover the whole and have no common part, they still must have something else in common, namely,

---

[66]"Porro ad continuum duo requiruntur, unum ut duae quaevis ejus partes totum aequantes habeant aliquid commune, quod adeo pars non est; alterum ut in continuo sint partes extra partes, ut vulgo loquuntur, id est ut duae ejus partes assumi possint (sed non aequantes), quibus nihil insit commune, ne minimum quidem" (GM v, p. 184).

[67]This seems to be, however, a consequence of the definition: for in intensive quantities (such as forces) two parts always overlap in a further part (i.e., the smaller of the two), and therefore they would never be said to be continuous. In this sense, the *partes extra partes* clause may have also been seen by Leibniz as an opening toward a different definition of continuity that was to apply to intensive quantities—since he surely wanted to declare forces or speed to be continuous, and his mereological definition did not allow this.

[68]This is explicitly stated after the definition in the *In Euclidis* πρῶτα. Two angles of different width would coincide in a common part (i.e. the entire smaller angle) and therefore should not be considered to be continuous, and much less a magnitude: "Magnitudo est continuum, quod habet situm. Angulus autem continuum non est" (GM v, p. 184).

[69]Note also that in general Leibniz still defines quantity in the classical way as something that has parts. In this sense, he had probably not seen that a purely mereological definition was more general than a definition recurring to measure. Moreover, the definition of *section* hinted at above still refers to the comparison on quantities.

their boundary. We may attempt a modern, symbolic expression of the Leibnizian definition of continuity in order to better clarify some of the most important aspects of it. Let be X a geometrical magnitude; $\mathcal{P}(X)$ the collection of its parts, where a part is in the Leibnizian sense a proper subset of the whole, homogeneous to it; and call ∂p the boundary of a part p (irrespective of the fact that such a part contains or not its own boundary—whether it is a closed set or not).[70] Then we may say that

X is *continuous* iff $\forall p_1 p_2 \in \mathcal{P}(X)$, $(p_1 \cup p_2 = X) \rightarrow ((p_1 \cap p_2 \notin \mathcal{P}(X)) \rightarrow (\exists b \subset X: b = \partial p_1 = \partial p_2))$

It is easy to understand Leibniz's *intent* with this definition if we apply it to a segment and its intervals (taken as its parts). In fact, where we give a numerical representation of such a segment, it even becomes apparent that Leibniz's definition

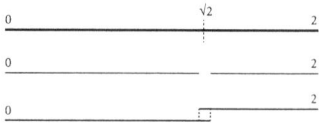

of continuity approaches Dedekind's notion of completeness given by the idea of a cut. If we take, for instance, the interval of rational numbers between 0 and 2 (which is dense, but not complete), we are not able to partition it as Leibniz's definition of continuity requires. If we consider, for example, the cut at number $\sqrt{2}$, which is between 0 and 2 but irrational, we can easily see that we may *either* find a partition of the interval in two sets that have *no element* in common, for their common boundary does not exist in the set (e.g. the sets of rational numbers strictly greater than $\sqrt{2}$, and all rational numbers strictly lesser than $\sqrt{2}$), *or* a covering of the interval in two parts which have *an entire part* in common (e.g. all rational numbers equal or greater than a certain $a < \sqrt{2}$, and all rational numbers equal or less than a certain rational number $b > \sqrt{2}$; the common part of which is the set of rational numbers between a and b). In either case, the whole would not be continuous according to Leibniz's definition: the first choice has an empty intersection; the second choice forces the sets to have a common part, rather than a boundary. It is only the existence of $\sqrt{2}$ itself that may allow us to divide the interval into the two closed sets of numbers $[0, \sqrt{2}]$ and $[\sqrt{2}, 2]$, the only common element of which is $\sqrt{2}$, which is not a part (but a section) of the intervals. The same reasoning may be repeated for all irrational numbers in the interval, and therefore the only complete interval is that composed by all the *real* numbers between 1 and 2. It is *continuous* according to Leibniz's definition.[71]

---

[70] That is to say, $\mathcal{P}(X)$ is a subset of the standard, set-theoretical Poweset of X. I have chosen this formulation of the Leibnizian definition, relying on the explicit reference to boundaries that we find in the *Specimen* (while it seems impossible to characterize the boundary simply stating $b \notin \mathcal{P}(X)$). I have also stressed the existence of such a boundary, in order to better show the similarity with Dedekind's definition. It seems clear to me, from Leibniz's examples, that he had in mind some sort of existential statement. I may add that the formula applies differently to topological spaces and subspaces (for boundaries may be differently considered in the two cases), even though this difference is far beyond Leibniz's "seminal" topological awareness.

[71] See DE RISI (2007, pp. 196–98), where I had offered a slightly different modern reformulation of Leibniz's definition. The fairly intuitive example given above, of course, may be complicated a bit if we mean the rationals as a subspace of the reals (rather than a topological space in its own

With such a definition, therefore, Leibniz was able to give an exact mathematical notion of continuity which approximates to our modern definition of completeness and might be employed as the foundation of continuity in geometry. There are some striking similarity with Dedekind's own definition, which also cuts the set in two parts, and asks for the existence of a separating individual; such a role is here played by the boundary of the parts, which has to be in common between them. We must be careful, of course, about attributing to Leibniz a fully-fledged definition of completeness or making him a forerunner of Dedekind. There is no doubt that the gap between the two formulations of completeness is enormous and that Leibniz's still sloppy definition, which does not apply without adjustments to its intended object, still falls short of that definitional exactitude that is Dedekind's greatest merit.

Of course, the main difference between Leibniz's and Dedekind's approaches is to be found in Dedekind's vigorous rejection of a geometrical foundation for completeness and his insistence that this notion should be grounded rather on a consideration of numbers. This is an important nineteenth-century view to which nothing could have exactly corresponded in the thought of Leibniz or his contemporaries. In the seventeenth century, most mathematicians still endorsed the Greek view (clearly expounded by Aristotle) that numbers are discrete quantities— that is to say that numbers are just natural numbers ($\mathbb{N}$). The status of real numbers was still very much ambiguous in the early modern age, and the general view was that irrational magnitudes such as $\sqrt{2}$ were to be considered primarily geometrical objects. Irrational numbers might, indeed, be employed in order to do calculations involving such geometrical magnitudes and some kind of correspondence between real numbers and the points of a line was surely envisaged (as it is obviously necessary in the foundations of Cartesian geometry). Nevertheless, irrational numbers were still regarded as an *image of the continuum* (in Wallis's words), or a *sign of magnitude* (in Barrows's words), rather than as anything forming a continuous whole in itself. This being the case, the set of all real numbers was not generally considered to be a proper mathematical object (but just a *set of tools*), and $\mathbb{R}$ was not investigated in any structural way. In this context, it remained doubtful whether real numbers should be taken to be *continuous* in any proper sense. But even if they may have been considered so from time to time, there is no doubt that their continuity was parasitic on the continuity of geometrical magnitudes rather than primary or original: while Dedekind's theory claimed the opposite. In this connection, we may note that Leibniz went further than his contemporaries in disentangling the consideration of numbers from that of geometrical magnitudes, and he always insisted on some ontological and epistemological independence of the former from the latter. Numbers are for him essentially *relations* and in this respect they have some ideal subsistence independent of space or extension. A natural number is sometimes defined by Leibniz (following Euclid) as a collection

---

right). In this case, the boundary of the rationals are the reals themselves, and the two parts also have boundaries in 0 and 2.

of unities. A number in general (including rational and irrational numbers) is also defined by him as anything which is *homogeneous* to a unity, where homogeneity seems to be taken in the sense of the Archimedean Axiom.[72] On the other hand, he always denied that infinite collections of numbers may be true wholes (they are rather syncategorematic aggregates), and in this respect he, like his contemporaries, never seems to consider or investigate $\mathbb{N}$, $\mathbb{Q}$, or $\mathbb{R}$ and their properties (including their continuity, or lack thereof).[73] In any case, as far as I know Leibniz never attempted to apply his definition of continuity to real numbers.

We have also noted that there is a certain ambiguity in Leibniz's writings on the necessity that a whole should always have its boundaries. In fact, Leibniz seemed concerned solely, in his foundational studies on continuity, with excluding "gaps" from the *continuum* and he did not explicitly discuss the case of missing boundaries. It is possible that he had considered the *open real* interval (0,1), which is Dedekind-incomplete, to be a continuous whole after all.

Speaking more generally, it does not seem that mereology (or topology) is the right setting in which to express Dedekind's notion of completeness. It is true, however, that Leibniz's mereological structure is enriched by a more robust definition of parthood (namely, one in terms of homogeneity and closure) which adds a further structure that, aptly interpreted, *might* be able to capture some more aspects of completeness.[74] It is still remarkable that Leibniz's definition of continuity employs an explicit quantification on the set of parts (which was not a feature of the ancient definitions of continuity). This clearly hints in the direction of a "second order property" as far as completeness is concerned. In this respect, his ability to distinguish from one another completeness and density (a first order property, which does not call into play the set of all parts) should definitely be considered a logical breakthrough.

The modern notion of completeness is, however, still insufficient to exhaust all that Leibniz meant when he evoked the notion of continuity. Since, in fact, Leibniz's definition is just a refinement of the old Aristotelian notion, it also inherits the mathematical merits of this latter. In particular, it seems apt to capture the topological notion of *connectedness*. We have already seen that connectedness, which is a much simpler property than completeness, could perhaps already have been hinted at in the Aristotelian definition.[75] But whereas reading a clear

---

[72]Note that the definition through homogeneity seems not be able to entirely replace the more restricted one in terms of natural numbers as aggregates of unities, since the Archimedean property seems to require natural numbers to be formulated. On Leibniz's notion of numbers, see the recent paper by SEREDA (2015).

[73]We may note that Leibniz attributed parts and homogeneity to each single number (which is in some respect an aggregate, or a set), not to the set of numbers. He might have stated that 7 is homogeneous with $\sqrt{2}$, but he did not claim that the set of numbers [0,1] is a homogeneous part of $\mathbb{R}$.

[74]The notion of boundary and closure may be defined, in fact, in a purely mereological setting. The question of homogeneity is more complicated, as Leibniz's definition of it remained a bit uncertain.

[75]See above, *note* 20.

mathematical intent into Aristotle's quite physical definition requires a certain stretch, Leibniz's purely mathematical treatment of the notion seems indeed to provide a viable definition of connectedness. In particular, since connectedness may be defined as the impossibility of representing the whole as the union of disjoint closed sets, Leibniz's explicit statement that parts must have boundaries (a statement that is not to be found in Aristotle), together with the possibility of conceiving of the set of all possible parts of a whole (likewise lacking in Aristotle), seems to constitute a real step forward toward an exact definition of connectedness. A set of real numbers such as $[0,1]\cup[2,3]$, which is Dedekind-complete but disconnected, would therefore count for Leibniz as a discontinuous set. The same is true for those topologically discrete sets, such as $\mathbb{Z}$, which are disconnected but complete: they would be discontinuous according to Leibniz's definition. It should be remarked that the flexibility of Leibniz's notion of continuity, which encompasses both completeness and connectedness, is rather artificial and is probably grounded on some ambiguity in the formulation of the principle.[76] There is no doubt, however, that such was Leibniz's intent and that, therefore, Leibniz's concept of continuity should be considered to be a hybrid topological/mereological notion combining completeness and connectedness.

We may also note that Leibniz's definition of continuity does not seem to entail density. In principle, it seems be possible to start (for example) with extended elements endowed with a boundary and homogeneous with the whole (i.e. atomic *parts*) and to make a continuous whole out of these latter by connecting them in their boundaries. Leibniz, however, does not seem to ever seriously consider this possibility, since, of course, he believes that he possesses important (and classical) arguments in favor of the infinite divisibility of geometrical extension. All geometrical applications of Leibniz's definition of continuity, therefore, assume density, and thus an atomless mereology. The parts involved in the definition of continuity are infinite sets of points and these are always further divisible. Nevertheless, the points themselves, as we have seen, are not parts of the whole, so that a point is not a mereological atom.

Something similar may be said about the Archimedean Axiom. Even though it may be employed in defining what a part is (since it may be involved in the notion of homogeneity), nothing seems to exclude the possibility that, in a continuous whole, there might also be non-Archimedean elements (which would not themselves be parts of it, of course).

---

[76]Connectedness may be defined as the impossibility of representing the whole topological space as the disjoint union of two closed sets. The above-sketched "Leibnizian" formula recovers connectedness by asking that in a connected space any partition in two (also non-closed) sets must have some element in common when we consider their boundaries (i.e. their closure). Alternatively, one could *define* parts as closed sets, and then define continuity simply as following: X is *continuous* iff $\forall p_1 p_2 \in \mathcal{P}(X)$, $(p_1 \cup p_2 = X) \rightarrow (p_1 \cap p_2 \neq \varnothing)$, which is good rendering of Leibniz's statement if we forget the (quite complex) "which is not a part" clause. This may approach a bit more the simpler Aristotelian setting.

It may not be surprising that the inventor of the Calculus felt the need to explore the deepest questions regarding continuity and there is no doubt that a similar definition would have been almost impossible prior to the development of infinitesimal analysis, along with the subtle discussions on density, closure, or homogeneity that it engendered. Nonetheless, Leibniz was able to achieve a degree of clarity in this definition that was to remain lacking in the work of most mathematicians active in the centuries between him and Dedekind.

4.6 A Gapless Proof of *Elements* I, 1

Leibniz's notion of continuity was first conceived of and developed for elementary geometry and it was here that it found its first and foremost applications. In fact, a new demonstration of *Elements* I, 1, is attempted in the *Specimen geometriae luciferae* in the lines immediately following the discussion of continuity. This is also an important innovation, since Leibniz's previous approaches to this Euclidean proposition seemed simply to *assume* the possibility of giving a mathematically viable definition of continuity, without ever attempting to do so.

Leibniz's first encounter with the foundational issues in *Elements* I, 1 may well have been in 1679, when he first attempted a *characteristica geometrica* and began to seriously work at his project on *analysis situs*. At the time, he immediately detected the gap in the Eucliedean proof, and complained that Clavius had not noticed it.[77]

We have a short essay entirely dedicated to this problem dating from those years, which begins by stating that:

> Euclid wanted to keep the greatest rigor in demonstrations: to this effect, he even proves that two sides of a triangle taken together are greater than the third, something that was mocked by the Epicureans, saying that this is known even by an ass. Nonetheless he did not prove other things that seem to need much more proof: and for instance in *Elements* I, 1, where he teaches how to construct an equilateral triangle on a given basis, he assumes that

---

[77] See Leibniz's letter to Vagetius, from December 12th, 1679, stating that "Clavius cum primam primi logice resolvere vellet, nihil egit; neque enim animadvertit vel ipse vel alius interpres demonstrationem Euclidis indigere hac propositione, quod duo circuli illic descripti alicubi sese secent. Quod aut probandum aut saltem assumendum erat inter principia" (A II, 1, n. 218, p. 768). A similar statement recurred still in the *Recommandation pour instituer la science generale* from 1686: "A quoy je trouve qu'Euclide, tout exact qu'il est, a manqué quelques fois, et quoyque Clavius y ait souvent suppléé par sa diligence, il y a des endroits, où il n'y a pas pris garde, dont un des plus remarquables et des moins remarqués se rencontre d'abord dans la demonstration de la premiere proposition du premier livre, où il suppose tacitement que les deux cercles qui servent à la construction d'un triangle equilatère, se doivent rencontrer quelque part, quoyqu'on sçache que quelques cercles ne se sçauroient jamais rencontrer" (A VI, 4A, n. 161, p. 705). In a passage of his *Demonstrationes Euclideas* Leibniz complained that even Clavius's reduction of Euclid into syllogisms did not elicited in him a doubt about the existence of the intersection point: "Hoc demonstrare quod circuli se secent Euclides et interpretes neglixere, etiam cum in Syllogismos redegere demonstrationem" (still unpublished in LH XXXV, I, 3, Bl. 2r).

two circles drawn from the endpoints of the straight line (taken as the base) and with radius equal to the straight line, cut each other. This is not evident at all.[78]

Yet, in the following discussion of the matter, Leibniz says that in order to prove that the two circles meet, one has to rigorously prove:

1) That the circumference of each circle is partly inside and partly outside the other circle (for if this is established, the circumference would necessarily intersect the other circumference, given the definition of section).
2) That the radius of the circle, extended, will intersect the circumference in two points only.[79]

Leibniz deals with the topic at length and his considerations are rich in reflections about order and the reciprocal positions of points. Basically, he attempts to employ mereological considerations in order to recover principles of order (by exploiting, as we would say today, the intrinsic partial ordering of any mereological structure). In fact, in the same group of essays (though not in this one) he was to arrive at a very neat formulation of Pasch's axiom, which is the most important principle ruling the ordering of points in the plane.[80] But however interesting these considerations may be for the birth of a geometry of space and situational (and ordering) relations, it is remarkable that in this early essay Leibniz never attempted to give any further explanation of the existence of the meeting points. The parenthetical remark about continuity in the above-mentioned quotation seems to be the only thing he has to say: by the definition of "section" (which he never gave), one should be able directly to show that the circles actually do intersect. This is restated at the end of the essay, the last words of which are:

---

[78]This and the following quotations are from an unpublished paper in LH XXXV, I, 2, Bl. 6–7, which has been transcribed in Echeverría's doctoral dissertation: see ECHEVERRÍA (1980, vol. 2, pp. 43–51). The original latin reads: "*Euclides* summum rigorum observare voluit in demonstrando: ideo probat quod in triangulo duo latera sint tertio majora, quod, ut Epicurei ridentes ajebant, etiam asinus novit. Interea non probat alia quae magis probatione indigere videntur exempli causa in prop. 1 lib.1 ubi triangulum aequilaterum super data basi constituere docet, assumit duos circulos ex rectae (quae pro basi sumitur) extremis et intervallo ipsius rectae descriptos, sese secare. Quod non equidem ita manifestum est".

[79]"*Primum*: Circumferentia in unius (DB) esse partim in altero partim extra alterum, EA (hoc enim posito illa hujus ambitum *secet* necesse est per sectionis definitionem). *Deinde*: Radium productum secare circumferentiam in duobus tantum punctis" (*ibid.*).

[80]The essay is unpublished in LH XXXV, I, 2, Bl. 14r; transcribed in ECHEVERRÍA (1980, vol. 2, pp. 78–79). I may mention that Pasch's Axiom, as it appears also in Hilbert's *Grundlagen* (as Axiom II, 4), may be taken as a Line-Line intersection principle, as it states that two segments (a side of a triangle and a straight line) meet in a point. As it should be conceived as an order axiom, though, it may be reformulated in such a way not to imply the existence of the point of intersection, but just the reciprocal position of the lines: and such a modified axiom would not change the underlying model of Hilbert's *Grundlagen*. In other words, the "continuity" part of Pasch's Axiom is redundant and, given a weaker form of Pasch's Axiom (i.e. a statement that does not require an actual intersection), it could be derived from the remaining Hilbert's axioms. I thank Victor Pambuccian for an enlightening discussion on this topic.

...and therefore the circumference is partly inside and partly outside the other circle, and thus it intersects its boundary and thus meets it.[81]

In a related essay, Leibniz attempted to further belabor the point, and moved decidedly towards some "topological" consideration of the intersection, by stating that if a point is inside (resp. outside) a given figure, that an entire part (say, a neighborhood) is inside (resp. outside) the figure. He then attempted to prove the existence of the point of intersection, by speculating on the fact that two parts of a continuous figure, which has a point inside and a point outside another closed figure, must meet the latter's boundary. Such texts are really remarkable from a proto-topological point of view, but in 1679 Leibniz still lacked a definition of continuity such as that of the *Specimen*, and the proof does not seem to conclude.[82]

A few years later, Leibniz sketched a very similar proof, that he enriched with a demonstration that the point of intersection of the two circles does not fall on the given segment (the base of the equlateral triangle), and that the points of intersection among circles cannot be more than two (*Elements* III, 10). He still concluded appealing to the general principle that.

> If a *continuum* has a part inside and a part outside a figure, it intersects the boundary of such figure.[83]

In the same years, Leibniz continued to work at the demonstration of *Elements* I, 1, and attempted to reshape it into a semi-formal demonstration in his important essay *Specimen analyseos figuratae*. The remarkable breakthough of this essay seems to be Leibniz's clear awareness of the necessity of an *existential statement* (such as the existence of the boundary, in his mature definition of continuity) in order to prove the theorem. As a matter of fact, Leibniz offered a formal derivation that the two circles *cross* one another, which is not very different from those already elaborated in 1679, which we have mentioned above. Then, however, rather than embarking in providing a definition of continuity, and without even mentioning the continuity of the circles, he introduced a *postulate* stating that if two things cross, their point of crossing (i.e. the intersection) *exists*. From an epistemological point of view, recurring to a postulate could not solve the issue for Leibniz (who wanted to derive the entire mathematics from definitions), and yet this formal essay

---

[81] "Est ergo circumferentia BD partim in circulo ABE partim extra, adeoque ejus ambitum secat seu ac proinde ei occurrit" (*ibid.*). The same line of reasoning was held by Leibniz in another essay dating from the same 1679, in which he attempted to give a further "situational" argument of the fact that the circle on the right has a point inside and a point outside the circle on the left: "ergo peripheria dextra sinistra occurrit" (LH XXXV, I, 11, Bl. 55, also published in ECHEVERRÍA (1980, vol. 2, pp. 295–301). A similar statement is still repeated in Leibniz's *Demonstrationes Euclideae* from the early 1690s, still unpublished in LH XXXV, I, 3, Bl. 3.

[82] The most remarkable text in this respect has been published in ECHEVERRÍA, PARMENTIER (1995, pp. 266–70).

[83] This can be read in an unpublished fragment in LH XXXV, XII, 1, Bl. 67r: "...sequitur ex hac generali propostione: Si continuum partem habet intra figuram partem extra, occurrit circumferentiae ejus aliquo puncto".

of the 1680s can be seen as an important step forward in the uderstanding of the difference between a merely situational property (the crossing) and a further existential statement (the intersection), which cannot be given by situation alone.[84]

In the same months, apparently, Leibniz penned down a further fragment of proof of Elements I, 1, which seems to cash out most of the work done up to this point, and which grounds the existence of the point of intersection on a continuity axiom (*axioma de continuis*), which is spelled out in terms that are not very far away from the final definition of continuity in the *Specimen geometriae luciferae*, but still seems to lack full generality.[85]

Ten years later, when Leibniz was able to actually give a mathematically viable definition of continuity, the discussion of *Elements* I, 1 that he engages in is still somehow similar and still based on a proof that some points of a circle are inside and some outside of the other circle. But at this point, in the *Specimen geometriae luciferae*, Leibniz is able to call upon his new definition of continuity and to *prove*, rather than assume, the existence of the intersection point (assuming, of course, the continuity of the two circumferences). Because, if the point of intersection were missing, there would not be a decomposition in parts of a circumference that would exhaust the whole circumference and yet intersect the first part at a boundary (a point) only. This is basically a derivation of the Circle-Circle intersection axiom.[86]

The *In Euclidis* πρῶτα, on the other hand, applies in a rather abridged way the notion of continuity to prove that the diameter of the circle does cut the circumference itself, thus fulfilling the second requirement of Leibniz's early discussion of *Elements* I, 1:

> Moreover, it follows from the nature of continuity that every *continuum* which is partly inside and partly outside a figure falls on its boundary. In fact, any two parts of a *continuum* that together make the whole have something in common even though they have no part in common. Let there be, then, two parts of a straight line, one inside and the other outside the circle. They have a common point. This point is also common to the circle, for it is in the part of the plane falling inside the circle and also in the part of the plane in which the straight line falling outside the circle lies. But anything that is common to the two parts of the plane, is in their common section, that is, the circumference.[87]

---

[84]The *Specimen analyseos figuratae* is a very remarkable essay still unpublished in LH XXXV, I, 14, Bl. 21–22. The postulate sounds: "Postul. 2: Datis occurrentibus habetur occursus". I thank Javier Echeverría for providing me with a transcription of this text.

[85]The latter fragment is still unpublished in LH XXXV, XII, 1, Bl. 69r. The dating of such fragments is still uncertain.

[86]The new proof of *Elements* I, 1, is to be found in GM VII, p. 284.

[87]"Porro sequitur ex natura continuitatis omne continuum, quod est partim intra partim extra figuram, cadere in ejus terminum. Nam *continui* duae partes quaevis totum aequantes habent aliquid commune, etsi partem communem non habeant. Sint ergo duae partes rectae, una intra circulum, altera extra circulum. Hae habent punctum commune. Id punctum etiam commune est tum circulo, quia est in parte intra circulum cadente, tum parti plani rectam continentis extra circulum jacenti, quia est in parte extra circulum jacente. Quicquid autem duabus plani hujus partibus est commune, id in communi earum sectione est, nempe in Peripheria" (GM V, p. 196).

A few lines below, a symbolic proof employing the machinery of the *characteristica situs* is also performed. In the following passage, $\overline{X}$ means, as before, the collection of all X's:

> It is worth arranging this demonstration in terms of the *Calculus of Situation*, so that we can familiarize ourselves a little with this latter. A plane is divided by the circumference of a circle into two parts, $\overline{X}$ and $\overline{Y}$, the first $\overline{X}$ being inside the circle and the second $\overline{Y}$ outside it. The circumference will, therefore, be $\overline{X}$ and $\overline{Y}$, that is to say, the *locus* of all points in common that are both X and Y. Let the straight line be extended from a point $\overline{Z}$, then the part of it which is inside the circle will be $\overline{Z}$ and $\overline{X}$, while the part outside of it will be $\overline{Z}$ and $\overline{Y}$. Their common point then (following the nature of continuity) is Z and X and Y; thus, it is X and Y; therefore, it is in $\overline{X}$ and $\overline{Y}$, that is to say, in the circumference.[88]

This must count, of course, as a symbolic proof of the Line-Circle intersection axiom. In both cases Leibniz was able to recover the *ad hoc* continuity principles which had been introduced by previous mathematicians in order to perfect Euclid's proofs using his new notion of continuity as completeness. He was able, in fact, to demonstrate these axioms as special cases of his more powerful principle and in the *Specimen* he explicitly generalized this latter so as to make it apply to all lines and surfaces:

> And in general, if any continuous line whatsoever is in any surface whatsoever, and it is partly inside and partly outside of a part of said surface, it will somewhere intersect the boundary of this part. And if any continuous surface whatsoever is partly inside and partly outside of any solid whatsoever, it will necessarily intersect somewhere the boundary of said solid.[89]

On the following page of the *Specimen*, Leibniz generalized his results even further by producing a symbolic expression for sets with common parts and common sections which easily extends the notion of continuity to any possible class of situational elements. It is an implementation of Leibniz's *characteristica geometrica* deploying a proper notion of continuity rather than special axioms for the intersections of lines and circles:

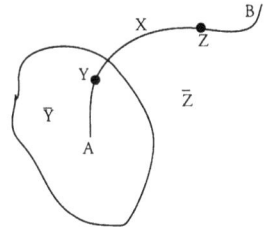

> We may express this also with a kind of calculus. Let's call $\overline{Y}$ a part of extension, and let us use the general name Y to refer to whatsoever point falls within this part $\overline{Y}$. Let us call all points of this extension that fall outside of this part by the general name Z, and call $\overline{Z}$ the whole extension outside of the part $\overline{Y}$. It is clear that the points falling within the boundary of the part $\overline{Y}$ are common to $\overline{Y}$ and $\overline{Z}$, and that they may partly be called either Y or Z, that

---

[88] Here is the original passage, with some corrections made to Gerhardt's transcription: "Operae autem pretium erit, hanc demonstrationem *Calculo situs* nonnihil accomodare, ut ei paulatim assuescamus. Planum per peripheriam circuli dividitur in duas partes $\overline{X}$ et $\overline{Y}$, unam $\overline{X}$ [intra] circulum, alteram $\overline{Y}$ [extra] circulum. Peripheria autem erit $\overline{X}$ et $\overline{Y}$ seu locus omnium punctorum, quae simul sunt X et Y. Recta autem ab uno termino producta sit $\overline{Z}$, ejus una pars, quae intra, circulum, est $\overline{Z}$ et $\overline{X}$, quae extra circulum, est $\overline{Z}$ et $\overline{Y}$. Punctum ergo utrique commune (ob naturam continuitatis) est Z et X et Y; ergo est X et Y; ergo est in $\overline{X}$ et $\overline{Y}$ seu in peripheria" (GM v, p. 197).

[89] "Et in genere, si linea aliqua continua erit in aliqua superficie, sitque partim intra partim extra ejus superficiei partem, hujus partis peripheriam alicubi secabit. Et si superficies aliqua continua sit partim intra solidum aliquod partim extra, necessario ambitum solidi alicubi secabit" (GM VII, p. 284).

is to say, that some Y is Z, and some Z is Y. The whole extension, on the other hand, is composed by $\overline{Y}$ and $\overline{Z}$ together, and it is $\overline{Y} \oplus \overline{Z}$. All its points are either Y or Z, and some of them are both Y and Z. Assume now that a new extension is given, namely AXB, which exists in the previous extension $\overline{Y} \oplus \overline{Z}$, and let us call this new extension by the general name $\overline{X}$, so that all its points, regardless of which, is X. It is immediately clear that every X is either Y or Z. If it is established from our data that some X is Y (e.g. A, which falls within $\overline{Y}$) and also that some X is Z (e.g. B which falls outside of $\overline{Y}$, and therefore in $\overline{Z}$), it follows that some X is both Y and Z. Thus, even if, generally speaking, nothing would follow from these premises, nonetheless in a *continuum* all this may be inferred from the premises thanks to the peculiar nature of continuity.[90]

This rather abstract and almost set-theoretical presentation of continuity was Leibniz's final accomplishment in this field of research. He continued throughout his life to restate that Euclid's proof of *Elements* I, 1, was in need of a continuity principle.[91]

4.7 The Composition of the Geometrical Continuum

I would say that the most important *logical* progress made by Leibniz with his new definition of continuity consists in the fact that this latter is a fully *structural* definition. We have seen that in the early modern era continuity was often simply assumed to be a monadic property that bodies or figures may or may not possess— a useless notion in mathematics, which in many respects assumed continuity as a primitive, unanalyzable concept. The original Aristotelian conception of continuity was relational but it was not structural. The idea that continuity (or even the structure of magnitude) may arise from the system of all possible parts of a whole and their reciprocal relations was never explored by Aristotle and he was therefore not able to

---

[90]"Hoc autem aliquo calculi genere etiam exprimere possumus, ut si alicujus extensi pars sit $\overline{Y}$ et unumquodque punctum cadens in hanc partem $\overline{Y}$ vocetur uno generali nomine Y, omne autem punctum ejusdem extensi cadens extra eam partem vocetur uno generali nomine Z, adeoque totum extensum extra illam partemn $\overline{Y}$ sumtum vocetur $\overline{Z}$. patet puncta in ambitum partis $\overline{Y}$ cadentia esse communia ipsi $\overline{Y}$ et ipsi $\overline{Z}$ seu partim posse appellari Y et Z, hoc est dici posse aliqua Y esse Z et aliqua Z esse Y. Totum autem extensum utique ex ipsis $\overline{Y}$ el $\overline{Z}$ simul componitur seu est $\overline{Y} \oplus \overline{Z}$, ut omne ejus punctum sit vel Y vel Z, licet aliqua sint et Y et Z. Ponamus jam aliud dari extensum novum, verbi gratia AXB existens in extenso proposito $\overline{Y} \oplus \overline{Z}$, et extensum hoc novum vocamus generaliter $\overline{X}$, ita ut quodlibet punctum sit X, patet ante omnia omne X esse vel Y vel Z. Si vero ex datis constet aliquod X esse Y (verbi gratia A quod cadit intra Y) et rursus aliquod X esse Z (verbi gratia B quod cadit extra $\overline{Y}$ adeoque in $\overline{Z}$), sequitur aliquod X esse simul et Y el Z. Unde cum alias in genere ex particularibus hoc modo nihil sequatur, tamen in continuo ex iis tale quid colligitur ob peculiarem continuitatis naturam" (GM VII, pp. 284–85).

[91]See Leibniz letter to Schenk of November 21st, 1712: "J'ai trouvé encor d'autres defauts dans les demonstrations d'Euclide: par exemple dans la prop. première du premier livre, il suppose que les deux cercles se rencontrent, sans l'avoir prouvé, ou sans en avoir fait un axiome" (DE RISI 2007, p. 621).

(nor did he want to) express the notion of completeness. This is rather to be found in Leibniz's definition, where continuity is neither a property nor a relation but rather a *system of relations* (Leibniz would say: *un ordre de relations*) existing among all the (infinite, possible) parts of a whole. Continuity is a structure and a structure is something which lies, logically speaking, in an altogether different realm from that of objects or relations.

There can be no question but that Leibniz's structural definition of continuity emerged out of his general views on mathematics, and these were themselves views which leaned toward structuralism in several fields—and in geometry in particular. In this respect, his celebrated definition of space as *an order of (situational) relations* opened up the way for him to an account of continuity along the same lines. The focus on the positional properties of things in space made it possible, for the first time, to conceive of continuity not as a monadic predicate or as a binary relation but rather as a system of such relations. I would, therefore, contend that the birth of such a notion of continuity constitutes one of the important consequences of the thematization of space as the object of geometry in the seventeenth century and in Leibniz's work in particular.

Nevertheless, it is very important to emphasize that Leibniz's definition of continuity, though indeed articulated in terms of a system of relations, is not articulated in terms of a system of *situational* relations. The system of relations in question here is rather a system of *mereological* relations and the notion of *situs* never appears in the definitions of continuity found in the *Specimen geometriae luciferae* or the *In Euclidis* πρῶτα. In particular, relations of order are generally subsumed by Leibniz under the general heading of "situation" and his definition of continuity, in contrast to Dedekind's, is not based on ordering relations.[92]

The main reason for this definitional strategy was probably Leibniz's strong commitment to the traditional Aristotelian definition of continuity, which was itself mereological. This, in its turn, was grounded in Aristotle's and Leibniz's further commitment to a definition of quantity as a mereological structure (quantity is for them something which has parts). Most of all, however, Leibniz's mereological definition of continuity is rooted in his own deeper agreement with Aristotle regarding the composition of the *continuum*.

Aristotle had denied that there can be a *continuity of points* (given his definition of continuity) and Leibniz himself cannot but endorse such a view. Leibniz's points are not homogeneous with the whole, and are therefore not *parts* of this latter. Thus, they cannot engender any mereological structure and a simple collection of points (without any further structure) cannot be continuous according to Leibniz's definition of the term. It should be noted, in fact, that Leibniz's definition of continuity is mereological all the way down, in the sense that it is expressed by a system of relations among parts, which are themselves continuous and (given *density*) composed of other parts, on and on to infinity. Such a definition is neither

---

[92] See for instance Leibniz's *Fifth Letter* to Clarke, §54, in which Leibniz puts together the two notions (ROBINET 1957, pp. 150–51). The idea is recurrent also in his mathematical papers.

inconsistent nor circular and the fact that the parts are continuous (as a *consequence* of the continuity of the whole) only means that a *continuum* is not reducible to simpler, non-continuous elements. One has to begin, so to speak, with a continuous whole. This was the classical Aristotelian way out of the *labyrinthus continui*: namely, stating that one cannot compose any extension from unextended things, and that points and other unextended elements are just abstractive limits (boundaries) of an originally given continuous extension.

In the overwhelming majority of his writings, both public and private, and running from his young years up to the very last months of his life, Leibniz simply remained faithful to the classical idea that unextended elements cannot compose extension, and that a set of points cannot be continuous. He was therefore convinced that continuity could not be constituted by a situational structure such as the one expressed in the above-mentioned definition of a circle. He did not attempt to reduce the elements of geometry to points and situational relations alone, taking continuity to be a derivative notion that could be defined starting from these latter. Rather, he permitted extension itself, or continuity, to count among the basic elements of geometry: on a par with situational relations. For example:

> When one meditates on geometry, two concepts come first of all to mind: namely, absolute space itself, in which nothing can be considered but extension; and the point, in which nothing can be considered but situation. Space has no situation, and a point has no extension. Space is infinite, and a point is indivisible. Space is the locus of all points. Midway between point and infinite space is the finite *extensum*, which has both extension and situation. A point outside of it can always be assigned. And it lies itself in infinite space, and bears a situation in respect to any other *extensum* . . . .[93]

Here, extension and situation (found respectively, in their pure forms, in space and in points) are the original elements of geometry. We may add that Leibniz's standard definition of extension is that of a "co-existent *continuum*" (as opposed to the successive *continuum* of time).[94] The notion that he really needs to supplement *situs* with, then, is that of continuity. This is clearly stated elsewhere:

> Extension indeed arises from situation but adds continuity to situation. Points have situation, but do not have continuity, nor do they compose it . . .[95]

---

[93]"Ordine meditanti de rebus geometricis ante omnia occurrent duo, *Spatium* scilicet ipsum absolutum, in quo per se nihil aliud considerari potest quàm extensio; et *punctum* in quo nihil aliud considerari potest, quàm situs. Spatium non habet situm et punctum non habet extensionem. Spatium est infinitum et punctum est indivisibile. Spatium est locus omnium punctorum. Medium inter punctum et spatium infinitum est extensum finitum, & id extensionem habet et situm. Assignari potest aliquod punctum extra ipsum. Ipsum inest spatio infinito, ad aliud quodlibet habet situm". From a paper from the mid-1680s, in LH XXXV, I, 5, Bl. 49, in DE RISI (2007, p. 624).

[94]This definition is widespread. It can be found, for instance, in § 17 of the *Characteristica geometrica* from 1679 (GM V, p. 147). It is discussed in strongly anti-Cartesian terms in the 1680s, for instance in the *Generales inquisitiones* (A VI, 4A, n. 165, p. 745), but it recurs also in Leibniz's late years (for example in the letter to De Volder from December 31st, 1700, in GP II, p. 221).

[95]Leibniz to Des Bosses, April 24th, 1709 (GP II, p. 370).

This dichotomy between situation and continuity recurs extremely frequently in Leibniz's writings, and has many ramifications outside the domain of geometry, extending into metaphysics itself. In a sense, the entire debate on corporeal substances in Leibniz's late metaphysics—namely, that on whether monads (and their structure) are themselves sufficient to compose bodies or need something further added—may be read from this perspective. We will not follow these developments here but it is important to see that the irreducibility of continuity to a situational structure is pervasive in Leibniz's thought.

Nonetheless, an important difficulty still lied ahead for an early modern mathematician, and for Leibniz in particular: a difficulty that stemmed from the enormous developments in mathematics that had occurred in the centuries elapsed since Aristotle and Euclid.

We have seen, in fact, a few excerpts from Leibniz's *characteristica geometrica* in which spheres, circles, planes and lines are defined as *collections of points* satisfying certain geometrical conditions. Leibniz had even introduced into his calculus a symbol (the short line over the letters) for a universal quantification that ranges over points. Even more, the general example given by Leibniz himself in the *Specimen geometriae luciferae* in order to formalize his newly found notion of continuity in terms of his geometrical characteristics also quantifies over points, rather than over parts. Here, as everywhere else in Leibniz symbolic treatment of geometry, geometrical figures are defined in a (naïve) set-theoretical way rather than in mereological form, with these collections of points nonetheless being claimed to have to be considered as *continuous*.

It may be remarked that Leibniz could have been in good company, because his views on the impossibility of composing the *continuum* out of points were quite widespread in the seventeenth century, and yet anyone working with Cartesian algebraic geometry (and then the geometry of transcendental curves) had to say that a curve is the collection of points expressed by the solutions of an equation. It should be remarked, however, that in early modern mathematical epistemology the connection between algebra and geometry was conceived of as quite extrinsic. Almost no one had yet dreamt of identifying an equation with a curve. Algebra was not a way of expressing geometry by symbols but rather a *tool* that a mathematician might use to solve a geometrical problem. In this respect no one took seriously (i.e. treated as a foundational statement) the apparent algebraic implication that curves were made of points. Curves were rather geometrical objects, which were not collections of solutions of equations and which received their continuity by their definition or by their generation through a continuous motion (for example), be it in imagination only or actually practically performed using compasses, strings and sliding rulers. Algebra, intended as a useful formalism to investigate certain properties of a curve, had nothing to say about the true nature of this latter.

Then again, Leibniz's *characteristica geometrica* was first conceived precisely in order to surpass the shortcomings of the application of algebra to geometry. It is a *characteristica propria* because it should be able to express the content of geometry without any extraneous element. The constant and variable letters do not range over numbers, as in Cartesian algebra, but straightforwardly over points, and the relations are those of congruence (or similarity), that is to say, basic *geometrical* relations. In fact, the definitions of curves offered by Leibniz's symbolic treatment (i.e. by taking the circle as $\overline{Y}$: A.B.C $\simeq$ A.B.Y), which employs only elementary situational relations, should even express the *essential definition* of said geometrical objects.[96] In this sense, such definitions should express the innermost essence of the geometrical figures, which therefore should actually be considered as a collections of points.

Here we see at work Leibniz's new structural conception of geometry. While it is true that a circle is a collection of points, it is also a collection of *situated* points (this is expressed in the above-mentioned formula by the dot between letters—meaning a relation of *situs*). In fact, a circle is a certain *structure* of points, which have an infinity of spatial relations to one another, as expressed by the formula. Once again, it is of paramount importance to realize how different such a conception of a geometrical figure was from the standard seventeenth-century notion of a geometrical object: the latter may well have relations with other objects and might even be analyzed in a finite set of components (as, for instance, in the case of the three straight lines forming a triangle), but, contrary to Leibniz's new conception of space and geometrical figures, it cannot be entirely resolved into a structured set of relations.

I think that this structural and almost set-theoretical conception of geometry that we find in Leibniz's writings may explain the fact that Leibniz could not be entirely satisfied with the classical Aristotelian views on the composition of the *continuum*, in so far as they entailed a multi-sorted, non-structured ontology of geometrical objects. As a matter of fact, we may spot another trend in Leibniz's mathematical writings, which is less represented but extremely important, in which he attempts different solutions to the old problem of the composition of the *continuum*.

We know that in his young years (around 1670), Leibniz was still (somewhat naively) attempting to conceive of extension as composed of indivisibles. He abandoned this view in the Parisian years, to embrace the above-mentioned classical solution to the problem of continuity, which is expressed by a stark dichotomy between situation and extension. It should be mentioned that still in the late 1670s he was still thinking about whether continuity could arise by some peculiar *situational* structure of points, so that extension might be considered to be composed of situated points after all. In his handwritten notes to Henry More's book on the *Immortality of the Soul*, in which More argues in the usual Aristotelian way about the impossibility of composing extension out of points, Leibniz remarks:

> This line of reasoning—namely, that extension is not composed of points because it would thereby be composed by nothings of extension—does not seem very certain. Because the points are something more than nothings, having a situation.[97]

---

[96] I have briefly investigated the notion of an essential definition in Leibniz's geometry in my DE RISI (2015, pp. 31–40).

[97] The passage most likely dates from the late 1670s and thus predates Leibniz's definition of continuity. Leibniz had already embarked, however, on his project of *analysis situs* and had already given his definition of space as a system of relations. The original text reads: "De plus le raisonnement ne me paroist pas trop asseuré, que l'étendue n'est pas composée de points, parce

The passage is quite elliptical, but it is hard not to see in it a hint that extension may be produced by points as a system of situational relations.

In the following years, Leibniz's conception of space as a system of situational relations was strengthened and developed. Since a point is defined (as we have seen) simply as an object the only property of which is to be considered as an element of a situational relation, it seems obvious to think of space as a system of situated points. In this respect, a close comparison can be drawn between Leibniz's mature definition of space and our present-day notion of a metric space, in which extension results not from a "material" accumulation of an infinity of unextended points but rather from a "formal" structure endowed by the system of all distances (i.e. situational relations) among its points.[98] Nevertheless, as the above-mentioned texts show, Leibniz seemed to deny that a simple situational structure might engender *continuity* and restated the divide between situation and extension.

In the 1690s, however, probably prompted by a renewed interest on his part in a mathematically sound definition of continuity, Leibniz expressed once again with great clarity the idea that extension should be considered as a structure of points (or of other lower-dimensional indivisibles) rather than as an unanalyzed "gunk". We have already mentioned the passage in the *Specimen geometriae luciferae* in which Leibniz directly declared to be "continuous" a figure composed of a collection of lower-dimensional objects arranged in a certain order (a spheroid made up of continuously varying ellipses). In a few further passages of the great *Dynamica* from 1689–1690 Leibniz dealt with similar examples in which a cylinder is seen as constituted by an infinity of concentric cylindrical surfaces (he even spelled out the singular point of such a foliation):

> I say, therefore, that the solid is constituted by these surfaces: not so much by reason of the fact that it is composed by them as by reason of the fact that these surfaces absorb all of its points and lines, with the possible exception of a few of them, which are determinate. The cylindrical surfaces inscribed in the cylinder absorb all of its points and lines, with the exception of the axis.[99]

Elsewhere in the same work, Leibniz considered a surface in motion to be made up of an infinity of lines sharing a common motion and added once again that such a collection of lines did not *compose* the surface but nonetheless (in some way still to be explained) *constituted* it: "*constitui, dico, non componi*". He further claimed that

---

qu'elle seroit composée de riens d'étendue. Car les points sont un peu plus que riens, ayants une situation" (A VI, 4B, n. 331, p. 1678). We may add that the standard objection here—namely, that a collection of points cannot compose extension, since adding a point to other points would be "adding a nothing to a nothing" (*velut si nihil nulli iungas*)—is to be found in Boethius (*De institutione arithmetica*, II, 4; ed. Friedlein, p. 87).

[98] I have briefly described Leibniz's constitution of a metrical space (without mentioning continuity) in my DE RISI (2019).

[99] "His autem superficiebus solidum constitui dico, non quod componant, sed quia omnia ejus puncta absorbent, forte quibusdam determinatis, ubi superficies cylindricae cylindro inscriptae absumunt omnia ejus puncta et lineas, praeter axem" (GM VI, p. 317; cf. also p. 319 for a similar use of "*constituere*").

bodies are constituted by points, that is to say, they are nothing but sums of points ("*nihil aliud sunt quam summae punctorum*").[100]

We may note that in these passages situation is no longer mentioned as the structural element needed to enforce continuity and, in fact, Leibniz did not explicitly propose any foundation for the operation of *constitution*. Leibniz's studies on geometrical continuity ought, therefore, to have explained which further principle, besides situation, was needed in order to obtain a proper *constitution of the continuum*—something that needed to be clearly distinguished from the standard Aristotelian discussion on the *composition of the continuum*.

While, however, the *Dynamica* had been written a few years before the *Specimen geometriae luciferae*, I would argue that after the geometrical breakthrough of the mid-1690s it is clear that such *constitution* may be characterized by the *mereological* structure added to the situational structure. In this respect the mutual irreducibility of situation and continuity is now revealed to be grounded in the interaction of two different structures: a situational structure on the one hand, which is often represented as a system of distances between points and that in any case has *points* as its basic elements and by which we may define geometrical objects and shapes; and a mereotopological structure on the other, by which continuity is defined and the elements of which are (extended) *parts* rather than points. Leibniz's claim that geometry is only possible by means of extension and situation together hereby becomes the idea that two different geometrical structures interact with one another. His claim of the irreducibility of extension to situation may be further explicated as the impossibility of accounting for the mereological structure as emergent from the underlying situational (i.e. metric) structure. In general, situational relations obtain among points and are, in this sense, first order relations. Even when Leibniz refers the situation of an extended figure to that of another extended figure, this generally means that there are certain situational relations between the *points* of the two figures in question: that is to say, a situational relation between figures is *reducible* to situational relations between individual points. On the other hand, continuity is defined by an irreducible appeal to parts and therefore needs to refer to (and quantify on) the elements of a further, second-order, structure: namely, the mereological one, constituted by sets of points rather than points. In this respect, while it is true that we can describe space and figures in purely situational terms, the facts that they have parts, which are collections of points, and that these parts themselves have a structure (a mereological one) is something that is irreducible

---

[100]The expression "constitui, dico, non componi" comes from GM VI, p. 370, where Leibniz discussed the surface constituted by lines and then repeated the above-mentioned example with the cylinder. Further on in the same work (GM VI, p. 399), Leibniz also mentioned a pyramid constituted of smaller and smaller rectangles. A similar passage, in which rectangles constitute a more complex solid figure, is to be found in GM VI, p. 415. Finally, he also mentions bodies constituted by points in GM VI, p. 498: "et quod de punctis duobus, id de quibuscunque veram esse ostendimus, adeoque et de mobilibus quibuscunque quae per puncta constituuntur, seu nihil aliud sunt quam summae punctorum, hoc est corporum sufficientis ad evitandum errorem dato minorem parvitatis". I thank Andrea Costa for bringing to my attention some of these passages.

to situational relations. In modern terms, Leibniz's ontology of space needs both elements (i.e. points) and sets (which give rise to parts).[101]

We may also add that parts must be regarded as *actually infinite* collections of points. There is no doubt, in fact, that before the nineteenth century Leibniz was one of the very few philosophers who envisaged the possibility of treating the actual infinite in mathematical terms. The very fact that he devised a peculiar notion of an actual and yet syncategorematic infinite in order to deal with this notion, shows the importance of the actual infinite in Leibniz's mathematical studies. The present issue seems to be a high point of Leibniz's new approach to infinite collections of elements, that had few matches before Dedekind and Cantor.[102]

Here is the real divide between Leibniz's conception and the classical, Aristotelian one. For Aristotle, an extended whole is just a different kind of being than a point (or a boundary in general) and one has to accept a two-sorted (or many-sorted) ontology of extended and unextended things. By contrast, there is a sense in which Leibniz's extended wholes are "constituted" by points, since the parts required to possess continuity are nothing but collections of points. In the same way, we have seen in a quotation from the *Specimen geometriae luciferae* that a solid body may be conceived of by Leibniz as constituted by a continuous family of surfaces. The continuity of such a family is not given by the reciprocal situation of the elements, and Leibniz is adamant that the set formed by these latter is continuous even without its having to be geometrically structured into a solid figure, or placed in space. Continuity is rather given by a further condition on the possibility of putting them into a regulated correspondence with the points of a continuous segment. There is a continuous map between a family of surfaces constituting a body, and the collection of points constituting a segment. The segment itself (and therefore also the body) is continuous thanks to its mereological structure, according to the definition of continuity given in the *Specimen*.[103]

Leibniz's ontology, then, is, in a sense, as two-sorted as the Aristotelian one: there are points, endowed with a situational structure; and parts (or sets of points),

---

[101] There might by some exceptions to this rule in a few geometrical writings, in which the situational relation cannot but apply to the figure as a whole. I have given a few examples of this attitude in my *Leibniz on the Parallel Postulate*, and will not discuss it further here.

[102] On the topic, see Arthur's fictional dialogue between Leibniz and Cantor in this volume.

[103] We will not deal here with the most difficult problem of the continuity of *time* in Leibniz's philosophy. We may notice, however, that a mereological approach to continuity as the one sketched above might in fact allow Leibniz to claim that time is continuous, and that it is nonetheless a collection of instants. The latter statement recurs in Leibniz's writings, and its formulation is especially striking in a draft from 1695—that might have been composed, therefore, in the same months of the *Specimen*. In it, Leibniz simply recognized the problem: "Tempus vero videtur necessario componi ex instantibus, quia duo instantia simul existere non possunt; itaque existit tantum instans praesens; [passatum] extitit, futurum existet. Video tamen nondum hinc sequi quod linea componatur ex infinitis punctis, sed tantum quod infinita puncta sunt in linea. Verum ecce difficultatem" (LH I, I, Bl. 24r., edited by MUGNAI 2000, p. 135). We may notice that a *situational* definition of the *continuum*, would not work in order to ground the continuity of time, since instants have no *situs* (cf. C 541).

endowed with a mereological structure. Aristotle's "substantialistic" view on the composition of the *continuum* has thus been transformed into a (similarly expressed) structural approach to the same problem. In this sense, *extension* (as a coexistent *continuum*, according to Leibniz's definition) has been dissolved into a system of parts.[104]

We may now attempt to offer an explanation of Leibniz's thought regarding continuity in his *characteristica geometrica* and in geometry in general. A formula such as $\overline{Y}$: A.B.C $\simeq$ A.B.Y, expressing the "equation" of a circle does not, and cannot, express the continuity of this latter. It only states the situational relations obtaining among the points that give rise to the figure of a circle (e.g., a set of points equidistant from a center). The formula, however, quantifies on the points of space and states that a circle is constituted by all these points. Whether the collection of all these points is rich enough to, and structured in such a way as to, guarantee continuity—this is an altogether different question, and depends *on space itself* as the collection of all points. The situation here is similar to that in ordinary algebra, where the circle equation $x^2 + y^2 = 1$ would produce Dedekind-complete or incomplete circles depending on whether the variables range over $\mathbb{R}$ or, say, the field of rational numbers $\mathbb{Q}$; and this is something that the equation itself cannot express.

Leibniz holds, therefore, that the *continuity of space* is the source of the continuity of the circle, which is defined as a collection of points. This is well expressed in the above-mentioned passage from the 1680s in which Leibniz immediately connects extension (and therefore continuity) to space, while situation is seen as an original property of points (or collections thereof). Space, however, is the source of the continuity of the circle not in the sense that it contains "points enough" to be continuous (i.e. complete), but in the sense that it has a mereotopological structure which makes it continuous. Such a structure is not and cannot be expressed as a system of situational relations among points, and therefore escapes the expressive power of a *characteristica geometrica*, by which we define the circle. Being a part of space, however, the circle inherits its mereotopological structure and acquires (so to speak) parts, boundaries, and situational relations between these latter. In this way, the circle acquires continuity. The whole process may be seen as a kind of embedding which induces a mereotopological structure on the subset.

We may note that such an embedding is only possible insofar as space itself is not *just* a mereotopological structure but *also* a situational system of points, and in particular the system of all situated points (otherwise a circle, defined as a system of situated points, could not embed in space). The interplay between the situational structure and the mereotopological structure, therefore, occurs *in space*.

---

[104] In the *Nouveaux essais*, I, III, § 6, Leibniz plainly states that the notion of extension is posterior to (and possibly generated by) the notions of whole and parts: "je crois plustost que l'idée de l'étendue est posterieure à celle du tout et de la partie" (A VI, 6, n. 2, p. 103).

4.8 The Continuity of Space

This latter, very important mathematical claim namely, that the continuity of a figure in space must be grounded in the continuity of space itself—was rather new, indeed almost unprecedented, in the seventeenth century, when geometry was still largely regarded as a science of figures rather than a science of space. It is nonetheless perfectly in line with the whole Leibnizian project of an *analysis situs*, that is to say, a new geometrical science which conceives of space as a *structure* of situational relations (an *ordre de situations*), and takes it to be the proper object of geometrical investigations.

Leibniz's epistemological attitude towards space and figures may be briefly sketched as follows.[105] In geometry, we are free to begin our enquiry by giving any arbitrary definition of a given figure. We may, for instance, follow Euclid in defining the circle as the curve equidistant from a given point (the center). After giving such a *nominal* definition, however, we should prove the possibility of the defined object (i.e. the internal consistency of the notion), and therefore transform the nominal definition into a *real* definition which can be actually employed in mathematical reasoning (since a merely nominal definition, which might possibly be inconsistent, would endanger the whole process of reasoning). Such real possibility may be proven, in principle, using any means might be available to us; in practice, however, most mathematicians in the early modern age, including Leibniz, thought that the easiest way to provide evidence of the possibility of a geometrical object was to *generate* it by giving a construction rule for it. In the case of the circle, this would amount, for example, to saying that it is generated by the rotation of a segment around one of its endpoints. Leibniz, however, here asks something more: the circular motion of the segment itself should be proven to be possible; otherwise we are just assuming, rather than proving, the possibility of the circle itself. The possibility of such a motion, though, may only be grounded, for Leibniz, in a property of space itself. Commenting on Euclid's Third Postulate, he wrote in the *In Euclidis* πρῶτα:

> Postulate 3: To describe a circle with any center and radius. This can be done in a plane by the motion of the radius while one of its endpoints remains still. But that a straight line can be moved while one of its endpoints stands still can be derived from the fact that the plane, or space, is uniform....[106]

---

[105]I have offered a lengthier treatment of the subject in my DE RISI (2015).

[106]The original passage reads: "Postulat. III. Quovis centro et intervallo circulum describere. Id in plano efficit motus radii uno puncto immoto. Posse autem moveri rectam uno puncto immoto ex eo colligitur, quod spatium planumve uniforme est, et quod versus unam est plagam, potest etiam versus aliam sumi quamcunque" (GM V, p. 206). Similar statements recur, however, frequently in Leibniz's works on geometry and can be found, for instance, in his paper from 1676 on *Generatio quidem rectae et circuli* (in ECHEVERRÍA, PARMENTIER 1995, p. 66), his essay *Uniformis Locus* from 1692 (in DE RISI 2007, pp. 582–85), his reading notes to Arnauld's *Nouveaux Elémens de géométrie* from 1695 (in DE RISI 2015, pp. 146–51).

The *uniformity* of space, here mentioned just in passing, is a property that Leibniz characterized elsewhere as a sort of isotropy (in modern terms), being the one grounding free rigid motion and rotations in space. The important mathematical and epistemological point, in any case, is that the very possibility of a geometrical figure is to be grounded, in the last analysis, in a property of the space in which it is embedded, and in this sense the true object of investigation of a perfected *analysis situs* should be space itself and its properties.

It is now easy to see that, since *uniformity* of space is necessary in order to state the possibility of a figure the points of which are all equidistant from a given point, the *continuity* of such a figure—which is not, for Leibniz, entailed, as it was for many of his contemporaries, by the simple notion of motion—should be grounded in a further property of space itself, that is to say, in *its own* continuity. This means that a circle is, in the last analysis, only possible as a figure, and will only intersect with other circles when crossing them (as in *Elements* I, 1), thanks to the uniformity and continuity of space itself. Such properties are themselves defined in structural terms, since space is nothing more than a system of relations. We have already seen, however, that uniformity is a situational property in the strict sense, and that it must depend on the structure of *situs* relations (i.e. broadly, on the structure of space conceived of as a system of distances, or as a metric space in abstract terms), while continuity is defined rather in terms of parthood relations, and therefore depends on the (independent) mereological structure of the same space.

This latter geometrical analysis, which relates back all the properties of a figure to the properties of the underlying space, was not, however, to exhaust the aims of Leibniz's *analysis situs*. It is well known, in fact, that Leibniz also embarked on the grand epistemological project of a kind of *logicism* in geometry. He wanted to show that the whole of geometry, and all its theorems, could be proven by logical laws alone and were ultimately grounded in the Principle of Contradiction, so that any deviation from the theorems of Euclidean geometry would not count as the consequence of an alternative system of axioms but rather as a logical inconsistency.[107] Given the preceding discussion, it should be clear that Leibniz could hope to achieve such a goal by proving, using logical means, that *space itself* cannot but possess a number of geometrical properties, such as the above-mentioned uniformity and continuity, on which the whole of classical geometry could rest. This proof, in turn, had to be grounded exclusively in the *definition* of "absolute space" as the order of all possible situations, and on a few metaphysical considerations establishing the necessity of such a structure.

We will not follow here the details of this geometrical and philosophical program, which was clearly doomed to fail. We may remark, however, that, if extracting such complex geometrical properties as uniformity (i.e. isotropy) or three-dimensionality out of a general notion of total situational order was surely a difficult (and in fact, an impossible) task, even greater difficulties stood in the way of attempting to prove

---

[107]For a more nuanced and complex account of Leibniz's logicism, see the essay by Valérie Debuiche and David Rabouin in this volume.

continuity from such a definition. Because, as we have noted, the mereological structure needed to define continuity cannot, in principle, be reduced to the situational structure by which space is logically defined. In fact, however, Leibniz's "logicism" in geometry had a hybrid nature and rested not only on logical laws but also on several metaphysical and phenomenological considerations. Leibniz claimed that space is somehow "ideal" and, while the exact import of the latter notion is difficult to establish, this certainly means that space is, in a certain way, a product of the mind (as an abstraction from physical extension, or perhaps as a form of intuition in an almost Kantian sense). This, in turn, implies that several properties of space may be derived not just from its logical definition (the order of all possible situations), but also from the operations of the mind that are necessary to produce it. In this way, Leibniz opened the way to a phenomenological foundation of geometry itself.[108]

The philosophical discourse of phenomenology enters into the foundations of geometry especially in relation to continuity. This can hardly be surprising, since the mereological structure cannot have a purely logical source in the notion of a situational order. In particular, Leibniz defines sensible (as opposed to intellectual) knowledge by stating that the former is a *confused* apprehension of reality. Confusion, in turn, is tantamount to a certain degree of indeterminacy, which causes us to perceive as homogeneous things that are not in fact so. For instance, we do not perceive all the infinite, actual divisions of matter and our sense-experience ceases once we have arrived at a certain degree of detail, leaving an undifferentiated extension beyond that point. The actually divided *parts* of matter (or physical space) are confused with one another, becoming an undifferentiated extension that has only indeterminate parts. In short, and without entering into the difficult and complex details of Leibniz's phenomenological theory, such (necessary) indeterminacy of perception must form the ground of the continuity of space. Geometrical space is a kind of idealization of the sensible *medium* (there would be no space without *sensibility*), and therefore retains traces of the structure of sensibility itself as the organ of confused perception. Ideal space is simply totally undetermined and homogeneous, since sensibility is the faculty producing such indetermination. The indeterminacy of space, in particular, is manifested (as we have seen above) in the indeterminacy of its parts. We remember, on the other hand, that Leibniz's "metaphysical" definition of continuity was grounded simply on the indeterminacy of parts. Therefore, "absolute space", i.e. the one, original, ideal geometrical space, is *continuous*.[109]

---

[108]I have offered a detailed treatment of Leibniz's phenomenology in relation to geometry in my DE RISI 2007.

[109]For further details I cannot but refer again to *Geometry and Monadology*. I may add here a nice quotation by Leibniz from an early essay on *Divisio terminorum*: "*Extensionem* vocamus quicquid omnibus simul perceptis commune observamus; et extensum vocamus cujus perceptione plura percipere possumus simul; idque indefinita quadam ratione, unde extensum est totum continuum cujus partes sunt simul et habent situm inter se, ipsumque totum rursus eodem modo se habet tanquam pars respectu alterius. Totum continuum est cujus partes sunt indefinitae, tale est ipsum

We do not need to enter into the details of Leibniz's phenomenology in order to understand that the proof that he was attempting to give fell short of its goal. In fact, even if we were to follow Leibniz through his complex epistemological theory of sensibility and idealization, we would thereby only be grounding the continuity of space on a definition of continuity as indeterminacy of parts—which was abandoned by Leibniz himself in his geometrical writings and has no definite mathematical meaning. This definition was, in fact, just the standard (and much easier) metaphysical definition of continuity that the *Specimen geometriae luciferae* attempted to surpass.

Leibniz also attempted other phenomenological deductions of the continuity of space. In his *Characteristica geometrica* from 1679, he made the following remarks:

(60) Between any two congruent objects, infinite other congruent ones can be taken; in fact an object cannot move to the place of another and still preserve its own shape except by passing through other congruents.
(61) Therefore, a line can be drawn from any point to any other. In fact, any point is congruent with any other point.

. . .

(108) If two objects are perceived simultaneously in space, then, by this very fact, the path from the one to the other is also perceived. And by virtue of the very fact that they are congruent, the path from the one to the place of the other is also thought. . . . Or, to put it yet another way: we perceive that a certain object can move, and thus come to be either in one place or in another one; but, since it can neither exist in a plurality of places at the same time nor move in just one instant, we perceive this place as continuous.[110]

Leibniz restated similar geometrical-phenomenological claims about continuity in his late years.[111] In such passages, it seems that continuity of space is simply tantamount to arcwise connectedness, since Leibniz discusses the possibility of tracing a line (in other texts, a straight line—geodesic connectedness) from any point to any other point. Therefore, however fascinating we may find such texts, and however sound we may judge Leibniz's phenomenological deduction of such a

---

spatium abstrahendo animum ab his quae sunt in ipso. Hinc tale continuum est infinitum, ut tempus et spatium. Cum enim ubique sibi simile sit, quodlibet totum erit pars" (A VI, 4A, n. 132, p. 565).

[110]The first two sections have been published in GM V, p. 161: "(60) Inter duo quaevis congrua assumi possunt infinita alia congrua, nam unum in locum alterius servata forma sua transire non posset, nisi per congrua. (61) Hinc a quolibet puncto ad quodlibet punctum duci potest linea. Nam punctum puncto congruum est". The latter sections, however, appear to be grounded only at the end of the essay, in §108. Since, however, this section is more metaphysical than geometrical, Leibniz himself had suggested (in the margins of the manuscript) excising §108 from the mathematical essay, and this was actually done by the nineteenth century editor of GM V. The last section was eventually published in ECHEVERRÍA, PARMENTIER (1995, pp. 228–30): "Cum duo simul in spatio esse percipiuntur, eo ipso percipitur via ab uno ad aliud. Et cum sint congrua, eo ipso concipitur via unius in alterius locum. Sunt autem duo puncta congrua. Itaque quod percipitur, duobus punctis simul perceptis, est Linea, seu via puncti. . . . Sive quod idem est, posse moveri, sive posse tam in uno loco quàm in alio esse, et quia non potest simul esse in pluribus locis, nec moveri in instanti, ideò locum illum percipimus ut continuum".

[111]In the *Initia rerum mathematicarum metaphysica* from 1715, in GM VII, p. 25.

property to be, it is not a matter here of continuity as Dedekind-completeness, and it is not useful for grounding Euclid's results.

Lastly, in some other texts Leibniz stressed the fact that the very notion of an "absolute" or "universal" space, meant as the system of all possible situations, implies that such a space contains "all points" (since a point is just an abstract situated element): *Spatium Universum est locus omnium punctorum*. This last statement is also sometimes reshaped into phenomenological terms, so that Leibniz can attempt to *prove* that absolute space encompasses the totality of situational relations by stating that the perception of any situational relations entails the perception of any other. In any case, Leibniz stated explicitly that such a universal or absolute space had necessarily to be *continuous*:

> Every place is in the same absolute space. Absolute space is continuous, otherwise some place could be interposed which would not be in it.[112]

This notion of continuous space as the totality of all points has in fact some relation to the notion of completeness, which also asserts something about the existence of a certain totality of ordered points. Nevertheless, the whole difficulty consists in stating what exactly "all" means here and it was to solve this difficulty that Dedekind's definition was first conceived. On the other hand, Leibniz's definition may bear a somewhat closer resemblance to the (quite problematic) definition of Linear Completeness given by Hilbert in the *Grundlagen*. Once again, though, Leibniz's philosophical proof is too general, and too vague, to provide an exact meaning for this notion.

In short, we may say that Leibniz's logicist program in geometry required a proof that there is just one possible space and that this space is continuous. In turn, the continuity of space, which could not be proven (in Leibniz's terms) from the logical definition of space, forced Leibniz to venture into a different domain, and to extend his logicist program in order to include a few phenomenological considerations. By such considerations Leibniz may have believed he had proven the continuity of space. But his notion of this latter's continuity had been formulated only with older, metaphysical notions (such as the indeterminacy of parts) or insufficient or vague mathematical claims (arcwise connectedness, totality of points). In the meantime, however, Leibniz had been able to achieve a remarkable mereotopological definition

---

[112]The passages in which Leibniz defines space as the totality of points are numerous, and I will only refer here to the *Scheda on situation and extension* from 1695, in which Leibniz explicitly connects this array of problems with continuity (to be found in DE RISI 2007, pp. 588–89; the Latin quote above is taken from there); and to the essay *Spatium absolutum . . .* from 1714, from which the longer quotation is taken: "*Spatium absolutum* interminatum est quod in situ amplissimum est. Ideo omnis locus est in eodem spatio absoluto. Spatium absolutum est continuum, alioqui locus aliquis interponi posset qui in ipso non esset" (DE RISI 2007, p. 609). We may note that Leibniz considered complex numbers to be internally inconsistent, and therefore there is a sense in which the maximal number field had been for him that of real numbers: complex numbers cannot be taken as coordinates of possible situational elements. Given that he believed that he could prove the three-dimensionality of absolute space (DE RISI 2007, pp. 205–215), the "totality" of its points had to be modeled on $\mathbb{R}^3$.

of continuity as completeness-connectedness which could be usefully employed in geometrical demonstrations. Such a definition, though, would have been difficult to apply to space itself using philosophical arguments: metaphysics and mathematics failed to weld together.

In this respect, the program of Leibniz's *analysis situs* was a failure especially in relation to continuity and he could not claim to have proven the logical necessity of the meeting of the two circles in *Elements* I, 1. Leibniz had perfected Euclid's proof by giving a viable mathematical definition of continuity, by which he could prove in a rigorous way that the two circles meet—*provided that they are continuous*. This was Leibniz's great accomplishment in the foundations of mathematics and may count as the first (although surely imperfect) grounding of a general theory of intersections. From our modern standpoint, no further discussion would be required, and we would be prompt to assume the continuity of the circles (or the continuity of the underlying space) as an axiom. Leibniz's epistemology, however, brought him to ask for more: to ask for a proof of the continuity of space itself that could eliminate the need for an axiom on continuity. Such a proof was a failure, for it could not be achieved, and exposed Leibniz's line of reasoning to a number of incursions in the muddy waters of metaphysics, and to a few paralogisms attempting to connect together a phenomenological definition of continuity (or a bunch thereof) with his different, and quite good, mathematical definition of the same notion. Leibniz himself could not have been fully satisfied by his own efforts towards an *analysis situs* grounding a theory of continuity. But we, disenchanted about the possibility of a logicist program in geometry, may still appreciate Leibniz's outstanding results in defining continuity.

Leibniz's definitions and demonstrations were lacking the exactness, simplcity, or power that we find in similar (but much later) attempts by Bolzano and Dedekind. They are still remarkable, though, as they are probably the only ones in the modern age to ground continuity in a *purely geometrical* way. Leibniz lived in an age in which the foudation of mathematics were not found in number theory or analysis, but rather in geometry, and the *Elements* were considered to be the groundwork of the whole mathematics. Yet, we have seen that no other early modern mathematician ever attempted at a geometrical definition of continuity, and the latter had to wait for the deep transformation of mathematics that occurred in the nineteenth century, and eventually produced Dedekind's arithmetical analysis of cotinuity. In this respect, Leibniz's unstable mediation between the Euclidean tradition in the proper sense and the modern need for rigour remained an isolated, and soon forgotten, attempt at maintaining continuity between two ages of mathematics.

Acknowledgements I thank the Max Planck Institute for Mathematics in the Sciences and his director Jürgen Jost for providing me with the possibility of writing a good part of this paper while I was based in Leipzig, and more in general for having fostered important occasions of discussion with other Leibniz scholars. I also thank Richard Arthur, Eberhard Knobloch, Julien Narboux and Erich Reck, who have discussed with me this paper in several occasions; and Andrea Costa, Siegmund Probst and Javier Echeverría for sharing with me some unpublished papers by Leibniz.

Bibliography

Alcoba, M.L. 1996. Leibniz: Geschichte des Kontinuumproblems. *Studia Leibnitiana* 28: 183–198.
Anapolitanos, D.A. 1990. Leibniz on Density and Sequentiae or Cauchy Completeness. In *Greek Studies in the Philosophy and History of Science*, ed. P. Nicolacopulos, 361–373. Dordrecht: Springer.
Arthur, R.T.W., ed. 2001. *Leibniz: The Labyrinth of the Continuum.* New York.
———. 2013. Leibniz's syncategorematic infinitesimals. *Archive for the History of Exact Sciences* 67: 533–593.
———. 2018. *Monads, Composition, and Force: Ariadnean reads through Leibniz's Labyrinth.* Oxford: OUP.
Beeley, P. 1996. Kontinuität und Mechanismus. Zur Philosophie des jungen Leibniz in ihrem ideengeschichtlichen Kontext. *Studia Leibnitiana Supplementa* 30.
Bennett, B. and I. Düntsch. 2007. *Axioms, Algebras and Topology*, in Handbook of Spatial Logics, M. Aiello, I. Pratt-Hartmann, J. van Benthem, Dordrecht, Springer
Boethius, S. 1867. *De institutione arithmetica*, ed. Friedlein. Leipzig: Teubner.
Bonitz, H. 1848. *Aristotelis metaphysica.* Bonn: Marcus.
Breger, H. 1986. *Leibniz, Weyl und das Kontinuum. Studia Leibnitiana Supplementa* 26: 316–330.
Calosi, C., and P. Graziani, eds. 2014. *Mereology and the Sciences. Parts and Wholes in the Contemporary Scientific Context.* Heidelberg: Springer.
Casati, R., and A. Varzi. 1999. *Parts and Places. The Structures of Spatial Representation.* Cambridge: MIT Press.
Caveing, M. 1982. Quelques remarques sur le traitement du continu dans les Éléments d'Euclide et la Physique d'Aristote. In *Penser les mathématiques*, ed. F. Guénard, G. Lelièvre, and Seuil Paris, 145–166.
Cohn, A.G., and A. Varzi. 2003. Mereotopological Connection. *Journal of Philosophical Logic* 32: 357–390.
Crockett, T. 1999. Continuity in Leibniz's mature metaphysics. *Philosophical Studies* 94: 119–138.
Dedekind, R. 1932. *Gesammelte Werke*, ed. R. Fricke, E. Noether, and O. Ore, Braunschweig: Bieweg.
De Risi, V. 2007. *Geometry and Monadology. Leibniz's Analysis Situs and Philosophy of Space.* Basel/Boston: Birkhäuser.
———. 2015. *Leibniz on the Parallel Postulate and the Foundations of Geometry.* Basel/Boston: Birkhäuser.
———. 2016. The Development of Euclidean Axiomatics. The systems of principles and the foundations of mathematics in editions of the Elements from Antiquity to the Eighteenth Century. *Archive for History of Exact Sciences* 70: 591–676.
———. 2019. Analysis Situs, the Foundations of Mathematics and a Geometry of Space. In *The Oxford Handbook of Leibniz*, ed. M.R. Antognazza, 247–258. Oxford: OUP.
De Risi, V. forthcoming a. Gapless Lines and Gapless Proofs. Intersections and Continuity in Euclid's Elements. *Apeiron.*
De Risi, V. forthcoming b. *Has Euclid proven Elements I, 1? The early modern debate on intersections and continuity.* In *Reading Mathematics in the Early Modern World*, ed. P. Beeley, Y. Nasifoglu, B. Wardhaugh. London: Routledge.
Echeverría, J. 1980. *La caractéristique géométrique de Leibniz en 1679.* Paris: Sorbonne. PhD dissertation.
———. 1990. Infini et continu dans les fragments géométriques de Leibniz. In *L'infinito in Leibniz*, ed. A. Lamarra, 69–79. Roma: Ateneo.
Echeverría, J., and M. Parmentier, eds. 1995. *G.W. Leibniz, La caractéristique géométrique.* Paris: Vrin.
Friedman, M. 1992. *Kant and the Exact Sciences.* Cambridge: Harvard University Press.
Froidmont, L. 1631. *Labyrinthus sive de compositione continui.* Antwerp: Moretus.

Garber, D. 2015. Leibniz's Transcendental Aesthetic. In *Mathematizing Space*, ed. V. De Risi, 231–254. Basel.

Giovannini, E. 2013. Completitud y continuidad en "Fundamentos de la geometría" de Hilbert: acerca del "Vollständigkeitsaxiom". *Theoria* 28: 139–163.

Giusti, E. 1990. *Immagini del continuo*. In *L'infinito in Leibniz*, ed. A. Lamarra, 3–32. Roma: Ateneo.

Glezer, T. 2017. *Kant on reality, Cause, and Force: From the Early Modern Tradition to the Critical Philosophy*. Cambridge: CUP.

Graßmann, H. 1847. *Geometrische Analyse geknüpft an die von Leibniz erfundene geometrische Charakteristik*. Leipzig: Weidmann. Now in *Gesammelte mathematische und physikalische Werke*, I, 1. Leipzig: Teubner 1896–1911.

Hartshorne, R. 2000. *Geometry: Euclid and Beyond*. New York: Springer.

Heath, T.L., ed. 1925. *Euclid. The Thirteen Books of the Elements*. Cambridge: CUP.

Hellman, G., and S. Shapiro. 2013. The Classical Continuum without Points. *The Review of Symbolic Logic* 6: 488–512.

Hellmann, G., and S. Shapiro. 2015. Regions-Based Two Dimensional Continua: the Euclidean Case. In *Logic and Logical Philosophy*, vol. 24, 499–534.

Hellman, G., and S. Shapiro. 2018. *Varieties of Continua. From Regions to points and back*. Oxford: OUP.

Hilbert, D. 1968. *Grundlagen der Geometrie*. Leipzig: Teubner (first ed. Leipzig 1899).

Jesseph, D. 2015. *Leibniz on The Elimination of Infinitesimals*. In *G.W. Leibniz, Interrelations Between Mathematics and Philosophy*, ed. P. Beeley, N. Goethe, and D. Rabouin, 189–205. Dordrecht: Springer.

Jullien, V., ed. 1996. *Eléments de géométrie de G.P. de Roberval*. Paris: Vrin.

Leibniz, G.W. 1846. In *Historia et origo calculi differentialis*, ed. C.I. Gerhardt. Hannover: Hahn 1846.

Leibniz, G.W. [GM]. *Mathematische Schriften*, ed. C.I. Gerhardt. Berlin/Halle 1849–1863.

Leibniz, G.W. [BW]. *Briefwechsel zwischen Leibniz und Ch. Wolff*, ed. C.I. Gerhardt. Halle 1860.

Leibniz, G.W. [GP]. *Die philosophischen Schriften*, ed. C.I. Gerhardt. Berlin: Weidmann 1875–1890.

Leibniz, G.W. [C]. *Opuscules et fragments inédits*, ed. L. Couturat. Paris: Alcan 1903.

Leibniz, G.W. [A]. *Sämliche Schriften und Briefe*, Darmstadt/Leipzig/Berlin: De Gruyter 1923.

Levey, S. 1998. Leibniz on Mathematics and the Actually Infinite division of matter. *The Philosophical Review* 107: 49–96.

———. 1999. *Matter and two concepts of continuity in Leibniz*. *Philosophical studies* 94: 81–118.

———. 2003. The Interval of Motion in Leibniz's Pacidius Philalethi. *Nous* (3): 371–416.

Lewis, D. 1991. *Parts of Classes*. Oxford: Blackwell.

Linnebo, Ø., S. Shapiro, and G. Hellman. 2015. Aristotelian Continua. *Philosophia Mathematica* 24: 214–246.

Mugnai, M. 2000. Two Leibniz Texts with Translations: LH IV 1, 9 r and LH IV 1, Bl. 24 r. *The Leibniz Review* 10: 135.

Palmerino, C.R. 2015. *Libertus Fromondus' Escape from the Labyrinth of the Continuum (1631)*. Vol. 42, 3–36. Lias.

———. 2016. Geschichte des Kontinuumproblems or Notes on Fromondus's Labyrinthus? On the true nature of LH XXXVII, IV, 57r-58v. *The Leibniz Review* 26: 63–98.

Panza, M. 1992. De la Continuité comme Concept au Continu comme Objet. In *Le labyrinthe du continu*, ed. J.-M. Salanskis and H. Sinaceur, 16–30. Berlin: Springer.

Pratt-Hartmann, I. 2007. First-Order Mereotopology. In *Handbook of Spatial Logics*, ed. M. Aiello, I. Pratt-Hartmann, and J. van Benthem, 13–98. Dordrecht: Springer.

Rabouin, D., ed. 2018. *G.W. Leibniz. Mathesis Universalis. Écrits sur la Mathématique Universelle*. Paris: Vrin.

Robert, A. 2009. William Crathorn's mereotopological atomism. In *Atomism in Late Medieval Philosophy*, ed. C. Grellard and A. Robert. Leyden: Brill.

Robinet, A., ed. 1957. *Correspondance Leibniz-Clarke*. Paris: PUF.

Roeper, P. 2006. The Aristotelian Continuum. A Formal Characterization. *Notre Dame Journal of Formal Logic* 47: 211–232.

Sereda, K. 2015. Leibniz's Relational Conception of Number. *The Leibniz Review* 25: 31–54.

Sylla, E.D. 1982. Infinite Indivisibles and Continuity in Fourteenth-Century Theories of Alteration. In *Infinity and Continuity in Ancient and Medieval Thought*, ed. N. Kretzmann, 231–257. Ithaca: Cornell University Press.

Varzi, A. 1994. On the Boundary Between Mereology and Topology. In *Philosophy and the Cognitive Sciences. Proceedings of the 16th International Wittgenstein Symposium*, ed. R. Casati, B. Smith, and G. White, 423–442. Vienna: Hölder.

Vitali, G. 1923. Della continuità nella geometria elementare. In *Questioni riguardanti le matematiche elementari*, ed. F. Enriques, vol. 1, 193–230. Bologna: Zanichelli.

White, M.J. 1988. *On Continuity: Aristotle versus Topology? History and Philosophy of Logic* 9: 1–12.

———. 1992. *The Continuous and the Discrete: Ancient Physical Theories From a Contemporary Perspective*. Oxford: OUP.

Wilder, R.L. 1978. Evolution of the Topological Concept of "Connected". *The American Mathematical Monthly* 85: 720–726.

Wyclif, J. 1893. In *Tractatus de logica*, ed. M.H. Dziewicki. London: The Wyclif Society.

Chapter 5
On the Plurality of Spaces in Leibniz

Valérie Debuiche and David Rabouin

Abstract According to a famous Leibnizian dictum, space is nothing but "an order of situations, or an order according to which situations are disposed". This so called "spatial relationism" is sometimes understood as a form of relativism (space is relative to the possible dispositions of bodies), which seems to entail the possibility of a spatial pluralism (possible worlds with non-equivalent dispositions of bodies might have non-equivalent spatial structures). An overly exclusive focus on the exchange between Leibniz and Clarke, considered to be the *locus classicus* for understanding Leibniz's views on space, did much to lend credence to such a conception. In addition, in several passages, Leibniz talked of other possible worlds as lacking this or that spatial property. In the exchange with De Volder, for example, he explained that God could have been pleased with a phenomenal world with gaps in it. In this description, as in other passages, it seems that we can imagine without contradiction a world with a geometrical structure different from ours. This stance was supported, as we will see, by influential scholars who concluded, without much trouble, that a plurality of spaces was conceivable for Leibniz. Yet this idea runs counter to two other no less famous Leibnizian dicta: first, that geometrical truths are absolutely necessary (one cannot deny them without contradiction); second, that geometry should be considered as the science of space and that we can specify various properties of geometrical space (such as tridimensionality, homogeneity, isotropy or continuity). Combined together, these last two claims tend to prevent any form of pluralism: space is the object of a science which describes its essential properties and these properties are truths which are absolutely necessary. These eternal truths apply to all possible worlds. Our goal in this paper is to confront this tension in Leibniz's description of space.

V. Debuiche
Aix Marseille Univ, CNRS, Centre Gilles-Gaston Granger, Aix-en-Provence, France

D. Rabouin (✉)
Université de Paris, Laboratoire SPHERE, UMR 7219, CNRS, Paris, France

© Springer Nature Switzerland AG 2019
V. De Risi (eds.), *Leibniz and the Structure of Sciences*, Boston Studies in
the Philosophy and History of Science 337,
https://doi.org/10.1007/978-3-030-25572-5_5

171

5.1 Introduction

According to a famous Leibnizian dictum, space is nothing but "an order of situations, or an order according to which situations are disposed"[1]. This so called "spatial relationism" is sometimes understood as a form of relativism (space is relative to the possible dispositions of bodies), which seems to entail the possibility of a spatial pluralism (possible worlds with non-equivalent dispositions of bodies might have non-equivalent spatial structures). An overly exclusive focus on the exchange between Leibniz and Clarke, considered to be the *locus classicus* for understanding Leibniz's views on space, did much to lend credence to such a conception. In addition, in several passages, Leibniz talked of other possible worlds as lacking this or that spatial property. In the exchange with De Volder, for example, he explained that God could have been pleased with a phenomenal world with gaps in it[2]. In this description, as in other passages, it seems that we can imagine without contradiction a world with a geometrical structure different from ours. This stance was supported, as we will see, by influential scholars who concluded, without much trouble, that a plurality of spaces was conceivable for Leibniz. Yet this idea runs counter to two other no less famous Leibnizian dicta: first, that geometrical truths are absolutely necessary (one cannot deny them without contradiction); second, that geometry should be considered as the science of space and that we can specify various properties of geometrical space (such as tridimensionality, homogeneity, isotropy or continuity)[3]. Combined together, these last two claims tend to prevent any form of pluralism: space is the object of a science which describes its essential properties and these properties are truths which are absolutely necessary. These eternal truths apply to all possible worlds. Our goal in this paper is to confront this tension in Leibniz's description of space.

One easy way out of the difficulty, which we study in the first section, consists in distinguishing between an "abstract" and a "real" space. The first would act as a mathematical framework, identical for all possible worlds, whereas the second would be related to the variable occupations of this framework by real bodies (some of them being "better" than others). We study the various forms that this widespread strategy has been able to take in the literature and show that it is either incompatible with Leibniz's relationism, or, in the best case scenario, neutral as regards the question of pluralism. Another solution would be to conceive less strictly the absolute character of geometrical truths and to consider whether such truths might be compatible with a plurality of spaces. This entails studying very closely the specific kind of necessity involved in the description of geometrical space. We recall first that Leibniz relied, on several occasions, on architectonic principles to characterize certain geometrical features such as the unicity of geodesics or, later,

---

[1] "Leibniz's Fifth Paper", §104 (GP VII, 415). Translated by Leroy E. Loemker, in Loemker (1989, 714).

[2] To De Volder, 24 March/3 April 1699 (A II 3, 545).

[3] See De Risi (2007).

the parallel postulate. In addition, he sometimes spoke as if God can create certain mathematical entities belonging to a certain world. These two claims, combined with the fact that he was well aware of the possibility of various metrics not only on surfaces but also in three dimensional space, lead us to conclude that it is possible to reassess the compatibility of spatial pluralism with the principle of the absolute necessity of mathematical truths. We do so in the last section, by putting particular emphasis on the *conditional* nature of mathematical truths.

5.2 Plurality or Unity of Space?

5.2.1 The Problem of the Plurality of Spaces and the Distinction Actual/Abstract

In section 10 of *Principes de la nature et de la grâce fondés en raison* (1714), one finds the following statement:

> It follows from the supreme perfection of God that he has chosen the best possible plan in producing the universe, a plan which combines the greatest variety together with the greatest order; **with terrain, place, and time arranged in the best way possible**; with the greatest effect produced by the simplest means; with the most power, the most knowledge, the greatest happiness and goodness in created things which the universe could allow[4].

Leibniz's claim can be interpreted in two different ways. Following the first line of interpretation, there are various spatio-temporal structures and God chooses one amongst them because it is arranged in the best possible way. Following the second interpretation, there is only one spatio-temporal structure and God can dispose things within it in various ways, one of them being better than the other. Although the first interpretation was endorsed by many Leibnizian scholars (see next section), it seems that only the second one is consistent with Leibniz's views on geometrical truth as *absolutely* necessary. Moreover, it seems to be supported by other famous passages such as that of the *De rerum originatione radicali* (1697) which gives a more precise content to the kind of "best" involved in the choice of a spatial disposition:

> Hence it is very clearly understood that out of the infinite combinations and series of possible things, one exists through which the greatest amount of essence or possibility is brought into existence. There is always a principle of determination in nature which must be sought by maxima and minima; namely, that a maximum effect should be achieved by a minimum outlay, so to speak. And at this point time and place, or in a word, the receptivity or capacity to the world, can be taken for the outlay or the terrain on which a building

---

[4]Loemker (1989, 639), GP VI, 603. Emphases in bold are ours, those in italic are Leibniz's. We have modified Loemker's translation. The French word "terrain" has been rendered by Loemker as "situation". This is confusing for the reader, since "situation" is usually used to render "*situs*". We have reestablished the word *terrain*, which is used by Leibniz in other texts.

is to be erected as commodiously as possible, the variety of forms corresponding to the spaciousness of the building and to the number and elegance of its chambers[5].

In this text, we see, so to speak, the two faces of Leibnizian space: the framework[6] within which things are arranged (the "receptivity or capacity") and the arrangement itself. There are, however, various difficulties involved in this distinction. First and foremost, "place", "time" and "terrain" are assigned to the side of the framework – by contrast to the "variety of forms". If we use this text to interpret the previous passage from *PNG*, we are led back to the first line of interpretation: if God can choose the best amongst various "terrains", this means that there are various *frameworks*, each of them able to "receive" various arrangements. This would be consistent with the metaphor of the architect who has to erect various buildings according to the variation of the terrains. However, it is possible to maintain the unicity of space in both descriptions by interpreting the expression from *PNG* as metonymic. By the terrain or the place "arranged in the best possible way" one might understand a (single) framework *endowed with* an optimal disposition of its parts.

This line of interpretation naturally leads to the idea that the two aspects of space could be interpreted as a distinction between an abstract (mathematical) space and a real (physical) space – the second corresponding to a certain disposition of bodies within the framework (satisfying, in addition, certain optimality conditions). Accordingly, relying on the distinction between real and abstract space, it might be possible to reconcile the idea of physical optimality, which seems to entail a plurality of spatial configurations, with the mathematical necessity requiring space's unicity. One abstract space for a multiplicity of real spaces.

The distinction between actual space and abstract space is relevant in Leibniz, since he explicitly made use of it. For instance, in the 5th letter addressed to Clarke, one can read:

27. The parts of time and place considered in themselves are ideal things, and therefore they perfectly resemble one another like two abstract units. But it is not the same with two concrete ones, or with two real times, or two spaces filled up, that is, truly actual[7].

In two other passages, he stated:

106. If there were no creatures, there would be neither time nor place, and consequently no **actual space**. [. . .] And therefore I don't admit what's here alleged, that if God existed

---

[5]Loemker (1989, 487), GP VII, 303.

[6]On the idea that space and time act as a "framework", see *Réponse aux reflexions contenues dans la seconde Edition du Dictionnaire Critique de M. Bayle, article Rorarius* (1702), GP IV, 568: "But space and time taken together constitute the order of possibilities of the one entire universe, so that these orders – space and time, that is – frame not only what actually is but also anything that could be put in its place, just as numbers are indifferent to anything that can be *res numerata*." (Loemker 1989, 583). We modify Loemker's translation, in particular we render the verb "quadrer" by its literal meaning, "to frame", instead of the overly vague "to relate".

[7]Loemker (1989, 700), GP VII, 395.

alone there would be time and space as there is now, whereas then, in my opinion, they would be only **in the ideas of God as mere possibilities**[8].

67. But space without things has nothing whereby it may be distinguished and, indeed, is nothing actual[9].

"Actual" is related to a certain disposition, or to a filling with material bodies, and "abstract" to space "without things", bodies or creatures, only existing in God's mind as an idea and defined as the order of possible situations[10]. Since the conceptual distinction between "abstract" and "concrete" (or "ideal" and "actual") appears to offer a solution to the tension involved in the idea of a plurality of spaces, it seems necessary that we improve our understanding of this distinction. In particular, we need to study how it is traditionally related to the question of plurality or, rather, how it might possibly make the question disappear. In order to do so, we will have to return to look at the writings that Leibniz addressed to Clarke and to their interpretation in the literature.

5.2.2 Abstract and Actual Space in the Literature

It is noteworthy that the texts referred to above and, in general, the correspondence with Clarke, are often read according to a line of interpretation focusing rather on the question of the reality of space. Leibniz is seen as denying actuality to absolute space, since actual space is only the space of actual things. From that claim, many commentators (Vailati, Khamara, Earman[11]) conclude that Leibniz defended the reality of a relative space. Leibniz's main purpose would then be to justify the reality of something which is not absolute and which is only dependent on relations. Thus, being relative, spaces might also possibly be multiple.

In 1976, a discussion was directly initiated between Nicolas Rescher and Yvon Belaval about the question of the plurality of spaces[12]. This debate is a rare and stimulating instance of the addressing of our problem and it is interesting to see how it was solved. Both Rescher and Belaval defended the claim that the nature of Leibnizian space, as an order of situations, allowed for the possibility of different

---

[8]Loemker (1989, 714), GP VII, 415.

[9]Loemker(1989, 708), GP VII, 407.

[10]Note, however, that these texts do not state that space "with things" in it is "real", and that this reality has to be related to material bodies. They rather affirm that there is no proper actuality or reality of space in and of itself (against Newton's reality of absolute space), and that space can only be said to be "actual" when it is considered as that in which actual things are. In this sense, talking about the actuality of space does not amount to claiming either that space is actual in the sense of being a material or physical entity or that space depends totally upon a certain disposition of actual bodies and has to change in some way with every change in the said disposition.

[11]Vailati (1997); Khamara (2006, namely "Chapter 3. Leibnizian relativism"); Earman (1979).

[12]The discussion began during a conference in Bern in 1976 and led to two articles: Rescher (1977), Belaval (1978).

worlds being endowed with different spatial structures. According to them, the very fact that Leibniz denied the reality of absolute space and conceived of space rather as an order relative to coexisting things was sufficient to ensure that there cannot be a unique form of it[13]. Among the spatial frameworks corresponding to other worlds than ours there might even be some which would be geometrically different from the one we know. In particular, one might envisage without contradiction non-Euclidean spaces. Rescher's main evidence to support this view involved reference to Leibniz's *analysis situs*, which he saw as a forerunner of modern topology and thus as characterizing an abstract space which might be endowed with various metric determinations (giving birth, amongst other possibilities, to the "actual" space)[14].

In the same period, founding his analysis on Leibniz's correspondence with Clarke, John Earman also assumed the plurality of spaces on the basis of their relativity[15]. Since the set of things is contingent and might potentially be different from what it actually is, it follows that space itself could be different (space being nothing else than the order of coexisting things). As a matter of fact, as soon as Leibniz's relationism (*i.e.* the notion that space is an order of situations or of coexisting things) is assumed to be tantamount to a relativism (*i.e.* to the notion that space is dependent upon existing things), several consequences seem to follow automatically: 1. There can be a plurality of spaces, since there can be other dispositions than the one which applies in our world; 2. As something relative to existing things, space is "real" or "actual", its reality depending on and deriving from the reality of things themselves; 3. The ideality of space, as the abstract order of "possible" situations, is opposed to its actuality as the order of "real" things.

In the 1990s, with the correspondence with Clarke still being taken as the main basis of discussion, the issue of Leibnizian space was still mostly regarded as one centred on the problem of the conditions and nature of its reality. In order to solve this problem several commentators, for instance Ezio Vailati[16] and Edward Khamara[17], considered actual space not as the order of co-existing things but

---

[13]Explicitly employing Rescher's arguments, E. Vailati claims that the plurality of spaces is obvious in Leibniz's thought, in Vailati (1997, 116).

[14]This argument seems nowadays particularly weak: a better knowledge of the texts on the *Characteristica Geometrica* has shown that Leibniz's project of *analysis situs* was much more an attempt at generalizing Euclidean geometry than it was the invention of an alternative geometry. Interestingly enough, the position that space is only one amongst several other orderings of possible worlds has been defended nonetheless by one of the main editors of Leibniz' texts on *analysis situs*: Echeverría (1999, 430, and note 8, where he discusses Rescher and Belaval). We will come back to this question in the last section.

[15]Earman (1979).

[16]Vailati (1997).

[17]Khamara (2006, 40-41). The author surprisingly introduces the notion of "point-particles" to clarify Leibniz's paragraph 47 in the fifth letter to Clarke about the explanation of what it is for one thing to be at the "same place as" another thing. Such an introduction is necessary because the author systematically replaces "things" in space by material "bodies". But it becomes superfluous as soon as "things" are conceived of in the most general sense, including material bodies, abstract figures, and dimensionless points.

as the order of existing bodies. The difference may seem subtle but it is a very important one. Firstly, by replacing "things" with "bodies", one overestimates the role of physical or material things, although nothing prevents things in space from themselves being abstract, as is the case in geometry. Secondly, by using "existing" instead of "co-existing", one neglects the relational aspect of spatial determinations. Existence alone does not suffice in order for there to be space. Co-existence is required. Although the commentators mentioned above do not appear to perceive this unfortunate consequence, it appears that, on such a view, to talk about a "geometrical space" would simply be meaningless. Mathematical space can only be considered as an abstraction supervening upon the disposition of real bodies. It is not clear how it could form the proper subject of an abstract theory in and of itself.

It is not surprising that this kind of reading came to be challenged once a better knowledge of Leibniz's geometrical studies had been established. Indeed, after the publication of Leibniz's texts on a *characteristica geometrica* by Javier Echeverría and Marc Parmentier in 1995[18], which expounded a purely geometrical conception of space, and after Vincenzo De Risi's book on the relation between geometry and monadology in 2007[19], the relevance of geometry for the understanding of Leibniz's doctrine of space was more fully recognized[20]. One of the significant changes in the interpretation of Leibniz consisted in the new emphasis placed on a neglected, if not entirely overlooked, difficulty (already mentioned in a paper from 1988 by Glenn Hartz and J.A. Cover[21]): whereas ideal mathematical space is continuous, actual space (if considered in the manner previously described, i.e. not as the order of "co-existing things" but as the order of "existing bodies") has necessarily to be discrete, since it depends on material bodies or on singular substances. This supported a better distinction between space and world. Conceived of as related to bodies, "actuality" should be considered as more likely referring to the created world rather than to space. But a second relevant consequence was the weakening of the idea of the plurality of spaces. The more space is conceived of as independent of real bodies, the stronger becomes the idea of its unicity as object of a geometrical theory[22].

In 2013, in a very interesting paper dealing with Leibniz's conception of space, Richard Arthur[23], while acknowledging the role of *analysis situs* in the story, seemed to admit another form of plurality for spaces. Leibniz's space, according

---

[18] Echeverría, Parmentier (1995).

[19] De Risi (2007).

[20] See E. Slowik: "On the whole, De Risi and Arthur have contributed greatly to the cause of disassociating Leibniz from the traditional, reductive, and external relationism that most modern philosophers of space and time have tended to read into his philosophy" (Slowik 2019, 111, note 16).

[21] Hartz, Cover (1988).

[22] Futch (2008) considered the issue of the plurality of spaces. As his analyses did not refer to Leibniz's geometry, they are not relevant for our survey. Besides, he accorded the primary place to time, not to space. For this reason, the arguments he used to explain the unity of space are not consistent with an approach focused on space itself.

[23] Arthur (2013).

to Arthur, must necessarily be "continuously changing and becoming something different", so that one can see it as a series of "instantaneous space[s]"[24]. On this interpretation the plurality of spaces is no longer connected to the plurality of possible worlds. There remains, however, some plurality conceived of as the possibility of different concrete spatial dispositions and constituting a general spatial framework[25]. This derives from the importance given to bodies and relative motions of bodies. Following Arthur, Ori Belkind[26] likewise continues to place bodies at the core of his interpretation[27]. He attributes reality to relations, since, according to him, "abstract space is ideal, i.e., it is constructed in the mind based on the potential set of relations between bodies" whereas "actualized space is a set of 'real' relations between material substances"[28]. His paper has, like Arthur's, the notable advantage of presenting space as conceived of through its metric, but it does not deal with the question of the unity or plurality of such a metric. Both papers reveal a slight but nonetheless real shift towards new considerations regarding actual and ideal spaces, and, consequently, regarding the issue of the plurality of spaces. Some other papers go in the same direction.

For instance, in a paper from 2012 criticizing what we have called the "relativist" point of view ("reductive relationism" in this paper's own parlance), Edward Slowik recalls that this point of view naturally leads to a possible spatial pluralism, whereas Leibniz insisted on the fact that geometry is Euclidean[29]. Space being characterized by its metric (*i.e.* by the nature of distance between points), the necessity of

---

[24]Arthur (2013, 500-501): "Such a space is therefore a three-dimensional partition, or system of boundaries or figures, corresponding to any one of the actually infinite divisions or foldings of matter at any given instant. Insofar as space is regarded as perduring, on the other hand, it is a *phenomenon*, in that it is an accidental whole that is continuously changing and becoming something different. It can be represented mathematically by supposing some set of existents hypothetically (and counterfactually) to remain in a fixed mutual relation of situation, and gauging all subsequent situations in terms of transformations with respect to this initial set."

[25]*Ibid.*, p. 514: "Phenomenal space is universal space, which consists in a different partition of places from one instant to another. [...] viewed through time it is continuously changing and becoming something different, so that it is never the same thing from one instant to the next."

[26]Belkind (2013).

[27]Belkind (2013, 474): "A body occupies a place when it has a particular set of distance relations to other bodies that are not moving (in Leibniz's language bodies that are fixed existents). One problem that immediately comes to mind is whether or not this definition of place is circular given that motion is ordinarily understood as change of place. If distance relations are determined in relation to bodies in which there is no motion, and if place is defined using distance relations, then it is not clear whether place or motion is the more fundamental concept."

[28]Belkind (2013, 464).

[29]E. Slowik: "On a straightforward interpretation of reductive relationism, the extension within bodies and the relative configuration of the bodies can remain invariant, whereas the actual distance relations among bodies (i.e. the geometry) can vary significantly. Likewise, one could employ the observations of various rigid body motions as a means of determining the geometrical structure of space as a whole. Yet, in contrast, there are a number of discussions in Leibniz's late corpus that single out (infinite) Euclidean geometry as the only possible spatial structure" (Slowik 2012, 121; see also note 70 attached to this paragraph).

being Euclidean seems to entail its unicity[30]. However, Slowik also notes that non-Euclidean space remains here a theoretical possibility due to the possibility of a different metric's applying on a spherical surface[31]. We will return to this point in the next section, since it shows that the problem of pluralism can be posed from within geometry itself. In 2015, focusing on the continuity of abstract mathematical space and the contiguity of material bodies, Daniel Garber updated Hartz and Cover's paper from 1988 by trying to give a more precise account of the relation between the "world of real extended bodies" and "geometrical extension" in terms of the actualization of ideal space in real world[32]. Restoring the quality of ideality to space, Garber's interpretation ultimately disconnects the claim regarding the actuality of space from the claim regarding its plurality, since reality refers only to the world of substances and of bodies, and not to space. The question at issue in Garber's account becomes the examination of the application of the eternal truths of geometry to the real and contingent world and the issue of the plurality of spaces is left unresolved.

In conclusion, it appears that the consideration of Leibniz's geometrical work on space tends to detract from the (surely excessive) importance that has been accorded to the correspondence with Clarke. This perspective on Leibniz's thought favors the distinction between space, as a continuous extended locus or as the locus of all possible coexisting things, and the world, as a discrete set of actual, real, and even material things. In other words, it restrains the tendency to identify so-called "actual" space with the world itself and indicates the fragility of the solution to the problem of plurality which presents itself most easily and immediately (namely: one abstract space, several actual spaces corresponding to this space's occupation by bodies). More generally, it calls for a reappraisal of Leibniz's *relationism* in the context of our problem.

5.2.3   The Failure of the First Solution: The Conception of Space as Abstract Order

As we saw in the previous section, the distinction between an ideal and a concrete space is quite widespread in the literature, although it is interpreted in various different ways. At any rate, it leads more to a tension than to a solution to the problem of the plurality of spaces. On the one hand, it seems that many

---

[30]This is related to the logical conception of necessity as the impossibility of being something else. But, as we will see in Sect. 5.4, the conception of necessity requires some more subtle analyses than this.

[31]E. Slowik: "Interestingly, it would seem to follow that a limited material world with, say, a spherical shape would have a non-Euclidean metric on that surface, given his notion of relative distance" (Slowik 2012, 121).

[32]Garber (2015, 243).

interpretations of Leibniz's relationism naturally lead to spatial pluralism; but on the other hand, the attention paid to the geometrical considerations of ideal space naturally calls for a spatial monism. One way out of the difficulty would be to reconcile these two points of view by proposing a plurality of real spaces corresponding to just one ideal space. But if we take the "arrangement" previously involved to signify a spatial disposition (we will see later that this is not the only possible interpretation), we immediately run up against an obvious difficulty: such a view directly contradicts Leibniz's "relationism".

Indeed, if by "relationism" we mean a certain view of space as an order of situations, in the sense of the relative dispositions of bodies, there seems to be no ontological distinction between the framework and the dispositions. Moreover, the theoretical distinction can only be one between an abstract frame and a series of dispositions which are *equivalent* (that is to say, none better or worse than any other). Any non-equivalent spatial disposition would amount to another set of relations giving rise to another space. To make this point clear, let us recall some famous texts grounding Leibniz's "relationism", for instance the 4th Letter to Clarke (§41):

> The author contends that space does not depend upon the situation of bodies. I answer: 'Tis true, it does not depend upon such or such a situation of bodies, but it is that order which renders bodies capable of being situated, and by which they have a situation amongst themselves when they exist together, as time is that order with respect to their successive positions[33].

To be sure, space does not depend upon "such or such a situation of bodies". But this is true in the specific sense that a body can take the place of any other body without any change thereby occurring in space itself. Mathematical space, which is abstract, is defined as the system of these "places". A body is said to be "taking the place" of another when, by moving, it acquires the same situational relations to a set of other bodies, considered as fixed, as the latter body had previously had. Space is nothing else than the set of all the possible places that bodies may occupy. This grounds the famous argument brought in the 5th Letter to Clarke (§47) as an explanation for the very genesis of an abstract notion of space:

> I will here show how men come to form to themselves the notion of space. They consider that many things exist at once, and they observe in them a certain order of coexistence, according to which the relation of one thing to another is more or less simple. This order is their situation or distance. When it happens that one of those coexistent things changes its relation to a multitude of others which do not change their relations among themselves, and that another thing, newly come, acquires the same relation to the others as the former had, we then say it has come into the *place* of the former; and this change we call a *motion* in that body wherein is the immediate cause of the change. And though many, or even all, the coexistent things should change according to certain known rules of direction and swiftness, yet one may always determine the relation of situation which every coexistent acquires with respect to every other coexistent, and even that relation which any other coexistent would have to this, or which this would have to any other, if it had not changed or if it had changed

---

[33]Loemker (1989, 690), GP VII, 376.

any otherwise. And supposing or feigning that among those co-existents there is a sufficient number of them which have undergone no change, then we may say that those which have such a relation to those fixed existents as others had to them before have now the same place which those others had. And that which comprehends all those places is called *space*. Which shows that in order to have an idea of place, and consequently of space, it is sufficient to consider these relations and the rules of their changes, without needing to fancy any absolute reality out of the things whose situation we consider[34].

The fact that space is defined as a system of what we would now call equivalence classes is made clear through a reference to the Euclidean definition of proportion: "I have here done much like Euclid, who, not being able to make his readers well understand what *ratio* is absolutely in the sense of geometricians, defines what are the *same ratios*. Thus, in like manner, in order to explain what *place* is, I have been content to define what is the *same place*.[35]" Places and space are henceforth not conceived in terms of a particular situation of bodies (which, strictly speaking, makes no sense), but in terms of an equivalence relation between situational relations.

This is the very foundation of Leibniz's famous reply to the hypothesis that God could have placed things "the quite contrary way, for instance, by changing east into west". In such a new arrangement, which Clarke takes to be another disposition within the same absolute framework, each body has taken the place of another one and thus constitutes, for Leibniz, *the very same space*:

> But if space is nothing else but that order or relation, and is nothing at all without bodies but the possibility of placing them, then those two states, the one such as it now is, the other supposed to be the quite contrary way, would not at all differ from one another. Their difference therefore is only to be found in our chimerical supposition of the reality of space in itself. But in truth the one would exactly be the same thing as the other, they being absolutely indiscernible, and consequently there is no room to inquire after a reason of the preference of the one to the other[36].

Note that some of the interpretations mentioned in the previous section would have to conclude here, in contradiction to Leibniz's claim, that this situation corresponds rather to *different* actual spaces. But be this as it may, we are now in a position to better assess the manner of posing the problem that was previously referred to (Sect. 5.2.1). A direct consequence of what Leibniz claims is that, if one acknowledges the possibility of *non-equivalent* dispositions, this will necessarily lead to non-equivalent spatial frameworks. But what this means is that, contrary to what is sometimes thought, the distinction between "actual" and "abstract" space is *neutral* as regards the issue of pluralism. The answer cannot lie in the distinction between a framework and various ways of occupying it. At least it cannot do so unless one specifies precisely the exact relation existing within the model between the various forms of occupation. If they are equivalent, then we are just talking about one and the same space, be it actual or abstract. But the price to pay for this is that

---

[34]Loemker (1989, 703), GP VII, 400.

[35]*Ibid.*

[36]3rd letter to Clarke §5, Loemker (1989, 682), GP VII, 364.

no arrangement can then be said to be better than another one, at least from a purely spatial point of view. On the other hand, if they are non-equivalent, then we are talking of different spaces, by the very definition of what a space is. Any "better" arrangement is thus a "better" space.

The definition of abstract space as a system of equivalence classes[37] is of great importance for another reason. It shows that the "actual" disposition has, in fact, no meaning outside of the "abstract" framework. Indeed, Leibniz explicitly warned against the danger of reifying this distinction. Although abstract space corresponds to various (equivalent) concrete dispositions, and although Leibniz sometimes goes so far as to speak of an "actual" space to designate a particular instantiation of the system of places, it does not follow, as so many commentators have assumed, that there are two distinct spaces. The distinction should rather be understood as a distinction between two aspects of the same space, in the same way as the abstract number and the thing numbered are not two distinct numbers, but two aspects of one and the same number:

> Although it may be true that, by conceiving of a body, one conceives of something more than space, it does not thereby follow that there are two extensions, that of space and that of body; for it is as when, conceiving of several things at once, one conceives of something more than number, i.e. res numeratas, and yet there are not two multitudes, one abstract, that of number, the other concrete, that of enumerated things. One can say in the same way that one should not imagine two extensions, one abstract and of space, the other concrete and of the body, the concrete being so only through the abstract[38].

This remark is also crucial in order to grasp that the distinction between an abstract space and a specific disposition has nothing to do with the fact that the second is « concrete » or « physical » or « real ». In the same way that the res numerata can already be an abstract object, for example when we use numbers in geometry, one can easily have a mathematical disposition as "actual" space. In fact, we need this to be true if we want to make sense of the text from the De rerum originatione radicali from which we started. When it comes to explaining what he has in mind when talking of one disposition being better than another, in the sense of more "determinate", Leibniz gives the following series of examples:

> The case is like that of certain games in which all the places on the board are to be filled according to definite rules, but unless we use a certain device, we find ourselves blocked out, in the end, from the difficult spaces and compelled to leave more places vacant than we needed or wished to. Yet there is a definite rule by which a maximum number of spaces can be filled in the easiest way. Therefore, assuming that it is ordered that there shall be a triangle with no other further determining principle, the result is that an equilateral triangle is produced. And assuming that there is to be a motion from one point to another, without anything further determining the route, that path will be chosen which is easiest or

---

[37]See Arthur (1994, 237): "Thus the hypothesis of fixed existents allows us to define place in terms of an equivalence: it is the equivalence class of all things that bear the same situation to our (fictitious) fixed existents. And when we take all possible situations relative to these fixed existents, we have a manifold of places, or abstract space." On the characterization of situational order by means of the concept of the equivalence of mutual situations, see also Winterbourne (1982, 203).

[38]Nouveaux essais sur l'entendement humain, II, ii, § 5 (1704), A VI 6, 127. Our translation.

shortest. Similarly, once having assumed that being involves more perfection than nonbeing, or that there is a reason why something should come to exist rather than nothing, or that a transition from possibility to actuality must take place, it follows that even if there is no further determining principle, there does exist the greatest amount possible in proportion to the given capacity of time and place (or the possible order of existence), in the same way as tiles are laid so that as many as possible are contained in a surface[39].

All the examples used by Leibniz to illustrate the optimality of a certain disposition are *mathematical*, either directly as is the case for the triangle or the geodesic, or indirectly as is the case for the game of position. In other words, the two aspects of the divine game (the terrain and the building) find their motivation in examples in which all the entities involved are *abstract*. It is thus very difficult to understand the situation in terms of actual and ideal space, if by "actual" one means something attached to "concrete", "real" or "physical" things.

This last remark helps us to question another way of reading the passage from the *PNG*. Indeed, one might consider that the possibility of a "better" arrangement is not attached to a spatial disposition, but to the fact that God can specify *non-mathematical* conditions, sometimes called by Leibniz "architectonic". These conditions consist in the addition of some constraints for the optimal determination not of dispositions, but of *phenomena*. In this case, space would be said to be "better arranged" or "arranged in the best possible way," not because its mathematical disposition would be better than some other, but because, in a mathematical space determined by a system of possible mutual situations and also of relative modifications of these situations, some non-mathematical constraints determine these motions and corporeal changes in a better way than others or in the best possible way. The propagation of light provides a typical example of this situation[40]. In a Euclidean space in which many trajectories are possible under general constraints, only one has a privilege as soon as one specifies a condition of optimality. Only additional conditions can, therefore, explain and specify the real laws of refraction. This also indicates that an element of "under-determination" ("indetermination" in Leibniz's parlance) is inherent in the mathematical structure of space.

But, once again, this solution is neutral as regards our problem of the plurality of spaces. The fact that Leibniz gave mathematical examples leaves the question open as to whether the choice of the optimal situation operates from within various *mathematical* models or within a single mathematical framework with various *physical* realizations (which would be equivalent from a mathematical point of view). Moreover, the fact that Leibniz regularly mentioned the case of the unicity of the geodesic as an example of an optimal solution tends to establish a strong association between the specification of these conditions and the choice of a mathematical spatial structure. How are we to understand this close association between a purely geometrical example and the illustration of the idea of the best choice? Does this imply that the principle of determination also concerns geometry?

---

[39]Loemker (1989, 487, modified), GP VII, 303-304.

[40]See *Tentamen anagogicum* (1695), section 2.1.

And if it does, is this sufficient to assert the plurality of spaces? To answer these questions, we need to examine more closely the role of optimality in mathematics.

5.3 Mathematics and Metaphysical Optimality

5.3.1 Architectonic Determination in Mathematics

At first sight, the principle of optimality only concerns, like the other architectonic principles, the choice of the actual world and of the laws governing it. For instance, it determines the optical laws of light reflection. We have seen that it may be interpreted as concerning non-geometrical determinations operating within a single general mathematical framework. But the difficulty then consists in the fact that Leibniz mostly uses *geometrical* examples to illustrate the nature of these constraints. Let us recall a few texts indicating the problem.

In *Tentamen anagogicum. Essay Anagogique dans la recherche des causes* from 1695, Leibniz presented the case of the choice of a triangle:

> This principle of nature, that it acts in the most determined ways which we may use, is purely architectonic in fact, yet it never fails to be observed. Assume the case that nature were obliged in general to construct a triangle and that for this purpose only the perimeter or the sum of the sides were given, and nothing else; then nature would construct an equilateral triangle. This example shows the difference between architectonic and geometric determinations. Geometric determinations introduce an absolute necessity, the contrary of which implies a contradiction, but architectonic determinations introduce only a necessity of choice the contrary of which means imperfection (. . .) If nature were brutish, so to speak, that is, purely material or geometrical, the above case would be impossible, and unless something more determinative were given than merely the perimeter, nature would not produce a triangle. But since nature is governed architectonically, the half-determinations of geometry are sufficient for it to achieve its work; otherwise it would most often have been stopped[41].

Two elements might be noted here. First, the distinction is maintained between architectonic determinations and geometrical ones, but both of them involve a kind of necessity – we will address this topic in Sect. 5.4. Second, even if a triangle is a geometrical object, it is said to be "produced" by nature in the given situation – we will elaborate on this "production" of mathematical entities in Sect. 5.3.2.

A second example can be found in the passage already quoted from the *De rerum originatione radicali*[42]. Once again, it concerns a triangle, but also a straight line, a very interesting example for our purpose. The main idea is that, without any further geometrical determination, and because of the principle of sufficient reason, if a triangle or a line have to be produced, they will be produced as an equilateral triangle and as a straight line. Where what is demanded is a line between two given points or

---

[41]Loemker (1989, 484), GP VII, 278-279.
[42]See Sect. 5.2.3.

a figure with three sides for a given perimeter, only a straight line and an equilateral triangle offer a singular or "determinate" solution. In any other case, we would find ourselves confronted with a plurality of equivalent or "indiscernible" solutions between which it would be impossible to choose. Such a lack of determination is made up for by the necessity of a determining reason which calls for simplicity. As a matter of fact, two points being given, the straight line is the simplest line between them, just as the equilateral triangle is the simplest figure of three sides for a given perimeter.

A third relevant passage can be found in the *Origo veritatum contingentium ex processu in infinitum*, dated from around 1689:

> Any truth which is not identical receives a proof; one proves a necessary truth by showing that the contrary implies contradiction, a contingent one by showing than there is more reason for it to be than for its opposite. Thus, like the wise man, God's first decree or resolution is to act in everything according to supreme reason. This is why, if we suppose a situation in which there should exist a triangle for a given perimeter, without anything being stated in the data wherewith one might specify the triangle, one should conclude that God will produce an equilateral triangle, freely but without any doubt. Indeed, there is nothing in the data which would prevent any other triangle from existing, and this is why the equilateral triangle is not necessary. However, for no other triangle to be chosen, it suffices that there be no reason in other triangles, but only in this one, for them to be preferred to the rest. It is the same if one is asked to draw a line from a given point to a given point; without any data to determine the species or the magnitude of the line, a straight line will certainly be produced, but freely, since there is nothing to prevent any other curve, but also nothing to incline one toward choosing it[43].

This passage makes it clear that, for any two given points, a straight line is determined not by geometrical reasons but by the principle of sufficient reason. Above all, the idea reappears that geometrical objects are "produced" freely but without any doubt by God. The invocation of the principle of sufficient reason is explicitly related to the question of God's choice and of the actuality of a certain world (the best one), as well as to the determination of the nature of space and of geometrical objects. All this seems to assert a relation between the actuality of the world and the determination of geometrical objects which is still quite mysterious. What can be the nature of a relation thus said to obtain between things usually considered as belonging to the distinct realms of necessary truths on the one hand and of free decrees on the other?

5.3.2 Actuality of the World and Mathematical Truths

When discussing the plurality of spaces, Nicolas Rescher and Yvon Belaval both propose to solve the problem by adopting the view that the plurality of spaces might correspond to the different ways of actualizing mathematical truths in different

---

[43] A VI 4, 1664. Our translation.

worlds. The set of all mathematical truths would thus remain the set of absolutely necessary truths in God's understanding, but only a subset of this set would be realized in the actual world. The previous examples suggest that, in conformity with Rescher and Belaval's contention, the spatial structure of the world can vary (depending on whether the geodesic is the Euclidean straight line or not). The additional conditions for the choice of a world would delimit a subset of geometrical truths (within the set of all possible mathematical truths) thus determining the structure of a specific world: the best one, when conditions are optimal.

Needless to say, one cannot interpret this notion of "actualizing" mathematical truths to mean that God would cause mathematical objects to exist as He causes real entities to exist. Leibniz is always clear on that point: mathematical objects are abstract and incomplete notions. They cannot be said to "exist" as real objects exist, i.e. as entities endowed with a complete notion:

> It is true that perfectly uniform change, such as is required by the idea of motion given to us by mathematics, is never found in nature any more than are actual figures which possess in full force the properties which we learn of in geometry, because the actual world did not remain within the indifference of possibilities but has arisen from the actual divisions or pluralities the results of which are the phenomena which occur and differ from one another down to their smallest parts[44].

As we have already seen, Leibniz nonetheless sometimes mentioned "actual" space not in the sense of "real" or "physical" space but just in the sense that mathematical truths can be "actual". Space is not actualized in itself: it is a mathematical notion belonging to the realm of possible things. But it can be said to be "actualized" when considered along with the actually co-existing things that fill it. More generally, all the "actual" phenomena follow mathematical rules: "Yet the actual phenomena of nature are arranged, and must be so, in such a way that nothing ever happens by which the law of continuity [. . .] or any of the other most exact rules of mathematics is violated[45]."

This being the case, "actual" space is only the set of mutual dispositions of actual things but is not radically different from abstract space. In this sense, to actualize certain mathematical objects only means to actualize certain mathematical possibilities as *structures of phenomena*. This is why Leibniz had no problem talking about worlds in which God had created this or that mathematical entity. In a very interesting passage from the *Dialogue effectif sur la liberté de l'homme et l'origine du mal* (1695), he stated:

> B. Is it not true that, if the order of things or divine wisdom were to ask God to produce perfect squares, God, having decided to fulfil the request, could not avoid producing incommensurable lines, although these lines, being imperfect, cannot be exactly expressed? Because a square cannot be without a diagonal, which is the distance between the opposite angles. Let us pursue the comparison further, and compare the commensurable lines with

---

[44]*Réponse aux reflexions contenues dans la seconde Edition du Dictionnaire Critique de M. Bayle, article Rorarius, sur le systeme de l'Harmonie preétablie* (1702), Loemker (1989, 583, modified), GP IV, 568.

[45]Loemker (1989, 583, modified), GP IV, 568. On that topic, see Garber (2015, 243).

the spirits who maintain themselves in purity, and the incommensurable lines with the less regulated spirits who thereafter fall into sin. It is plain that the irregularity of the incommensurable lines comes from the very essence of figures and does not have to be imputed to God. It is also plain that such an incommensurability is not a positive evil, something which God could not produce. **It is also very true that God could have avoided it by creating neither figures nor continuous quantities, but only numbers or discrete quantities.** But the imperfection of incommensurables has been counterbalanced by much greater advantages, so that it was better to accord a place to them so as not to deprive the universe of figures[46].

The analogy rests on God's enjoying the possibility of choosing a world in which some mathematical truths are not "created". Leibniz even provided a precise model of a world in which all the mathematical relations between existing objects might be conceived of only by means of discrete quantities[47].

This analogy, if extrapolated, might improve our understanding of the relation between geometrical truths and the actuality of the world. To support such an extrapolation, one need only note that Leibniz quite often imagined alternative world-structures. We thus have only to look for alternatives entailing some geometrical oddities. For instance, in the 5th letter to Clarke (§30), Leibniz had no trouble saying that the world could have been finite: "Absolutely speaking, it appears that God can make the material universe finite in extension, but the contrary appears more agreeable to his wisdom[48]." Even more surprisingly, he explained to De Volder that a world with discontinuous changes, i.e. in which space has to be discontinuous, is conceivable. Indeed, the continuity of physical phenomena is said, once again, to be an effect only of God's wisdom:

> The axiom *in changes there are no leaps,* which I use, is like that. I believe that it follows from the law of order and depends on the same reasoning by which everyone recognizes that motion does not take place through leaps, i.e., that a body does not move from one place to another distinct place unless it goes through the intervening ones. I admit that once we have assumed that the author of things was pleased by continuity in motion, by this very fact leaps are excluded. But how do we prove that this pleased Him except through experience or the principle of order? For, since everything happens through the perpetual production of God and, as they say, by a continuous creation, why could He not have, so to speak, transcreated a body from one place to another at a distance from it, leaving behind a gap either in time or space, e.g., by producing the body at *A* and immediately thereafter at *B* etc.? Experience teaches that this does not happen, but the principle of order, which makes it the case that *the more things are analyzed the more they satisfy the intellect*, establishes the same thing[49].

---

[46]A II 3, 16-17. Our translation.

[47]Two questions then naturally follow: would such a world still be spatial? If not, what would be a non-spatial world? Both the idea of a world endowed with space but not with incommensurable magnitudes and the idea of a non-spatial world without any order of co-existence are intriguing ideas. But they do not relate to our topic, especially as Leibniz's use of such a possibility is mainly analogical and not to be considered in itself.

[48]Loemker (1989, 700), GP VII, 396.

[49]Leibniz to De Volder, 24th March/3rd April 1699, transl. Lodge (2013, 311), A II 3, 545. See also "I don't say that the vacuum, the atom, and other things of this sort are impossible, but only

The *a priori* argument for continuity has nothing to do here with absolute necessity and the reduction to non-contradiction. It is derived from the "principle of order" which satisfies the intelligence. In the same vein, Leibniz even went so far as to imagine a world containing only spherical beings:

> Whence it follows that there is a reason for eternal things too: if one would suppose a world existing from eternity and with only globes in it, one would have to give a reason why they are globes and not squares[50].

Such worlds are certainly conceivable but they are not the actual world and the question of the conditions of their existence did not much interest Leibniz. Moreover, if the texts cited above tend to affirm the theoretical possibility of a spherical or a discontinuous world, these possible worlds could still be understood as taking sense in a more general and absolute geometrical framework subsisting in God's mind. Much in the same way as I can conceive of a sphere in Euclidean geometry and pose questions regarding its geodesics, I can imagine a "globular" world which would be a realization of a subset of certain truths of Euclidean geometry[51]. In other words, these examples show that there can be some variation in the actualization of geometrical truths but they do not testify in and of themselves to a plurality of geometries and of spaces. On the other hand, since the nature of distance determines the spatial form, if we follow the example of the geodesic given by Leibniz it seems that the diversity of their determination involves the possibility of different spatial forms, at least from a theoretical point of view. This interpretation corresponds to the quite modern idea defended by Riemann: space can be characterized by its metric, and if different metrics are possible, so are different spaces[52]. We can now progress in our questioning: is this just an anachronistic reading? Could Leibniz himself have accepted such a variation in metrics?

---

that they are not in agreement with divine wisdom" (Letter to J. Bernoulli, Jan. 1699, GM III, 565; transl. Ariew, Garber (1989, 170).

[50] *Principia logico-metaphysica* (ca. 1689), A VI 4, 1645. Our translation.

[51] This world would play the same role as what we now call a "Euclidean model" of a non-Euclidean geometry. It does not suffice to ground an alternative geometrical science, but it certainly shows that a geometry in which the parallel postulate is not satisfied cannot be contradictory without Euclidean geometry's being so as well. See V. De Risi: "To push Leibniz up a road he had certainly not thought of taking, we may conclude by saying that, according to his logical and epistemological principles, non-Euclidean geometries would be viewed as coherent systems of axioms; for negating the parallel axiom does not in the least imply a contradiction" (De Risi 2016, 119).

[52] B. Riemann, *Sur les hypothèses qui servent de fondement à la géométrie*, in Riemann (1854, 200-299). Y. Belaval briefly referred to Riemann's work.

5.3.3   The Problem of the Theoretical Possibility of the Plurality of Spaces

Let us sum up the problem which we arrived at in the previous sections. On the one hand, one might hold that the diversity of geometrical properties in different possible worlds involves only a variation of the phenomenal structure occurring on non-geometrical grounds. But on the other hand, in order to illustrate this variation Leibniz regularly proposed *geometrical* examples. In these latter, architectonic principles form part of the geometrical determinations themselves. For example, as regards the nature of space, it seems that the determination of the nature of geodesics rests entirely upon the architectonic "principle of the simplicity" – just as the choice of an actual world depends upon the "principle of the best", both principles being expressions of the "principle of sufficient reason". All these elements tend to speak in favor of a plurality of different spaces for different worlds. In fact, since the application of architectonic principles is required in order to choose the nature of a geodesic just as it is required in order to choose the best possible world, and since the existence of a choice supposes a plurality of possibilities amongst which one is to be selected, it seems to follow that the notion of the plurality of (possible) spaces is as susceptible of being defended as is the notion of the plurality of (possible) worlds. For instance, it will be possible for God to create a "non-Euclidean" world in the sense of a world in which geodesics are not Euclidean straight lines.

But this conclusion strongly depends on attributing to Leibniz the anachronistic idea that space can be determined by the nature of its metric. To support the view that our geometrical examples are more than just a useful metaphor, we need therefore to further examine whether Leibniz considered the possibility of a plurality of distance relations and the role of architectonic principles in these geometrical questions. First and foremost, did Leibniz ever envisage the possibility that the notion of distance (and, accordingly, the notion of the shortest way, or the geodesic) might be relative to a certain set of data and, consequently, possibly plural? Secondly, if he did, did he consider that one particular distance might enjoy some privilege over all other possible distances? Thirdly, if he did, did he prove such a privilege by means of logical or by architectonical arguments?

Let us begin with the first question. As early as 1679, as he began to work on the geometry of situation, Leibniz envisaged geodesics on a spherical surface, even if he did not develop this notion in great details at the time[53]. He then worked on this notion in several papers and, twenty-five years later, he presented the results of these researches as a clear plurality of the distance relation:

> Ph. Space, considered with respect to the length separating two bodies, is called 'distance,' and with respect to the length, width, and depth, is called 'capacity'.
>
> Th. To speak more clearly, the *distance* between two situated things (punctual or extended) is the magnitude of the shortest line possible that can be drawn from one to the

---

[53] See Echeverría, Parmentier (1995, 272). On Leibniz's studies on geodesics, see De Risi (2007, Appendix 6 and 7, 592-595).

other. This distance may be considered either absolutely, or in a certain figure containing both distant things. For instance, a straight line is absolutely the distance between two points; but these two points being within the same spherical surface, the distance between them on this surface is the length of the smallest great-arc of a circle that can be drawn from one point to the other. It is also proper to note that distance is not only between bodies, but also between surfaces, lines, and points. It may be said that the *capacity*, or rather the *interval*, between two bodies or two other extended [things], or between an extended [thing] and a point, is the space constituted by all the shortest lines which may be drawn from one to the other. This interval is solid, except when the two situated things find themselves within the same surface and when the shortest lines between the points of the situated things must also fall within this surface or must be intentionally taken within it[54].

Leibniz distinguished between different distance relations: an "absolute" one, which is Euclidean, and a distance defined within a certain figure, such as geodesics on curved surfaces. He explicitly presented geodesics on a spherical surface (portions of great circles) as equivalent to straight lines on a plane surface. It was also by analogy with the distance determined between two points on a surface that he conceived of the "capacity" obtaining between points and surfaces, or between surfaces, or between surfaces and solids. All this patently concerns properties of space itself.

Some important and well-known consequences follow from this consideration. Firstly, it is possible – that is to say, non-contradictory – that there be not only one, but a multiplicity, and even an infinity of geodesics between two given points. For instance, there is an infinity of geodesics between two antipodal points on a sphere. Secondly, "parallel lines" (making alternate angles with a given line equal) can intersect and thus violate the fifth Euclidean postulate, just as the meridians on a sphere (which are at right angles with "parallels") intersect in both antipodal points. It appears, then, that Leibniz was entirely aware that the notion of distance, which is central to his geometry of situation, does not *logically* entail all the properties of Euclidean space[55]. Distance being conceived through the mutual relation between points, it can be Euclidean but can also not be. And, if we hold consistently to the generalization of distance to "capacity", the same will be the case regarding three-dimensional space.

But in order to fully answer our question we need to remind ourselves of the problem that such a conclusion would in turn entail. How is it possible to reconcile this notion of the plurality of distances with the nature of geometrical truths as *absolutely necessary*? Or to go still further, how are we to deal with the fact that some geometrical properties are not logically deducible from the basic geometrical notions, given that Leibniz himself constantly repeated that geometrical truths obey the principle of non-contradiction and are demonstrable by means of the simple logical analysis of their terms? These questions call for an examination of the very nature of geometrical truths. How are we to reconcile such a nature both with the

---

[54]*Nouveaux essais sur l'entendement humain*, II, xiii, § 3 (1704), A VI 6, 146-147. Our translation.

[55]In the same way, the notion of congruence can be defined on a sphere or a cylinder, see De Risi (2007, 181, n. 52), GM V, 189.

notion of an other than strictly logical relation between distances, straight lines, and the nature of space, and with the intervention of the principle of sufficient reason in geometry?

5.4 The problematic Issue of the Necessity of Geometrical Truths

5.4.1 The Necessity of a Euclidean Metric Grounded on Architectonic Principles?

The fact that Leibniz conceived of different kinds of distances does not imply, in and of itself, that he admitted the notion of a plurality of spaces. Indeed, the plurality of distance relations is compatible with what is nowadays called the intrinsic geometry of a surface. This geometry leaves undecided the question whether or not the surfaces it deals with are immersed in a larger and fixed "extrinsic" space (as was the case, for example, when Gauss first considered the intrinsic geometry of surfaces in three-dimensional Euclidean space). Even if Leibniz explicitly generalized his view to three-dimensional spaces, in the form of his notion of "capacity", this is still not enough for us to be certain that he did not have in view what we would now call sub-manifolds of Euclidean three-dimensional space alone. Moreover, as noted in the previous passage from the *Nouveaux essais*, the Euclidean distance is said to be "absolute".

Leibniz was undoubtedly convinced that space must necessarily be Euclidean. The principal testimony to this fact is that he attempted on various occasions to demonstrate the fifth Euclidean postulate regarding parallel lines – which does not hold true, as he knew perfectly well, on surfaces such as spheres. Recent works by Vincenzo de Risi[56] reveal, however, that, despite his efforts, Leibniz never managed to logically establish the truth of this postulate, whereas he had long since "demonstrated", or so he thought, the three-dimensionality of space[57]. Examining his attempts, it is astonishing and revealing to see that he was eventually forced to rely once again on architectonic principles. Around 1712, he wrote in the *In Euclidis* πρῶτα:

> (12) But it is necessary to show, if one straight line makes equal angles to two, then any straight line would make equal angles to them. **I have tried many things and I see that this indeed cannot be easily demonstrated**, that if from the straight line AB the equal perpendicular lines AC and BD are drawn out, then CD is a straight line equal to AB, and the angles in C and D are right angles. Even though we assume that the straight line from C towards D is drawn out with a right angle, it cannot easily be shown to intersect with

---

[56]De Risi (2016).

[57]*De primis Geometriae elementis* (1680), in Echeverría, Parmentier (1995, 281-283); *Essais de Théodicée*, § 351 (1710), GP VI, 307-308.

D with a right angle there. Since Euclid met a difficulty in demonstrating the properties of parallels, he assumed axiom 13, that when straight lines are drawn out from a straight line, with unequal angles (since the argument reduces to this), they come together and thus are not parallels; for he defined parallels, as those which do not come together in a plane. If he had chosen rather to define parallels, as those which make equal angles to a straight line, then it would have been necessary for him to assume this axiom: that such lines do not come together. And indeed the point is that the connection between these two should be demonstrated: making equal angles to a straight line, and not coming together. I demonstrated this a little before, if we assume that an equal angle is made to any straight line. **Therefore, it was necessary to show the latter assumption**.

(13) **It seems to me that the whole thing can follow from a higher principle, namely that of Determinant Reason**[58].

Leibniz was obliged, then, to admit that he had to refer to non-logical principles, such as that of determinant reason in order to complete his demonstration. In another text from the same period, the *Initia rerum mathematicarum metaphysica* (ca. 1714), he related to one another the "simplicity" of Euclidean distance, its optimal determination and the definition of space:

> *Space is the order of coexisting things*, or the order of existence for things which are simultaneous. In each of the two orders – that of time and that of space – we can judge relations of *nearer to* or *farther from* between its terms according as *more* or *less* middle terms are required to understand the order between them. Thus, two points are nearer if the maximally determined intervening forms arising from them produce a simpler configuration. Such an interval of maximum determination, that is, the minimum and at once the most conformal figure made by the intervening terms is the simplest path from one to the other; in the case of points this is the straight line, which is shorter between nearer points[59].

In another passage, Leibniz explicitly related optimal determination to the principle of sufficient reason:

> There is, moreover, a definite order in the transition of our perceptions when we pass from one to the other through intervening ones. This order, too, we can call a *path*. But since it can vary in infinite ways, we must necessarily conceive of one that is most simple, in which the order of proceeding through determinate intermediate states follows from the nature of the thing itself, that is, the intermediate stages are related in the simplest way to both extremes. If this were not the case, there would be **no order and no reason** for distinguishing among co-existing things, since one could pass from one given thing to

---

[58]Translated by De Risi (2016, 175-176, our emphasis).

[59]Loemker (1989, 666-667), GM VII, 18.

another by any path whatever. It is this minimal path from one thing to another whose magnitude is called *distance*[60].

If we lay emphasis on the first part of this passage, one could argue that Euclidean distance is one amongst infinitely many other possibilities and that God chooses it as the structure of our world according to the principle of sufficient reason; but where we lay emphasis rather on the ideas expressed in its latter part, it seems that Leibniz went a step further and posited the existence of a *via simplicissima* as a condition for the existence of space itself ("If this were not the case, there would be no order and no reason for distinguishing among co-existing things"). The first line of interpretation is the one which we have already encountered in the text *Origo veritatum contingentium ex processu in infinitum*. The second would imply that Leibniz was not describing the geometry of the best of all possible worlds, but rather the very structure of geometry as the science of space itself. Be that as it may, we stumble in both readings upon the same difficulty: how are we to reconcile the definition of geometrical truths as absolutely necessary with the unexpected appearance of architectonic principles in geometry?

5.4.2 A Fresh Look at Geometrical Necessity

It is remarkable that the texts cited above tend to modify the initial meaning ascribed to the notion of the necessity of mathematical truths in Leibniz's thought. A widespread conception of necessity, originating with Saul Kripke[61], consists in defining it as "truth in all possible worlds". Such a definition might sound typically Leibnizian, in particular if one considers texts like the following:

> We learn, then, that propositions which relate to essences are one thing and propositions which relate to the existence of things quite another. Essential, it is certain, are those [propositions] which can be demonstrated by the resolution of terms; that is to say, those which are necessary, i.e. virtually identical, and whose contraries are impossible or virtually contradictory. And those truths are eternal which not only will prevail as long as the World remains, but which would also have prevailed if God had created the World in accordance with another reason[62].

Essential truths are demonstrable by "the resolution of terms". They can be reduced to identical propositions (*A is A*, *A is not non-A*, *AB is B*, and so on) by means of a finite number of operations replacing the propositional terms with their definitions[63]. They also are independent of God's choice to create such and such a world. Since essential truths are also necessary truths, Kripke's definition seems

---

[60] Loemker (1989, 671), GM VII, 25.

[61] Kripke (1963).

[62] *De natura veritatis, contingentiae et indifferentaeque atque de libertate et praedeterminatione*, ca. 1685-1686; A VI 4, 1517. Our translation.

[63] See *Generales inquisitiones de analysi notionum et veritatum*, §56–61 (1686), A VI 4, 757–758.

in accordance with Leibniz's own characterizations of necessity. Nevertheless, we need to read this text very carefully. In *Dialogue effectif sur la liberté de l'homme et l'origine du mal*, Leibniz wrote that a world in which there would be no figures and no continuous quantities is conceivable. This assertion does not invalidate the passage just quoted, since such a world would not *contradict* the mathematical truths about figures and continuous quantities. These propositions remain true, whether they are actualized in the world or not. But the case presented in the *Dialogue effectif* highlights the fact that "necessary" or "true whether God has created the world or not" do not signify the same thing as "true *in* each possible worlds".

If we want better to assess the various forms of necessity in Leibniz, we need only to examine the passage immediately following the previous text:

> However, one should not consider that only singular propositions are contingent [...]; I even think that there are certain propositions which are entirely, universally true within the series of things and which cannot ever be violated even by a miracle, not because they could not be violated by God, but because, as He chose this series of things, He Himself decreed that He should observe them (as the specific properties of the series chosen). [...] In truth, one does not have to be disturbed by the fact that I have stated that the laws of this series of things are essential, whereas we previously said that these very laws are neither necessary nor essential, but contingent and existential. Indeed, since the existence of this series is contingent, depending as it does upon God's free decree, **the laws of this series will certainly be absolutely contingent, yet hypothetically necessary and essential once the series has been set up**[64].

In texts like this one, another kind of necessity makes its appearance. Although not reducible to identical propositions, some propositions can be considered as both essential and necessary. But their necessity is said to be "hypothetical", since it is anchored in a decision, lying ultimately with God, to create this particular series and not another one, i.e. this particular world and not another one. Unlike physical laws, which allow for the possibility of miracles, hypothetically necessary truths cannot be violated by God within the world that He has chosen, although these truths are not inviolable in any possible world. Accordingly, Kripke's reduction of necessity to "that which is true in all possible worlds" is not consistent with Leibniz's conception of this other form of necessity, no less universal and eternal than the necessity of absolutely necessary truths.

The issue then becomes that of determining how such a characterization of hypothetical necessity can be related, if it can be related at all, to geometrical truths[65]. Indeed, one of the main aspects of our problem regarding the plurality of spaces originates from the idea that geometrical truths have to be absolutely necessary, so that their opposite involves contradiction, and that they depend upon God's understanding alone and not upon His will. However, as already noted, Leibniz sometimes explained that, two points being given, God "freely" produces a straight line, since He does everything according to a "supreme reason", or that He

---

[64]*De natura veritatis, contingentiae et indifferentaeque atque de libertate et praedeterminatione* (ca. 1685–1686), A VI 4, 1518. Our translation.

[65]See Lin (2016).

could have created a world with gaps, or without geometrical figures. We also saw that the very definition of the straight line in Euclidean geometry depends upon the principle of sufficient reason, which is true of our world, but not of every possible world.

5.4.3   The Second Solution: The Conditional Nature of Geometrical Truths

In the *Nouveaux essais sur l'entendement humain*, Leibniz elaborated on the specific nature of mathematical truths:

> Concerning *eternal truths*, it must be observed that they are, at bottom, **all conditional** and say in effect: such a thing being posited, such another thing is (also); for instance, saying: *every figure having three sides will also have three angles*. I do not say anything else than, supposing that there is a figure with three sides, this same figure will have three angles; I say *this same*, and this is the respect in which Categorical propositions, which can be formulated without condition although they are, at bottom, conditional, differ from those called 'hypothetical', as would be the case for this proposition: *If a figure has three sides, its angles are equal to two right [angles],* wherein one clearly sees that the *antecedent* proposition (i.e. the three-sided figure) and the *consequent* (i.e. the angles of the three-sided figure are equal to two right [angles]) do not have the same subject, as they had in the former case where the antecedent was *this figure is three-sided* and the consequent, *the figure in question is three-angled.* [. . .] The Scholastics hotly discussed *de constantia subjecti*, as they called it, that is to say, how a proposition advanced with reference to a subject can possess real truth if the subject referred to does not exist. It can do so because **there is no truth but conditional truth and because what is said is really that, assuming the subject exists, it will be found to be as is proposed of it**[66].

It is not easy to understand what Leibniz has in mind when he claims that mathematical truths are "at bottom" ("*dans le fonds*") always conditional, especially as this point is almost entirely ignored in the literature[67]. We will not claim to give

---

[66]*Nouveaux essais sur l'entendement humain*, IV, xi, § 10 (1704), A VI 6, 446–447. (Our translation and our emphasis.) The following passage is also very interesting: "But one will still ask on what such a connection is based, since there is surely some reality in it which is not deceptive. The answer will be that it is based on the relation between ideas. But one will ask in reply where would these ideas be if no mind existed and what would then become of the real foundation for the certainty of eternal truths. This finally leads us to the ultimate foundation of truths, namely, to that Supreme and Universal Mind which cannot fail to exist and Whose understanding truly is the sphere of eternal truths, as St Augustine recognized and expressed in a very vivid way. And so that it may not be thought that there is no need to have recourse to this notion of a Supreme and Universal Mind, it must be remembered that **these necessary truths contain the determining reason and the regulating principle of existences themselves, or, in a word, the very laws of the Universe**. Thus, these necessary truths being prior to the Existences of contingent Beings, they must necessarily be grounded in the existence of a necessary substance." (Our translation and our emphasis.)

[67]On "conditional" truth and the problem *De constantia subjecti*, see Rauzy (2001, Chapter II, §6), where one can find a survey and a discussion of the literature. See also: Mates (1986); Adams

a complete account of it here, but focus only on its most obvious and clearest thesis. All mathematical truths are of the form: such or such a thing being given, such or such another thing is also given. It is possible, then, that the antecedent here may not exist (*"en cas que le sujet existe jamais"*)[68], without the proposition being thereby rendered false.

Characterizing mathematical truths as conditional is thus not inconsistent with these truths' absolute necessity and with the usual characterization of mathematical truths in terms of the reduction to identical propositions. For instance, *if there is a trigon, then it necessarily has three angles* is both a conditional truth and reducible to identical propositions. The demonstration of the triangularity of a trigon does not require any recourse to metaphysical principles such as the principle of the best, or of the sufficient, reason. The truth of the conditional proposition can be proven by a finite resolution of terms and by the reduction to identical propositions[69].

However, as previously explained, the invocation alone of the principle of contradiction and of the absolute nature of mathematical necessity is not enough to provide an "absolute" character to mathematical truths regarding the structure of space. It is entirely certain that Leibniz considered the world-structuring principle of Euclidean distance to be the privileged principle as regards the actual created world, but not in logical terms. As Vincenzo De Risi aptly remarks:

> Such is the fate of non-Euclidean geometries in Leibnizian geometry. They are eradicated by a couple of demonstrations which, beyond any defect that can possibly be found in them, present us with the magnificent oddity of resting on the principle of reason. Accustomed as we are to the philosopher's reiterated assertions that mathematics and geometry are entirely deducible from the principle of contradiction, we cannot but be impressed[70].

(1994). None of these authors deal in detail with the specific problems related to the application of this category of "conditional truth" to mathematical theorems.

[68] In his analysis of the question, Rauzy emphasizes that one should relate existence to the actuality of the thing and not to the concept alone (2001, 119-120). However, when commenting upon the passage quoted from the *Nouveaux essais* (2001, 125) and the fact that the subject of a conditional truth might be "non-existent", he follows another interpretation in terms of non-contradiction alone. This is at odds with the numerous passages in which Leibniz talked of mathematical truths as concerning not only existent things, but also possibilities (see *Nouveaux essais* II, xiv, §26 or II, xiii, §17). We shall thus stick to the interpretation of existence in terms of actuality – in accordance with the discussion above on the "actualization" of mathematical truths.

[69] This idea evocates what is called 'if-thenism' by contemporary philosophers of science. According to them, mathematical truths ultimately can be formulated as *if p, then q*. This thesis historically stemmed from the difficulty to conciliate the plurality of geometries with the logical nature of mathematics. On that topic, see A. Musgrave on Russellian logicism: "After he had adopted the logicist thesis, Russell sought a way to bring geometry into the sphere of logic. And he found it in what I shall call the *If-thenist manoeuvre*: the axioms of the various geometries do not follow from logical axioms (how could they, for they are mutually inconsistent?), nor do geometrical theorems; but the conditional statements linking axioms to theorems do follow from logical axioms. Hence geometry, viewed as a body of conditional statements, is derivable from logic after all." (Musgrave 1977, 110).

[70] De Risi (2007, 260).

The issue of the Euclidean nature of space is a peculiar one since it allows us to relate features usually associated with hypothetical necessity to the conditional nature of mathematical truths. Indeed, the choice of the straight line as a geodesic nourishes Leibniz's main illustrations of the choice of the best possible world. Just as the choice of actualizing some mathematical truths and not others does not depend on logical principles, the proof of the Euclidean nature of space requires the same architectonic principle of sufficient reason (or of best determination) to be satisfied. Or to put it another way, just as God can create a world without squares in which no diagonal would be expressible, even if the existence of diagonals follows logically from the nature of squares (as stated in the *Dialogue effectif*), it is possible, in the sense of non-contradictory, to conceive of a globular world in which all things would be spherical and in which there would exist an infinity of geodesics between certain points. In order to choose the most perfect world Leibniz resorted, just as he did in order to ground the unicity of a geodesic, to metaphysical principles.

To better understand the point, let us further develop the example of the *Dialogue effectif*. Although the existence of diagonals follows logically from the nature of squares, God could have created, so Leibniz argued, a world without any square and thus without any diagonal. Interestingly enough, in such a world, with only numbers and no continuous magnitudes, one could simulate a squared form by means of rational ordinates. However, the magnitude of the diagonal of such a "square" (which would not really be a square since there would be, according to Leibniz, no figure in such a world) would be inexpressible. Such a diagonal would have to be said to be impossible relative to the given, i.e. due to the choice of such a "discrete" world. But it would not be impossible in and of itself. Moreover, it would still remain possible to perform demonstrations and to affirm truths about it. For instance, the incommensurability of the diagonal would still be demonstrable in a world equipped only with numbers (just think of the classical demonstration by *reductio*, which is already mentioned by Aristotle). Thus, mathematical truths are both absolutely true, since they are demonstrable by the principle of contradiction (as is typically the case in demonstration by *reductio*), and endowed, nonetheless, with a kind of conditionality, since their actualization in a world depends on what is given in the world in question[71].

The situation would be the same if we imagined a world in which everything geometrical had to be constructed by ruler and compass: the trisection of an angle would not be possible in such a world, although it would be so in a world endowed with richer means of construction. In light of such a case, we see that the necessity attached to a mathematical truth does not lie in the proposition in and of itself, but rather in the manner in which a proposition depends on certain hypotheses. One can, therefore, at the same time argue that the trisection of an angle is possible in

---

[71]Leibniz's central argument in this text consists in explaining that our world is the best since it is the richest regarding phenomena, even if the incommensurability of some magnitudes or things can be regarded as a flaw, as an imperfection. This example is used by reason of its comparability to the existence of sin and evil in the world.

one world but not in another, *and* argue that there is an absolute truth underlying this fact: in every possible situation, it will remain absolutely true that the trisection of an angle is not constructible with ruler and compass alone. This provides the solution to our initial dilemma: the unicity of the geodesic is demonstrable under certain conditions which guarantee the Euclidian nature of space (say, the fifth postulate). This truth is absolutely necessary. But the condition itself has no logical necessity. We know that it holds in our world because it satisfies architectonic principles, but there is nothing contradictory about a world which would not satisfy it[72] – typically a spherical world.

5.5   Conclusion

To sum up, spatial pluralism appears to be a natural consequence of Leibniz's relationism, especially when this latter is read as a form of relativism (space is relative to the disposition of bodies). Space as the structure of the phenomenal world can vary from one world to another. Leibniz regularly talked as if other worlds than ours were endowed with different spatial structures, or asserted that God has chosen our world because it has the "best" spatio-temporal structure. However, this pluralism enters into tension with the fact that space is also the object of geometry and, as such, has various properties which are absolutely necessary, i.e. which cannot be denied without contradiction. In this second sense, geometrical properties are eternal truths, subsisting in God's mind before He chose to create the world. One easy way out of the difficulty is to postulate a distinction between an ideal space, mathematical and unique, and a real space, concrete and corresponding to its various occupations. But we have shown that this distinction between ideal and actual space is in fact neutral as regards our problem. Although it is true that Leibniz sometimes (though rarely) made such a distinction, the various occupations of the same framework are characterized by him as *equivalent* systems of places. Conversely, when it came to giving examples of how one arrangement could be "better" than another, Leibniz regularly gave *mathematical* examples. In the first case, the various occupations amount to one and the same space; in the second, the non-equivalent dispositions act on the level of the ideal space itself.

   All this being the case, we are obliged to elaborate another solution by high-lightening a few facts. Firstly, it makes sense for Leibniz to say that God can create or produce this or that mathematical entity. But this does not mean that God can make such an object exist *stricto sensu*, since all the mathematical truths exist in His mind, as in the sphere of eternal truths, and no mathematical object exists in the sense that "real" entities exist. But by creating such and such a world, God brings into existence a phenomenal structure which can correspond to only one single

---

[72]"Satisfaction" is taken here in the sense of "actuality", which we describe above as holding for mathematical entities.

subset of mathematical truths (typically a "globular" word). This corresponds to what Leibniz called "actual space". Moreover, we have shown that Leibniz regularly gave, as an example of properties which depend on God's choice, the unicity of the geodesic in Euclidean space (the fact that the straight line is the *via simplicissima*). At the same time, we have shown that Leibniz's geometrical studies indicate that he was well aware of the non-contradictory character of geometrical situations in which geodesics are not determined according to a single principle. As we have seen, he even went so far as to generalize this situation obtaining upon surfaces in Euclidean space (typically spheres) to three-dimensional space itself. Relying on these facts, we have proposed a different path than the distinction actual/ideal. This path emphasizes the *conditional* nature of mathematical truths, thus allowing the two aspects of Leibnizian space to be compatible with one another: mathematical truths are absolutely necessary in God's mind, but they subsist therein only as conditionals. When we turn to consider these truths' actualization, it may well be that some conditions are necessary for their existence which are not given in all possible worlds.

At this point, however, there remain two possible lines of interpretation and only one of these leads to a plurality of spaces. On the one hand, one might hold that Leibniz asserted, in the end, that the foundation of geometrical truths is not the principle of contradiction alone but also involved certain architectonic principles (of simplicity, determination, or sufficient reason). Indeed Leibniz sometimes claimed that necessary truths, just as much as contingent truths, follow from the combination of the principle of reason with the principle of identity[73]. The very title of the *Initia rerum mathematicarum metaphysica* from 1715 seems to support this idea: mathematical things *originate* from metaphysical principles. If such is the case, then the unicity of space in God's mind can be preserved, but the notion of the necessity of geometrical truths has to be reconsidered (it cannot be reduced merely to the absence of contradiction). This interpretation is also reinforced by Leibniz's claim, in the *Initia metaphysica,* that any indetermination of geodesic lines would amount to an absence of order and, consequently, of space. Thus, in order to defend the unicity of space, we need to deflate the logical conception of mathematical truths and accept the reintroduction of metaphysical principles into geometry. On the other hand, because of the relation obtaining between the conditional nature of mathematical truths and the possibility of actualizing just some, not all, of the conditions necessary for the validity of these latter, one might hold that the architectonic principles serve to *specify* the geometrical structure of a given world.

---

[73] See "Remarques sur le Livre sur l'origine du mal, publié depuis peu en Angleterre" (§14): "I have elsewhere made this remark, which is one of the most important in philosophy,drawing attention to the fact that there are *two Great Principles*: namely, *that of Identicals or of contradiction*, which states that, where there are two contradictory enunciations, one is true and the other is false; and that of Sufficient Reason, which states that there is no true enunciation the reason for which cannot be seen by who possesses all the knowledge required to perfectly conceive of it. Both principles occur not only in necessary truths but also in contingent ones." (*Essais de Théodicée*, 1710, GP IV, 413–414. Our translation.)

The principle of determination might then be seen as a means to selecting one object amongst others, for instance to specifying one single shortest line amongst many other possible lines. These other possibilities would then correspond to other possible spaces. It would not necessarily be interesting to study such spaces, just as it is not entirely interesting to develop just a rational geometry alone, even if such a geometry is possible, and true, in God's understanding. But in that case, the plurality of spaces is possible and fully compatible with the characterization of geometrical truths in terms solely of their logical consistency[74].

It is certainly difficult to choose between these interpretations, but the difficulty may well reflect Leibniz's own hesitations on a topic with regard to which, as he said: *he has tried many things*, recognizing at the end that *it cannot be easily demonstrated*.

Bibliography

Adams, R. 1994. *Leibniz. Determinist, Theist, Idealist*. New York: Oxford University Press.

Ariew, R., and D. Garber, eds. 1989. *G.W. Leibniz, Philosophical Essays*. Cambridge: Hackett.

Arthur, R. 1994. Space and Relativity in Newton and Leibniz. *British Journal for the Philosophy of Science* 45: 219–240.

———. 2013. Leibniz's Theory of Space. *Foundations of Science* 18 (3): 499–528.

Belaval, Y. 1978. Note sur la pluralité des espaces possibles d'après la philosophie de Leibniz. *Perspektiven der Philosophie* 4: 9–19.

Belkind, O. 2013. Leibniz and Newton on Space. *Foundations of Science* 18 (3): 467–597.

De Risi, V. 2007. *Geometry and Monadology. Leibniz's Analysis Situs and Philosophy of Space*. Basel: Birkhäuser.

———. 2016. *Leibniz on the Parallel Postulate and the Foundations of Geometry*. Basel: Birkhäuser.

Earman, J. 1979. Was Leibniz a relationist? *Midwest Studies in Philosophy* 4 (1): 263–276.

Echeverría, J. 1999. L'harmonie post-établie. In *L'actualité de Leibniz: les deux labyrinthes*, ed. D. Berlioz and F. Nef. Stuttgart: Steiner.

Echeverría, J., and M. Parmentier, eds. 1995. *G. W. Leibniz. La caractéristique géométrique*. Paris: Vrin.

Futch, M. 2008. *Leibniz's Metaphysics of Space and Time*. Dordrecht: Springer.

Garber, D. 2015. Leibniz's Transcendental Aesthetic. In *Mathematizing Space*, ed. V. De Risi, 231–254. Birkhäuser: Basel.

Hartz, G., and J.A. Cover. 1988. Space and Time in the Leibnizian Metaphysics. *Noûs*: 493–519.

Khamara, E.J. 2006. *Space, Time, and Theology in the Leibniz-Newton Controversy*. Frankfurt: Ontos Verlag.

Kripke, S. 1963. Semantical Considerations on Modal Logic. *Acta Philosophica Fennica* 16: 83–94.

Lin, M. 2016. Leibniz on the Modal Status of Absolute Space and Time. *Noûs* 50 (3): 447–464.

---

[74]Note however that this does not mean that there is no geometry true "in every world" since on this interpretation also there may be a set of geometrical truths acting as necessary conditions for the possibility of a spatio-temporal structure in and of itself. But the point is that this "absolute geometry", to take up a modern vocabulary, would have to hold prior to the intervention of architectonic principles, and thus prior to what, according to Leibniz, makes space Euclidean.

Lodge, P., ed. 2013. *The Leibniz-De Volder Correspondence*. Yale: Yale University Press.

Loemker, L., ed. 1989. *G.W. Leibniz, Philosophical Papers and Letters*, 2nd Edition. Trans. L. Loemker. Dordrecht: Kluwer.

Mates, B. 1986. *The Philosophy of Leibniz. Metaphysics and Language*. New York: Oxford University Press.

Musgrave, A. 1977. Logicism Revisited. *British Journal of Philosophy of Science* 38: 99–127.

Rauzy, J.-B. 2001. *La doctrine leibnizienne de la vérité. Aspects logiques et ontologiques*. Paris: Vrin.

Rescher, N. 1977. Leibniz and the Plurality of Space-Time Frameworks. *Rice University Studies* 63 (4): 97–106.

Riemann, B. 1854. *Œuvres Mathématiques*. Paris: Gauthier-Villars.

Slowik, E. 2012. The 'Properties' of Leibnizian Space: Whither Relationism? *Intellectual History Review* 22 (1): 107–129.

Vailati, E. 1997. *Leibniz and Clarke*. Oxford: Oxford University Press.

Winterbourne, A.T. 1982. On the Metaphysics of the Leibnizian Space and Time. *Studies in History and Philosophy of Science* 13 (3): 201–214.

Chapter 6
One String Attached: Geometrical Exactness in Leibniz's Parisian Manuscripts

Davide Crippa

Abstract In this paper, I shall discuss Leibniz's considerations on the problem of exactness, as they can be reconstructed from published and unpublished letters, notes, drafts and sketches he did while Paris between 1673 and 1676. Leibniz's critical target was Descartes' Géométrie (1637). Descartes had managed to include into geometry all algebraic curves on the ground that they can be constructed by one continuous motions. Few rogue elements, called "mechanical curves" did not comply with this criterion of exactness, and therefore were not allowed in solving geometrical problems: these were, for example, the Archimedean spiral, the quadratrix and the cycloid. From the beginning of his mathematical studies, in 1673, and throughout his career, Leibniz set the task to reform the Cartesian criterion of exactness. Leibniz objected that Descartes' criterion is not tenable, because there exist curves constructible by one continuous motion which cannot be associated to any algebraic equation. In this paper, I shall discuss several examples of such geometrical, non-algebraic curves: evolutes and involutes, the parabolic trochoid (the curve generated by the rolling of a parabola along one of its tangents) and the cycloid. The key to reform the Cartesian demarcation was, for Leibniz, to admit within geometry curve constructions obtained by strings that could be bent into arcs. In this way, the previously mentioned curves resulted geometrical insofar as they could be generated by one and continuous motion.

D. Crippa (✉)
Institute of philosophy, Centre for Science, Technology, and Society Studies, Czech Academy of Sciences, Prague, Czech Republic

© Springer Nature Switzerland AG 2019
V. De Risi (eds.), *Leibniz and the Structure of Sciences*, Boston Studies in the Philosophy and History of Science 337,
https://doi.org/10.1007/978-3-030-25572-5_6

6.1 Introduction

In this article,[1] I shall discuss Leibniz's considerations on the problem of exactness, as these latter can be reconstructed from published and unpublished letters, notes, drafts and sketches which belong to his period in Paris between 1673 and 1676. As has been highlighted in recent literature (Knobloch 2006; Probst 2012) Leibniz's considerations took their starting point from Descartes' *Géométrie*, and in particular the second book of this treatise, devoted to the nature of curves.[2]

By the term "exactness" we usually refer to a crucial methodological concern for early modern geometers, namely that of fixing appropriate norms for deciding which curves, understood as tools for solving problems, should be admitted in geometry and which should not.[3]

Ancient and early modern geometers relied on two fundamental modes for defining geometrical objects (see Molland 1976). One way was to define a mathematical entity, for instance a curve, by saying how it was constructed. This mode of identification may be called "specification by genesis" (Molland 1976, p. 23), or "description" (*descriptio*), as was customary in early modern period. Secondly, a mathematical entity, like a curve, could be identified through a property that the curve as a locus needed necessarily to possess: this is the so-called "specification by property" (Molland 1976, p. 23).

Ancient geometers saw curves primarily as a result of a construction: only when a curve had been generated through a certain rule, could one study its properties (Molland 1976, p. 33; Acerbi 2007, p. 468). The primacy of the specification by genesis in the geometric practice of ancient mathematicians can be also related to a classical model of knowledge, whose origins may be traced back to the Aristotelian conception of science (Detlefsen 2005, p. 237). This epistemological view is explicit in Proclus' account of the role of problems and constructions in the edifice of Euclid's geometry. According to the Neoplatonic scholar, whose influence was far-reaching in the early modern period too, geometrical constructions had a twofold purpose: on the one hand, they formed a sort of "bridge" which brought the reasoner from the givens of the problem to the sought-for result, and on the other they granted the existence of those objects that are themselves invoked in the course of a proof, or whose properties are made the object of a mathematical theorem, by producing them in the imagination (Detlefsen 2005, p. 242. See also Harari 2008; Rabouin 2009).

---

[1]I would like to thank Pietro Milici, Achim Trunk, Siegmund Probst, Vincenzo de Risi and Charlotte Wahl, whose help has been invaluable to bring this article to light. I would also like to thank Prof. Michael Kempe for having made my research stay at G. W. Leibniz Bibliothek in Hannover possible. This article was written with the support of the DAAD grant 57214227 (*Förderprogramm Forschungsstipendien – Kurzstipendien*, 2016), of ANR-17-CE27-0018-0 Mathesis (*Édition et commentaires de manuscrits mathématiques inédits de leibniz*, 2017-2021), and completed with the support of the GAČR Grant 19-03125Y (*Matematika v Českých zemích*).

[2]Descartes (1897–1913), vol. 6, p. 388.

[3]Cf. Bos (1984, 1988, 2001), and the recent Panza (2011), Mancosu (2007), and Mancosu and Arana (2010).

This mode of conceiving of geometrical objects was still prominent in Descartes' *Géométrie*. Indeed he also relied on specification by genesis as the fundamental means of defining new curves and ensuring their knowledge.[4]

Hence, when Descartes pondered which curves should be considered exact, and therefore geometrical, he adopted a constructive viewpoint. Curves are to be admitted in geometry:

> Provided that they can be conceived as described by a continuous motion or by several successive motions, each motion being completely determined by those which precede it; for in this way an exact knowledge of the magnitude of each is always obtainable.[5]

In the light of this criterion, Descartes managed to include within geometry a larger class of curves that extended the Euclidean means of construction (i.e. the circle and the straight line) relying on the same fundamental characteristics: unicity and continuity of the constructing motions. One of the central theses advanced in the *Géométrie* is that all and only those curves complying with this constructional criterion can be described by finite polynomial equations of the form $F(x, y) = 0$, where F is a polynomial of finite degree, with integer coefficients. In this way, Descartes was able to build a sound geometrical foundation for algebraic reasoning, and thus apply algebraic techniques in geometry (for instance, in the calculus of tangents).[6]

This criterion of exactness was sufficiently large to include most of the curves known to Descartes and his contemporaries, except for few a rogue elements, called "mechanical curves" in *La Géométrie*:

> the spiral, the quadratrix and similar curves, which really do belong only to mechanics, and are not among those curves that I think should be included here, since they must be conceived of as described by two separate movements, whose relation does not admit of exact determination.[7]

---

[4] See in particular Bos (1984, p. 322), and Panza (2011, p.74ff), where the constructional criterion grounding Descartes' ontology of geometry is discussed at length.

[5] Descartes (1952, p. 43). The original, full passage from *La Géométrie* says: "Mais il est, ce me semble, tres clair que, prenant, comme on fait, pour Geometrique ce qui est precis et exact, et pour Mechanique ce qui ne l' est pas; et considérant la Geometrie comme une science qui enseigne generalement a connoistre les mesures de tous les cors; on n' en doit pas plutost exclure les lignes les plus composées que les plus simples, pourvû quon les puisse imaginer estre descrites par un mouvement continu, ou par plusieurs qui s'entresuivent et dont les derniers soient entierement reglés par ceux qui les precedent: car, par ce moyen, on peut toujours avoir une connaissance exacte de leur mesure ..." (Descartes 1897–1913, vol. 6, p. 389).

[6] One must remark though that Descartes never affirmed nor denied the converse claim that any algebraic equation could represent a curve constructible by one continuous motion (cf. Itard 1984, p. 276, and Bos 2001, p. 404–405). A first proof that all algebraic curves are, at least locally, constructible via articulated devices, which include the linkages described by Descartes, was given by the mathematician A. Kempe, in 1876 (Bos 2001, p. 405).

[7] "...la Spirale, la Quadratrice, et femblables qui n'appartienent véritablement qu'aux Mechaniques et ne font point du nombre de celles que ie penfe deuoir icy estre receues, a cause qu'on les imagine descrites par deux mouuemens separées et qui n'ont entre eux aucun raport qu'on puisse mesurer exactement" (Descartes 1897–1913, vol. 6 p. 390).

Although recent historiography has seen a growth of hypotheses in order to explain the exclusion of the spiral and the quadratrix from geometry (see, among the most recent theses: Mancosu 2007; Panza 2011, p. 82), there are clear mathematical reasons which could have inspired Descartes.

Firstly, in his description of what he called "mechanical" curves, Descartes relied on the ancient representations for the generation of the Archimedean spiral and of the quadratrix.[8] Accordingly, in the case of the quadratrix, a radius of a given circle pivots with uniform velocity around the centre while another segment, equal in length to the radius, moves uniformly along the vertical direction, during the same time interval. Given a circumference and its centre, the spiral is generated by a mobile point which covers the radius of the circle while the radius rotates, both motions being uniform and occurring at the same time.

Hence both generations make appeal to two independent uniform motions which must be set in such a way that, while one motion translates a point or a segment along the radius of the circumference, the other rotates a segment around the centre of the circumference. This description was not considered geometrical enough by Descartes, since it could not translate into a purely constructional criterion until the reference to the speeds of the motions was eliminated from the definition of the curves (cf. Panza 2011, p. 81–82).This was not a task accomplishable in geometry, however: the only alternative would be to use the specification by property of the curve, which involved the comparison between segments of straight lines and circular arcs into one proportion. Perhaps relying on an ancient, Aristotelian tradition, Descartes denied outright that this operation was performable with exactness, and held that the proportion between segments and arcs was unknowable to the human understanding (Descartes 1897–1913, vol. 6, p. 412; Bos 2001, p. 342; Mancosu 1999, p. 77). As a consequence, mechanical curves were not expressible by polynomial algebra and could at most be traced by approximate methods.

Descartes supplemented the demarcation between geometrical and mechanical curves with a normative component: only geometrical curves constituted legitimate inferential steps in geometric problem solving, and only they, therefore, could lead to exact knowledge. According to him, this restriction was a virtuous one, since it purportedly allowed the geometer to solve, in principle, *all* problems of geometry, as is emphatically proclaimed at the outset of *La Géométrie*,[9] As we shall see below, this point was repeatedly criticised by Leibniz in the course of his career.

With hindsight, the divide between geometrical and mechanical curves, leaving its normative component aside, can be seen as the precursor of the fundamental dichotomy between algebraic and transcendental curves or functions.

The terms 'algebraic' and 'transcendental', when referred to curves, were effectively introduced by Leibniz in the second half of seventeenth century as part of his criticism of Cartesian geometry. From the beginning of his mathematical studies, in 1673, and throughout his career, Leibniz called "algebraic" (and before this, "analytical") those curves termed by Descartes' "geometrical", whereas he termed

---

[8]His main sources were indeed the third and fourth books of Pappus' *Collectio* (in Pappus 1876, vol. I, p. 234–235, 253), and Archimedes' treatise on *Spirals* (cf. Archimedes 1881, vol. II, p. 52).
[9]Descartes (1897–1913), vol. 6, p. 369.

"transcendental" those curves that could not be expressed by an algebraic equation, and coincided therefore with "mechanical" curves in Descartes' terminology.[10]

Such a terminological change can be seen as one of the effects of a lifelong effort, begun in Paris in 1673, to conceive and implement a general reform of the Cartesian construal of geometry, guided by two main goals.

The first goal was that of extending the method of analysis promoted by Descartes, consisting in reducing geometric problems to finite polynomial equations, into a new symbolic calculus admitting infinitary and infinitesimal objects, and therefore capable of dealing with "transcendental" problems, namely problems irreducible to finite algebra: in particular, quadrature and rectification problems and inverse tangent problems. This method became, eventually, the differential and integral calculus.[11]

The second goal of Leibniz's project was that of reshaping or wholly removing the normative aspect inherent in the Cartesian distinction between geometrical and mechanical curves, so as to include non-algebraic curves as legitimate tools in geometric problem-solving. It should be pointed out, in fact, that the family of mechanical curves, which numbered very few elements when Descartes published his geometry, acquired in the second half of the seventeenth century a central place in mathematical research and became one of the main subjects of study.

In several writings, Leibniz suggested that the overcoming of the strictures imposed by Descartes on the edifice of geometry was essentially made possible thanks to the adoption of the formalism of calculus which, together with finite algebra, would express both geometrical and mechanical curves by means of a single symbolism.[12]

---

[10]Cf., among the earliest occurrences of this terminology in Leibniz's mathematical practice: AVII3, 23, AVII3, 38_{12}, AVII5, p. XVIII. Here and in the following sections, I will use the letter 'A', followed by a roman and an arabic numeral, in order to refer to the edition of Leibniz's collected works published in the Academy Edition of Leibniz's miscellaneous works (Leibniz 1923–). Thus, 'AVII3' will refer to the third volume of the seventh tome of the Edition of the *Akademie der Wissenschaften*, and 'AVII3, 23' will refer to the text number 23 contained in that volume, where the expression "transcendental" effectively appeared for the first time. On the other hand, Leibniz maintained the word "geometrical" in order to denote both algebraic and transcendental curves (compare also Breger 1986, p. 122–123).

[11]The most complete, so far, introduction to Leibnizian calculus is Bos (1974). Other important contributions are: Giacardi (1995) and Roero (1995). See also the collective volume: Heinekamp (1986).

[12](Cf. Huygens 1888–1950, vol. 9, p. 451; Leibniz 2011, p. 54; Child 1920, p. 25; Vuillemin 1962, *Introduction*, in particular p. 33). The idea that infinitesimal calculus could overcome Descartes' normative separation between geometrical and mechanical curves can be gleaned, for example, from the following lines, where Leibniz explained to the British mathematician John Wallis the uselfulness of the calculus: "dum novo Calculi generis effeci ut etiam Algebram transcendentia, Analysis subiciantur, et curvas, quas Cartesius male geometria excluserat, suis quibusdam aequationibus explicare docui, unde eorum proprietates certo calculi filo deduci possunt. Exemplo Cycloeidis, cui aequationem ibidem assigno $y = \sqrt{2x - xx} + \int \frac{dx}{\sqrt{2x-xx}}$, ubi $\int$ significat summationem, et d differentiationem, x abscissa ex axe inde a vertice, et y ordinatam normalem" (AIII7, p. 350–351).

These remarks suggested the thesis (proposed for instance in Breger 2008b) that Leibniz gradually abandoned, at least in his mature production, the classical way of describing curves through their specification by genesis, and came to rely solely or mostly on the symbolic apparatus of calculus instead.[13]

In this way, Leibniz would be taken to have also inaugurated a shift that invested the whole of mathematical practice in the subsequent century, marked by a passage from an explicit representation of curves, based on their geometric construction and visualisation, to an implicit one in which curves were denoted by symbolic expressions (at first equations and afterwards functions).[14]

In this article, I will try to put this general view into perspective by reconstructing part of the attempts to reform the Cartesian demarcation between geometrical and mechanical curves that Leibniz initiated during his mathematical apprenticeship in Paris, between 1672 and 1675. At that time, the symbolism of the calculus did not yet exist or existed in its inchoate state, so that Leibniz's approach was grounded mostly on geometrical considerations. In Sect. 6.2, I shall survey the main tenets of the early criticism made by Leibniz of Cartesian geometry as a geometry restricted to finitary objects (polynomial equations), which cuts out the most interesting parts of mathematics, namely curvilinear problems that have to do with quadratures and rectifications. In Sect. 6.3, I shall survey a second criticism Leibniz moved to Descartes' exactness criteria. According to Leibniz, Descartes' theory of curves was praiseworthy, in so far as it extended the boundaries of ancient geometry beyond the circle and the straight line, yet it was marred by notable errors of classification, since curves that fully qualified for geometry were displaced by this theory rather into the sphere of mechanics. In particular, Leibniz objected that Descartes had wrongly treated as non-geometrical rolling curves like the cycloid or the parabolic trochoid, respectively generated by the rolling without slipping of a circle or a parabola over one of their tangents. I shall evaluate the rationale which pushed the young Leibniz to consider it to be a categorical error to group together curves that of grouping together curves like the cycloid or the parabolic trochoid with the Archimedean spiral, and I shall finally argue that the exactness criterion that underpinned Leibniz's early views of geometry represented the first step in a programme he would pursue throughout his life.

---

[13] As we read in Breger (2008b): "It proved to be difficult to show that all transcendental curves really could be arrived at from a few simple means of construction. In addition, it may have been the case that the mathematical community perhaps contrary to Leibniz's original expectation did not express any interest in further construction methods. So Leibniz and with him the mathematical community were finally content simply to use an equation as the basis for the legitimation of a curve" (Breger 2008b, p. 149–150).

[14] Bos (1988, p. 55). On the plane of the general evolution of philosophical thought, this historical movement seems to concord with the pushing back of the role of imagination, understood as the capacity to visualise shapes and figures through suitable diagrams, when it comes to geometrical matters. Concerning the retreat of imagination and its philosophical justification see Arana (2016) and Beeley (2008, p. 91, 93ff).

The main sources I shall analyse are constituted by a group of manuscripts and notes, redacted between the end of 1674 and the beginning of 1675 according to the analysis made by the editors of Leibniz's collected works, but unpublished in Leibniz's lifetime.[15] Since these volumes collect drafts, sketches and working notes not prepared by their author in view of a publication, they appear somehow repetitive, untidy, at times enigmatic or containing a certain amount of inconsistencies. They are nevertheless worthwhile documents, and their study will permit a more thorough account of Leibniz's almost daily struggle between the acquisition of new insights, new methods and ground-breaking mathematical discoveries on one hand, and on the other the need to legitimate the newly acquired mathematical knowledge against the background of received views about exactness, especially those propounded by Descartes in his *Géométrie*. It is this struggle that I am going to illustrate in the rest of the paper.[16]

6.2 The Rise of Archimedean Geometry

Leibniz started studying the Latin edition of Descartes' *Géométrie* in earnest as early as 1673.[17]

By that time new problems, derived either from the development of physics or emerging within mathematics itself, had promoted a significant shift in what could be considered the central and the peripheral questions in geometry. In particular, the traditional construction problems which had acted as the main driving force behind the conception and the genesis of Descartes *Géométrie* had lost their impact and became peripheral in the mathematical practice of the second half of the seventeenth

---

[15]Most of the manuscripts I have been able to peruse have been recently published in Leibniz's miscellaneous works (in particular, in volumes AVII, 3, 4, 5). A small number of manuscripts that I am using in this paper have been transcribed by Siegmund Probst by Uwe Meyer and by Achim Trunk with a view to their forthcoming publication in the seventh tome of the seventh volume of Leibniz's mathematical manuscripts. I am particularly indebted to Dr. Probst for having granted me the possibility of consulting his transcriptions.

[16]We could quote, as a general reminder for the importance of studying manuscripts, or at least different sources than published ones the following remark made by David Hilbert: "Die Wissenschaft wird auch mündlich übertragen, nur aus Büchern ist unfruchtbar – so etwa" (in Thiele 2009, p. 271).

[17]Cf. Mahnke (1926, p. 9). The two Latin editions of *La Géométrie*, from 1649 and 1659–1661, edited by Frans Van Schooten and enriched with important studies by De Beaune, Van Heuraet, Hudde, de Witt et al, enjoyed a larger circulation, during the second half of the seventeenth century, than the first, French edition. Leibniz's critical appraisal of Descartes' analytic geometry has been the object of an extensive literature. Among classical studies, see: Boutroux (1920), the already quoted Mahnke (1926), Belaval (1968), and Hofmann (1975). An important paper is also Breger (1986). Recently, the publication of volumes 3, 4 and 5 of Leibniz's mathematical manuscripts has made available a wealth of material from the years 1673–1676, which has stimulated and continues to stimulate further research into Leibniz's reception of Cartesian geometry. Among the most recent contributions, see for instance: Knobloch (2006, 2015) and Probst (2012).

century. Concomitantly, Cartesian geometry did not seem to provide sufficiently powerful tools to solve in a systematic way problems which had become crucial in the agendas of mathematicians, such as problems of quadrature and problems of constructing a curve, given its tangents (the inverse tangent problem).

Leibniz soon realized that Cartesian geometry suffered from these shortcomings and was not able to bring geometry to its own completion, contrary to what is declared in the final lines of the *La Géométrie* (cf. Descartes 1897–1913, vol. 6, p. 485).

Starting from 1673, the year in which he became acquainted with Descartes' geometrical works, Leibniz elaborated a conception of the rationale, the limits and the historical significance of Cartesian geometry that he maintained essentially unchanged throughout his career. Such a view is grounded in a conceptual history of geometry, that certainly reflects Leibniz's personal viewpoint although it is well-grounded in the practice of his time.

Accordingly, geometry should be divided into three realms on the ground of the domains of problems investigated and the methods involved: the Euclidean, the Apollonian and the Archimedean realms.[18]

The first kind of geometry, the Euclidean one, is essentially the geometry of the first six books of the *Elements*, in which problems are solved by the intersections of circular arcs and segments according to the first three postulates of *El.* Book I or, in the language of instruments, by the use of straightedge and compass.

Apollonian geometry aimed to increase the resolutory capacity of Euclidean geometry working in two main directions: it taught how to solve "rectilinear" problems, namely problems requiring the construction of segments by the intersection of higher curves (the so-called "determinate problems" in the language of Descartes' *Géométrie*), and problems requiring the construction of geometrical curves as loci of points (the so-called "indeterminate" problems). In both cases, Apollonian geometry extended the number of permissible solving methods beyond straight lines and circles.

Viète, Descartes and his followers, such as François de Sluse,[19] still adhered to the main goals of the Apollonian programme in geometry, although they revolutionized the classical methods of the ancients thanks to the introduction of symbolic algebra. As Leibniz declared in his *Dissertatio esoterica de usu geometriae* (Summer 1676)[20]:

> In our time, the use of algebra in geometry has been restored, more generally and evidently, thanks to new forms of characters. By all means, the great man François Viéte, by teaching

---

[18]One of the first evidence is offered by an interesting manuscript from Summer 1673, titled *Fines Geometriae* ("On the limits, or realms of geometry"): AVII4, 36. Cf. also AVII6, 7; 49$_1$. A related manuscript is: *De imperfectione analyseos*, from 1673 (AVI, 3, 40) As an evidence of the long-lasting presence of this perspective on the development of geometry, one can compare the manuscript: *Scientiarum diversos gradus nostra imbecillitas facit*, that should have constituted an introduction to the unfinished treatise *De scientia infiniti*. See Roero (1995, p. 356ff), in particular.

[19]Cf. AVII4, p. 594.

[20]Except when specified otherwise, all translations in this article are mine.

how to express quantities through letters and to compute with species or symbols as if they were numbers, cast the foundations of a certain new science that some call "Analysis", others "specious", and I'd rather call "Symbolics" ... but that the praise for having placed this beautiful science in broad light must be given to this excellent man, René Descartes.[21]

Descartes clearly saw that:

... Equations of two or three unknowns could exhibit the linear loci of the Ancients, and the loci with respect to a surface (...) and here I come back to something discovered by Descartes in the Geometry: without doubt, given the nature or the symptom of the curve, from the number of those which are determined by the relations and powers of certain straight lines, to find its equation, its construction and its tangents; and given a rectilinear problem, to find the curves that I have mentioned the intersections of which satisfy the problem. But we must confess that the honour is partly due to Fermat, for his having invented, already earlier, the beautiful method of maxima and minima by which, although in a different way from the Cartesian one, the tangents of curves are also found. But a general Method for constructing all rectilinear problems by the intersection of the said curves, is due to Descartes alone, as far as I can judge with good reason. This is truly a most splendid invention, by which he had himself rightly claimed to have opened an immense field of speculation.[22]

However in spite of Viète's *Logistica speciosa* or *Symbolica* and its subsequent development with Descartes' general method for solving (rectilinear) problems, some of the most interesting properties of the curves that were part of Apollonian or even of Euclidean geometry, starting from the areas of figures bounded by geometrical curves or their arc lengths, continued to lie beyond the power of Apollonian geometry.[23]

---

[21] "Nostro seculo novis characterum formis usus Algebrae in Geometria amplior et manifestior redditus est. Magnus profecto Vir Franciscus Vieta, cum omnes magnitudines literis exprimere, et speciebus sive symbolis perinde ac numeris computare docuisset; novae cujusdam scientiae quam alii Analysin, alii Speciosam vocant, ego Symbolicam malim, fundamenta jecit... Sed scientiae pulcherrimae in clara luce collocatae laus summo Viro, Renato Cartesio debebatur". AVII6, 49_1, p. 502.

[22] "... aequationes duarum triumve incognitarum exhibere veterum Locos Lineares; et ad superficiem (...) Et huc redit quicquid a Cartesio detectum est in Geometria: nimirum data natura vel proprietate Lineae ex earum numero quae per rectarum relationes potentiasque determinantur, invenire ejus aequationem, descriptionem, tangentes; et dato problemate rectilineari, curvas quales dixi invenire, quarum intersectionibus problemati satisfiat. Fatendum est tamen in partem honoris venisse Fermatium, pulcherrima illa de Maximis et Minimis Methodo jam ante inventa, a Cartesiana prorsus diversa, qua et curvarum Tangentes exhibentur. At Methodus generalis problemata rectilinearia omnia per curvarum quales dixi intersectiones construendi, uni Cartesio quantum judicare possum in solidum debetur. Inventum sane pulcherrimum, quo ut ipse recte ait immensus speculandi campus aperitur." AVII6, 49_1, p. 504. Leibniz alludes to Frans van Schooten's Commentary: Descartes (1659–1661), vol. I, p. 170.

[23] "Quod tamen non nisi de problematis illis verum est, quae a me rectiliniaria vocantur, quae scilicet non nisi rectarum linearum magnitudines relationesque quaerunt aut supponunt: nam cum linearum curvarum, aut spatiorum ipsis conclusorum magnitudo quaeritur (: quod saepius fit, quam ille forte crediderat, quod animum ad Galileanam Mechanicen, non satis applicuisset :) neque aequationes neque curvae Cartesianae nos expedire possunt ...". AVII6, 49_1, p. 504.

Leibniz claimed that problems of quadrature and rectification traditionally belonged to the third type of geometry, the "Archimedean" one, after the name of its most illustrious master. This type of geometry was cultivated during the first half of seventeenth century when old methods like the study of the centre of gravity or the method of indivisibles[24] were rediscovered and improved upon by Galileo, his disciples Cavalieri and Torricelli, and later on by Guldin, Roberval, Gregoire de St. Vincent, Pascal, Mercator, Gregory, Barrow, Wallis, and Huygens.

Certainly, Descartes could not have been unaware of the divide between the "Apollonian" and the "Archimedean" geometry, both historically and conceptually, even if he was not familiar with this terminology. Leibniz attacked, on the contrary, the normative component that Descartes and his followers had allegedly ascribed to this demarcation, according to which only the Euclidean and Apollonian fields deserved the name of "Geometry" because only they could be successfully studied by finite algebra:

> ...To cut curvilinear magnitudes in a given ratio, to transform them into rectilinear or into other curvilinear magnitudes and more generally, to exhibit quantities given in position, magnitude and shape, this is doubtless the highest peak of Geometry, but Descartes never aspired to it. He saw the path blocked for him; and his equations and his analysis were of no use in solving these problems. He knew that the discoveries of Archimedes and his disciples about the measure of curvilinear figures were limited, and had to be due to the acuteness of an individual, matched with a large dose of good luck. For this reason, he despaired about comparing geometrically the curved and the straight, and did not touch problems of quadrature; and left out, classifying it as falling within the sphere of what he called the "mechanical" (as if it were hopeless), whatever he could not bring under his analysis.[25]

Hence, Leibniz was convinced that Descartes had essentially failed to discover a method of solving the problems of Archimedean geometry, and had concealed this failure by presumptuously ruling out as "mechanical" what could not be reduced to algebraic equations. In this way, exactness would be confined within the domain of the Apollonian geometry without a sound justification, and leaving out the richest and most interesting part of geometrical research.

---

[24]Leibniz was convinced that Archimedes used indivisibles to discover theorems about the quadratures and cubatures of curvilinear figures or solids of revolution (AVII6, 49$_1$, p. 498: "Archimedem ego semper in tantum miratus sum, in quantum licet mortalem; usque adeo insignia ejus inventa, et profunda, et superioribus dissimilia, et in omnem posteritatem valida fuere. Indivisibilia certe, aut si mavis infinite parva, Geometriae sublimioris clavem, adhibuit primus, tecte licet, et ita, ut admiratio inventis, et rigor demonstrationibus constaret.)"

[25]"Magnitudines curvas in data ratione secare, in rectilineas vel alias curvilineas transform[are] et generalius, quantitates exhibere, positione, quae sint datae magnitudine et figura, fastigium est haud dubie Geometriae, sed ad quod Cartesius ne aspiravit quidem. Videbat aditum obstructum sibi; neque ullum esse in solvendis his problematis usum aequationum et analyticae suae. Agnoscebat quae ab Archimede et ejus discipulis circa Curvilineorum dimensiones inventa sunt, angusta esse, et sagacitati singulari cum bona quadam fortuna conjunctae deberi. Unde de comparatione Geometrica recti et curvi, desperavit, et problemata quadraturarum non attigit; et quae sub analysin cogere non poterat sub Mechanicorum nomine velut desperata reliquit" (*De descriptione geometric curvarum transcendentium*, AVII7, 49$_2$, p. 512, from January 1675).

For this reason, Descartes' belief that he had provided a solution to Viète's problem – or 'meta-problem' as we may call it – of leaving no problem unsolved was ultimately deceptive.[26]

In an open criticism of the "short-sightedness" of Cartesian geometers Leibniz invoked the need to search for a method that would represent for Archimedean geometry what Viète's and Descartes symbolic algebra had represented for Apollonian geometry:

> Indeed, insofar as we search for the magnitude of curved lines, or for the surfaces bounded by these lines (...) neither Cartesian equations nor Cartesian curves can help us. We need a completely new kind of equations, new constructions and new curves. Indeed a new calculus, not offered by anyone yet, of which now I could, if nothing else, give at least some marvellous specimens.[27]

Starting from this conviction, Leibniz sketched, while he was still in Paris, the programme leading to the differential and integral calculus.[28]

A beautiful summary of this programme can be found in a letter written in 1678 (hence, when Leibniz was already in Hannover) to Jean Gallois:

> However, it is not the algebra of Viète or Descartes which can reach the solution of all problems: because it only goes until the problems of rectilinear geometry ...whereas the hardest problems, and those who have most influence on mechanics do not reduce to any equation of a certain degree. They depend instead on some extraordinary equations, that I call 'transcendental', as they are of all the degrees at once, either together, or alternatively. New curves are necessary to construct them, and a new kind of Algebra to treat them with dignity...for this reason, one should not be surprised that 'Viète and even Descartes and their disciplines had not been able to do almost anything about such sorts of problems. And what others have done regarding them consisted in nothing but a few particular, fortunate and ingenious findings. On the contrary, I can see the means to treat all this analytically, and I have many valuable specimens of my method.[29]

---

[26]As Leibniz pointed out: "Enim ille [namely, Descartes] saepe loquitur splendidius sane quam verius; methodo sua geometriam ad perfectionem perductam esse, quanta ab homine optari possit; nullum esse problema, cuius non aut solutionem aut solvendi impossibilitatem monstret. Certas sibi rationes esse praescribendi limites intellectui, deniendique quicquid aliquando inveniri possit". AVII,4, p. 594–595. See also Mahnke (1926), p. 60.

[27]"nam cum linearum curvarum, aut spatiorum ipsis conclusorum magnitudo quaeritur (...) neque aequationes neque curvae Cartesianae nos expedire possunt; opusque est novi plane generis aequationibus, constructionibus curvisque novis; denique et calculo novo, nondum a quoquam tradito, cujus si nihil aliud saltem specimina quaedam, mira satis, jam nunc dare possem." VII6, p. 504.

[28]Cf. in particular the fundamental manuscripts: AVII5, 38, 40, 44, which contain the 'invention' of the calculus. See also Bos (1974).

[29]"Ce n'est pas pourtant l'Algebre de Viete ou de des Cartes qui puisse arriver a la solution de tous les problemes: puisqu'elle ne va qu'aux problemes de la Geometrie rectilineaire (...) au lieu que les problemes les plus difficiles, et qui ont le plus d'influence dans la mecanique ne se reduisent à aucune equation d'un certain degré. Ils dépendent de quelques equations extraordinaires, que j'appelle Transcendentes, par ce qu'elles sont de tous les dégrés tout à la fois ou conjointement, ou bien alternativement, il faut de nouvelles lignes courbes, pour les construire, et il faut une nouvelle espece d'Algebre pour les traiter dignement (...) c'est pourquoy il ne faut pas s'étonner si Viete des Cartes même et leurs disciples n'ont pû presque rien faire sur ces sortes de problemes. Et

The similarity in content with the passage reported above is further evidence that Leibniz's criticism of the limitations of Cartesian geometry followed a seamless trajectory from its origins to the later developments that brought Leibniz ultimately to the foundational articles on infinitesimal analysis. Let us point out, in particular, that Leibniz insists on the necessity of introducing a new algebra on a par with new curves in order to deal with transcendental problems. Leibniz's reflection on these new curves shall be the topic of the next section.

6.3 Geometrical Curves Beyond Cartesian Geometry

The criticism discussed above was not the only kind of criticism Leibniz made of the *Géométrie*. In his early manuscripts, Leibniz also attacked the structure of Cartesian geometry, and used against Descartes himself the very arguments that Descartes had employed to demarcate exact, geometrical from non-exact, or mechanical curves.

The core of Leibniz's objections can be summed up as follows. Descartes' criteria of exactness based on the equivalence between geometric constructibility via a unique and continuous motion on one hand, and on the possibility of expressing curves through finite polynomial equations on the other is not tenable, because there exist curves which are constructible by one continuous motion; thus these curves are geometrical, although they cannot be associated with any algebraic equation. Consequently, the very demarcation between geometrical and mechanical curves ought to be rethought.

In a draft of a letter to Oldenburg, composed in March 1675, Leibniz discussed this objection to the Cartesian criteria of exactness in precise terms:

> I call geometrical constructions the ones which are accomplished by a ruler and a compass, or by rulers and compasses joint one with another and moving or guiding one the other. These are those accomplished by curves, that Descartes had correctly taught that should be received in geometry (…) but it must be known that Descartes had only stumbled upon the class of these curves, forgetting about a great multitude of other ones. In fact, as he had only stumbled upon those which could be dealt with in his way, he did not even suspect the existence of other ones. But I will disclose a totally new world of curves (*novum plane orbem aperiam curvarum*), which can be described geometrically, by one continuous tracing (*uno ductu continuo*) and only by some rulers composed with other rulers or compasses, with no less simplicity than parabolas or hyperbolas, which nevertheless, it is impossible to deal with in Cartesian geometry.[30]

---

ce que les autres ont fait la dessus ne sont que de certaines rencontres particulieres, heureuses ou ingenieuses. Au lieu que je voy moyen de traitter tout cela analytiquement et j'ay beaucoup d'essais considerables de ma methode." (AIII2, 245, p. 567).

[30] AIII1, 46, p. 205: "Constructiones autem voco Geometricas, quae regula aut circino, vel regulis aut circinis inter se compositis seque compellentibus aut ducentibus perficiuntur. Tales sunt quae per curvas fiunt, quas Cartesius recte docuit recipiendas esse in geometriam (…) sciendum est tamen Cartesium in earum speciem tantum incidisse et immensae aliarum multitudinis oblitus esse. Nam ille, cum in eas tantum incidisset, quas suo more tantum tractari poterat, de caeteris ne suspicatus est quidem. Ego verum novum plane orbem aperiam curvarum, quae geometrice

The beginning of the passage reported here is a correct summary of Descartes' canon of exactness in geometry, as this is presented in the second book of *La Géométrie*. As we have seen in the introductory section, the first criterion for geometricity was, in Descartes' view, the possibility of constructing a curve by one continuous motion. These constructions are ensured, in the *Géométrie*, by a set of configurations of moving figures, or articulated compasses, that we can call "geometrical linkages" (following Panza 2011, for instance) and which generalize the constructional possibilities of ruler-and-compass in Euclidean geometry.

Descartes characterized geometric linkages by a set of necessary requirements, which can be thus summarized (see Bos 2001, chapter 24, and especially Panza 2011, p. 74ff.):

- Geometric linkages can be formed by configurations of moving rulers, or segments, subject to translations and rotations, such that the motions of any of these parts are dependent on a principal motion (a rotation or a translation).
- A geometric linkage traces a curve as a trajectory uniquely determined by the continuous motion of one of its points (we can imagine the curve traced by a pen fixed upon one of its points).
- Moreover, the curves constructed by geometric linkages are independent of the speeds of the tracing motions.
- As soon as a curve has been traced by a system of geometric linkages, it can become itself a component of the system, and thus trace other curves, which in their turn can become parts of more articulated devices.

The first three conditions above ensure the continuity and unicity of the tracing motion and the fact that the motion is speed-independent, whereas the fourth condition grants that geometrical linkages have a compositional nature: if a curve has been constructed by a certain linkage, it can enter another linkage as its component (Panza 2011, p. 88).

As an example, let us consider how Descartes constructed, via a geometric linkage, the conchoid of Nicomedes (Fig. 6.1), i.e. the "first conchoid of the ancients" as Descartes called it (Descartes 1897–1913, vol. 6, p. 395). The conchoid was employed by the ancient Greek in order to solve solid problems like the duplication of the cube and the trisection of an angle, and was also familiar to seventeenth century geometers through Pappus' *Mathematical Collection* and Eutocius' commentary on Archimedes's treatise On the *Sphere and the Cylinder*.[31]

The *Géométrie* contains a sketchy description of a linkage for the tracing of this curve. Starting from Descartes' brief remarks, we can reconstruct it as follows. This geometric linkage is formed by a circle with centre E and radius EH, hinged to the pivoting ruler AEH. When AEH rotates around a fixed point A, the circle

---

uno ductu continuo exacte ac solis regulis circinisve compositis non minore quam parabola aut hyperbola aliaeve altiores Cartesianae simplicitate describi possunt; quas tamen impossibile sit tractare per Cartesii Geometriam."

[31] See Pappus (1876), vol. 1, p. 243–44; Archimedes (1881), vol 3, p. 267.

translates along the vertical BE and point H traces a curve HH'. By construction, EH remains constant during the motion. The curve traced in this way is a conchoid, according to the characteristic property of this curve.

Descartes' apparatus for the conchoid, according to the plausible reconstruction given above, has all the necessary requirements to qualify as geometrical linkage. Firstly, the local motions of any of its components depends on a principal motion, the rotation of the segment AEH. Secondly, the trajectories of all the motions generated by the conchoid-linkage are independent of the variations in the speed of the rotating segment. Finally, all the components of a conchoid-linkage are geometric curves in their turns: a certain number of segments and a given circle.

Lastly, the conchoid can be easily associated with a finite algebraic equation. Indeed, taking as a reference Fig. 6.1, let us name $AB = a$, $EH = b$, while θ is the angle formed by the pivoting radius with the axis ABH'. The curve can be described by the polar equation: $\rho = a\sec\theta + b$, as can be easily verified by looking at the figure. By choosing a suitable framework, transforming the previous expression into a Cartesian equation is easy and will yield the following: $(x^2 + y^2)(x - a)^2 = b^2 x^2$.

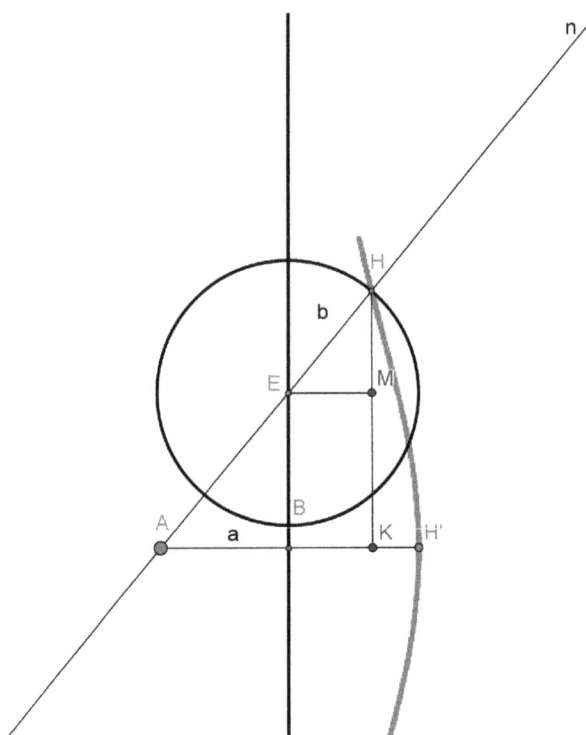

Fig. 6.1 The conchoid of Nicomedes generated by a Cartesian linkage. By putting $AK = x$ and $KH = y$, and considering the right-angled triangles AKH and EMH, which remain similar during the motion, we can find a finite polynomial equation describing the curve

Descartes was certainly able to obtain the same, or a similar equation describing the curve via a geometrical reasoning (sketched in Fig. 6.1) and could therefore associate the conchoid to a quartic equation.

By contriving the generation of the conchoid of the ancients through a geometric linkage, Descartes also succeeded in showing that a curve traditionally ranged with the spiral and the quadratrix,[32] ought to be more properly grouped with algebraic curves like the circle and the conic sections.

But in what did Descartes' exactness standard fall short, according to Leibniz's criticism? As we have just seen, geometrical linkages serve a twofold theoretical purpose. On the one hand, they ensure the unicity and continuity of the motion which engenders a geometric curve, on the other, they also ensure the exact knowledge of its magnitude (or "mesure", following the original French edition), namely the possibility of associating a curve with a finite algebraic equation in two unknowns. According to Leibniz's criticism, choosing geometric linkages as the sole way to implement unique and continuous motion is also an unnecessary limitation of the boundaries of geometry, since there are constructions that do not violate the requirements of unicity and continuity of motion and can engender a whole new realm of curves (*novum orbem... curvarum*), not necessarily algebraic.

Leibniz did not give examples of such curves in the letter to Oldenburg. Although I could not find any passage where Leibniz discussed any transcendental curve constructed "by some rulers composed with other rulers or compasses", as stated in the letter to Oldenburg, I will argue in the remaining part of the article that he had in mind at least two main examples of alleged geometrical curves which cannot be dealt with Descartes' geometry: the curves constructed by evolution (particularly the involute of the circle) and the rolling curves (in particular the parabolic trochoid and the cycloid).

6.3.1 The Involute of the Circle and the Curve of Bertet as Geometrical Curves

Leibniz came to know the theory of evolutions through the third book of Huygens' *Horologium Oscillatorium*, published in 1673 (Leibniz's annotations to Huygen's *Horologium* dated back to Spring 1673, as shown in AVII4, n. 2).

The dual notions of evolute and involute (see Fig.6.2) were introduced by Huygens by considering a smooth curve, namely a curve without cusps or changes in its concavity, around which a taut string is wrapped and unrolled, so as to trace a companion curve, that we shall call the "involute" of the given curve.[33]

---

[32] According to Pappus, the conchoid should be ranged together with the Archimedean spiral and the quadratrix into the same class of "linear" loci (Pappus 1876, vol. 1, p. 271).

[33] In Huygens' terminology, the involute is the curve "described by evolution" (*descriptaex evolutione*). The curve around which the string is wrapped is called "evolute" (this corresponds to definition IV in the *Horologium Oscillatorium*, Huygens 1888–1950, vol. XVIII, p. 189. See also Yoder 1988, p. 6). A given evolute can be associated with an infinite number of companion curves,

Fig. 6.2 From *Horologium*
Oscillatorium, Book III,
definitions 2 and 3. A string is
applied to the evolute ABC
and then tightly unrolled
starting from the free end D
which, originally coinciding
with A, traces the involute
ADE. The length of the arc
AB is equal to the length of
the tangent BD to the circle
at point B

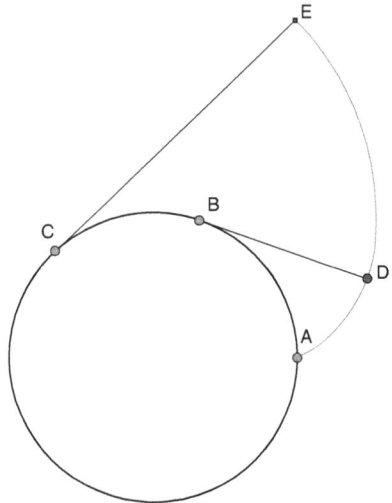

The use of a mathematical string to generate the companion of a given curve implements, by virtue of its way of operating, a unique and continuous motion, thus it obeys to the constructional criterion of exactness defined as in Descartes' geometry. In certain cases, the involute of a given algebraic curve is an algebraic curve in its turn[34]: in this way, Descartes' global ideal of geometrical exactness, based on the equivalence between the unique-motion constructional criterion and the algebraic criterion, would be preserved.

However, it is also possible to associate, by evolution, with a given algebraic curve a non-algebraic one.

The case of the circle is a particularly salient one in this respect (see Fig. 6.2). Indeed its involute could be described by Huygens only "qualitatively" as the result of a geometrical construction, but not analytically via an equation (See Yoder 1988, p. 93). The reason, is that the companion curve of the circle is not an algebraic curve.

The example of the involute of the circle first appeared, among Leibniz's writings, at the beginning of 1673 (in the tract AVII4, n. 10, p. 141) and subsequently in a draft of a letter, possibly addressed to Huygens in the beginning of 1675.[35]

---

since Huygens assumed that the string has an arbitrary length, and its extremity can coincide with any point on the initial curve (Huygens 1888–1950, vol. XVIII, p. 189).

[34] For instance, the involute of an algebraic curve like the semi-cubical parabola is another algebraic curve, namely a parabola (Huygens 1888–1950, vol. XVIII, p. 207–209). Using a geometrical reasoning, Huygens also discovered that the involute of a cycloid (a mechanical curve according to Descartes) is a similar cycloid. This result had a fundamental importance for the practical problem of constructing the pendulum clock.

[35] See AIII, 1, 29. Hofmann dates the letter to Summer 1674, although an examination of the watermark reveals that the latter was more likely from a later period, possibly January 1675 (I am indebted to Siegmund Probst for this suggestion).

In the first manuscript from 1673, Leibniz conjectured that the involute of the circle could not be expressed by a finite polynomial equation. Leibniz's sketchy argument proceeds by contradiction. In brief, the supposition that an involute is an algebraic curve would lead us to the absurd conclusion that this curve should be associated with an equation of higher and higher degree, hence with an equation of a form not envisioned by Descartes. The reason, according to Leibniz, is that the involute of the circle could be used to rectify arbitrary arcs of a given circle, and via this property, established thanks to the use of strings in the generation of the curve, it could be used to find the ratio between any two given arcs. Since Leibniz knew, through Viète, that this problem could not be associated with a single, finite polynomial equation with a fixed degree, he concluded that the solving curve could not be associated with a polynomial equation in a fixed finite degree either. Hence, it could not be associated with an equation belonging to Descartes' algebra.[36]

This argument of impossibility was merely sketched in 1673, but was refined by Leibniz in the subsequent years. Leibniz would in fact use the very relation between the rectification problem and the universal section of the angle to prove that there is not a finite algebraic formula in order to express relation between an arbitrary circular arc and its corresponding tangent (cf. AVII6, 51, proposition LI; see also below).

In the tract AVII4, 50, from Autumn 1673, and again in his AVII5, 45, from November 1675, Leibniz discussed a second curve, similar to the involute of the circle, called "the curve imagined by Bertet", or the *linea Berthetiana*, by the name of its discoverer. This curve belonged to a family of curves, called by Leibniz: "curvae protensae".[37]

In a letter to Bertet himself from November 1675 Leibniz gave the following description of the curve, by means of a pointwise construction. Let BD be an arc of a circle with centre A. On the arc BD, let a point C be chosen at will. Let the radius AC be extended up to a point E, such that EC equals the arc BC. Let another point

---

[36] AVII4, 10_1, p. 141: "Illud ergo restabat quaerere, an data curva geometrica dari possit alia geometrica, quae datae evolutione describatur, seu, quae ita comparata sit, ut curvae datae tangentes sint quaesitae perpendiculares[,] ut si evoluta circuli describi posset geometrice per aequationes licet altioris gradus haberemus quadraturam; licet ea problema altioris gradus futura sit, ut in mesolabo. An forte malum in eo est, quod in quolibet puncto opus foret aequatione nova, ut patet in circulo. Nam si haberetur figura cuius puncta geometrice determinarentur aequatione quadam, quae evolutione circuli describeretur, haberetur cuilibet arcui circuli aequalis recta, quae describeretur scilicet illa evolutione, et cum illa recta data futura sit, daretur et ratio eius ad aliam rectam toti circumferentiae aequalem, ergo et arcus in quo circulum tangit, ita haberetur sectio angulorum universalis. At ea haberi non potest, nisi per aequationes alias atque alias continuo altioris gradus. Ergo pro initio aequationis opus foret aequatione graduum infinitorum, et ita semper regrediendo. Quod est impossibile. Res tum accuratius consideranda".

[37] AVII5, 45, p. 318. Leibniz had come to know this curve in 1673 through Ozanam, but attributed his discovery to the Jesuit mathematician Jean Bertet (1622–1681). Since the manuscript AVII4, 10 dates from 1673, it is possible that Leibniz had written down these notes as a result of his discussions. In the tract AVII4, 50, Leibniz remarked that the curve was originally proposed by Bertet to Ozanam (AVII4, 50, p. 808).

(C) be chosen, and let the same operation be repeated for this point. Continuing in this way, we shall obtain a net of points that belongs to the locus of a curve.[38]

In the manuscripts relating to the curve of Bertet, Leibniz supplemented two constructions by continuous motions. A first construction (see, for instance, AVII4, 50, p. 810) occurred with the aid of two motions, using a string in order to measure the successive portions of the arc, while the radius moves around the radius of the circle, as in Fig. 6.3 (on the left). To this, Leibniz added a second construction by *uno tractu*, i.e. by one tracing, or one motion (AVII5, 45, p. 319; see Fig. 6.3, right).

The curve of Bertet differs from the involute of the circle, as it can be ascertained by considering the expression of the elements of arcs for the respective curves, with the given circle taken as a reference (see Figs. 6.4 and 6.5). Even so, the same argument advanced by Leibniz in order to claim the non-algebraic nature of the involute holds for the curves of Bertet too, since this argument depends only on the property of rectifying circular arcs, which is shared by both curves on a par.

On the other hand, it should be pointed out that Descartes, would certainly have been dissatisfied with their construction via strings. In a passage of the second book of the *Géométrie*, in fact, Descartes made it clear that constructions

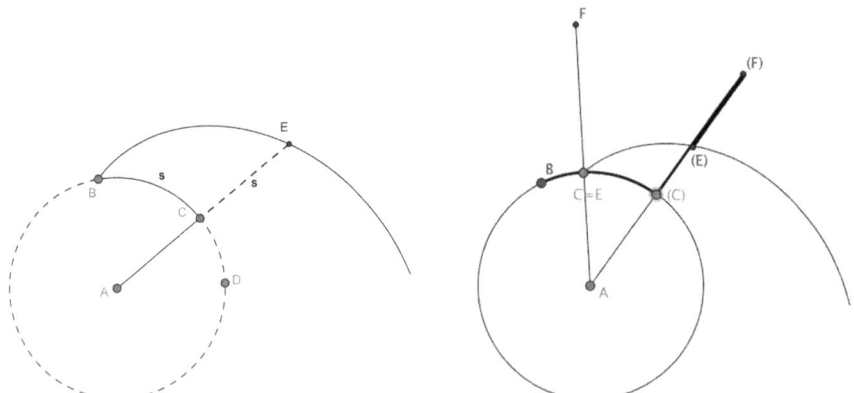

Fig. 6.3 The curves of Bertet (AIII, 1, 68, p. 309). On the left, its construction with two motions: while the radius AE rotates around the center A of the circle from B to C, the arc BC is rectified along AE with the aid of a string. On the right: the same construction, performed by one continuous motion or "one tracing" (*uno tractu*). Let us take a circle with centre A and radius AB, and a ruler AF which rotates around A. In the initial position, let the ruler AF cut the circle in C (this point is chosen at will). A string is fastened to B, and wrapped around the arc BC. At C we can fix an eyelet or a small ring, so that the string goes through it, and it revolves around AF, passing through a second eyelet in F, until it reaches the point C. The string must be left free at one end in C. In this manner, when the ruler AF rotates and goes, for example, from F to (F), the string covers the circular arc $C(C)$, and the free end (originally at C) moves along AF of a distance equal to the length of the arc $C(C)$. While AF rotates, the free end of the string traces the arc $C(E)$

---

[38]AVII5, 45, p. 317, AIII1, 68, p. 308–309. See also Hofmann (1975), p. 195–196.

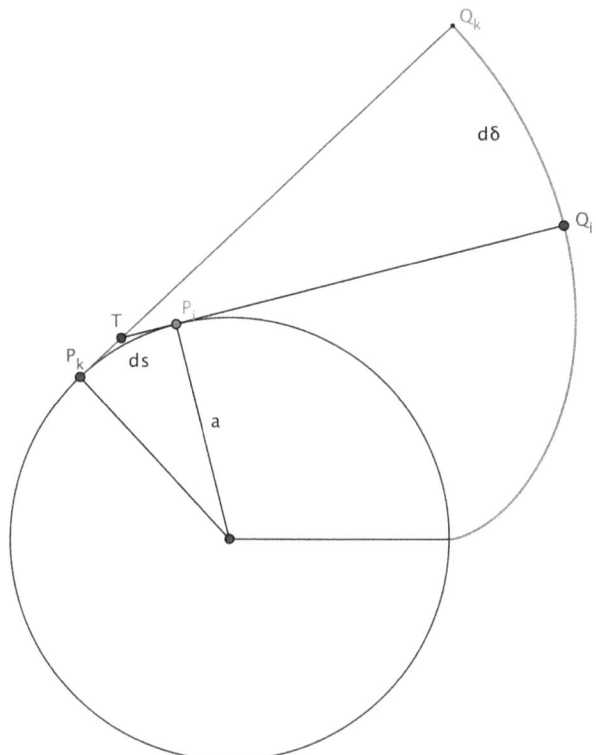

Fig. 6.4 From AIII, 1, 29, p. 116. The infinitesimal triangles $M P_k P_i$ and $T Q_i Q_k$ are approximately similar, so that $Q_i Q_k = d\sigma \approx \frac{a\,ds}{s}$

Fig. 6.5 From AIII, 1, 68, p. 309. The element $E(E)$ of the curve can be found by applying Pythagoras' theorem to the infinitesimal triangle $E(E)(G)$. By construction, $(E)(G) \approx C(C)$. Since triangles $AC(C)$ and $AE(G)$, having one side infinitely small, are similar, putting $(E)(G) = \Delta s$, we shall have that $E(G) \approx \frac{(a+s)\Delta s}{a}$. In conclusion:
$$E(E) \approx \frac{\Delta s}{a}\sqrt{a^2 + (a+s)^2}$$

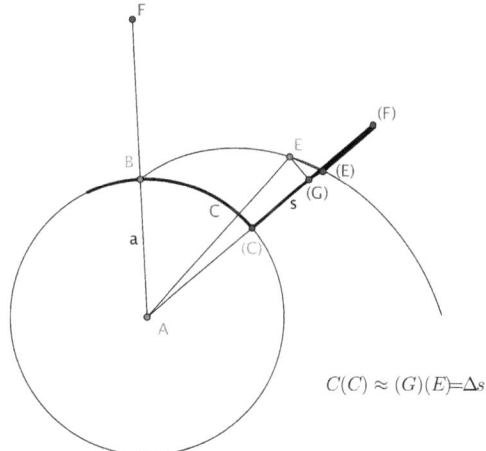

$C(C) \approx (G)(E) = \Delta s$

using strings were legitimate only in cases where the strings were employed to measure segments or distances, as in the constructions of the conic sections by the gardener's method (Descartes 1897–1913, vol. 6, p. 412). On the other hand, any construction in which a string was bent to adapt to curvilinear magnitudes should be treated as ungeometrical. When he ruled out constructions obtained by bending strings, Descartes was perhaps considering special methods for the generation of the Archimedean spiral, that I shall refer to also below (see Bos 2001, p. 348), but he had more likely in mind the general operation of comparing straight lines (segments) and curvilinear arcs, which in his view did not meet the requirements of exactness.[39]

In summary, when Leibniz wrote to Oldenburg, in March 1675, he must have realised that the involute of the circle and Bertet 's curve violated the Cartesian standard of exactness, on whose basis they ought to be considered mechanical. Nevertheless, the very use of strings involved in their construction assured their generation by a single and continuous motion (let us recall that Leibniz explicitly claimed the constructibility by one "tractus" of Bertet's curve in Autumn 1675, but he must have been aware of the constructibility of the involute upon his reading of Huygens' *Horologium*). Leibniz was certainly aware that a tension existed between Descartes' firm rejection of strings from geometry and their successful employment to implement the constructions of curves via a single and continuous motion, and solved such a tension by opting for an unconditional acceptance of the one-continuous-motion criterion as a criterion of geometricity, even if such acceptance involve breaking the equivalence between geometrical and algebraic curves, one of the pillars of Descartes' geometry.

It should be added that, during his Paris sojourn, Leibniz came to know about another fruitful use of strings in curve constructions, which became very important in the subsequent evolution of his mathematical thought (see below, the conclusion of this article). According to a story told, many years after the events, by Leibniz himself, the architect Claude Perrault introduced him to a curve traced by a chain clock dragged on a plane.[40] Because of the friction of the clock on the plane, the chain "in traction" is always tangent to the traced curve, called "tractrix" (Fig. 6.6). The role of traction in the first instrumental method of generating a curve given some

---

[39]In fact Descartes rejected those lines which: "semblent a des chordes, qui deviennent tantost droites et tantost courbes, a cause que, la proportion qui est entre les droites et les courbes n'estant pas connue et mesme, je crois, ne le pouvant estre par les hommes, on ne pouvait rien conclure de là qui fust exact et assuré". Descartes (1897–1913), vol. 6, p. 412.

[40]Leibniz recounted to Huygens his first meeting with the curve as follows: "Vous estes tombé de vous meme sur une idée, que j'avois deja, mais que j'ay apprise d'un autre. C'est de feu Mons. Perraut le Medecin qui me proposa de trouver quelle ligne se produit en menant une extrminté du fil le long d'un règle, pendant que l'autre extremité tire un poids par le plan horizontal dans le quel la règle tombe. Je trouvay bientost que c'est la quadratrice de la figure des tangentes canoniques du cercle, et par consequent dependante de la quadrature de l'Hyperbole. Je croyois d'avoir seul cette application de ce mouvement " (letter from 1/11 October 1693, AIII5, 191, p, 646). In the published article *Supplementum geometriae dimensoriae* Leibniz told a similar story about how he had come to know the curve in Paris, through the architect and editor of Vitruvius Claude Perrault (Leibniz 2011, p. 135).

Fig. 6.6 The tractrix,
according to Leibniz's draft.
Figure from AVII6, 21, p. 259

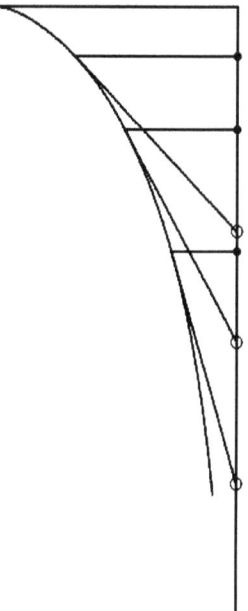

tangent conditions led to the general name of "tractional motion". In articles and correspondence from the 1990s (see below), Leibniz employed tractional motion (generalized to other curves than the tractrix) to provide a general solution to the inverse tangent problem, and thereby justify the exact nature of both algebraic and non-algebraic (or transcendental) curves in a purely geometrical way.

It is ascertained that Leibniz knew about the tractrix in Paris, since he drew at least a picture of it, reproduced in Fig. 6.6 above (AVII6, 21 p. 259). The manuscript, dating from 1676, does not contain any text. Leibniz must have noted, as he would recall many years later (see Bos 1988, p. 9; Tournes 2009, p. 14ff.), that the chord in tension is constantly tangent to the traced curve, but the absence of any other document related to the tractrix, up to 1676, does not allow us to draw further conclusions regarding the role played by this curve during his stay in Paris.

6.3.2   Enter Rolling Curves

Manuscript sources from 1674 and the beginning of 1675 reveal that Leibniz was thinking of other non-algebraic curves that he viewed nevertheless as geometrical. This is the case, for instance, of the so-called trochoids or rolling curves, that are discussed in a series of tracts (38_{11}, 38_{12}, 38_{13}) from October 1674, taken from a long inquiry formed by fifteen sheets, *De serierum summis et de quadraturis plagulae quindecim*.[41]

The first trochoid mentioned in Leibniz's mathematical notes (see, among the earliest evidence: AVII4, 2, 7, both written in early 1673) is the cycloid. Leibniz probably came to know this curve via three main sources: Honoré Fabri's *Synopsis geometrica* (1669); Blaise Pascal's *Lettres de Dettonville* (1658) and *Histoire de la Roulett* (1659) (Pascal 1913, pp. 195–209 and 334–384 respectively); and Huygens' treatise *Horologium Oscillatorium* (Paris, 1673), read and annotated in April or May 1673 (AVII4, 2). His research into rolling curves was also stimulated by the contacts with Danish mathematician Ole Roemer, who was in Paris in 1674, studying the epicycloid of a circle and its use in the construction of gear wheels[42]; and by his acquaintance with Philippe de La Hire, who in the middle of the 1970s was also studying the subject of trochoids.[43]

The cycloid was initially identified as the trajectory of a point attached to the rim of a wheel, or another round object, when the wheel rolls on the ground.[44] Abstracting from this physical situation, the curve can be conceived of as the locus described by a point on a rolling circle.[45]

But whereas it is intuitive to refer to the motion of a wheel in order to construct a cycloid (the point is made in Pascal 1913, p. 7), this description is not satisfactory for a geometrical inquiry. In fact real wheels can move in more than one way and sometimes unpredictably (for instance, a wheel over an icy surface does not move in a same way as a wheel over the ground), hence taking a moving wheel as a paradigm of rolling motion might lead to confusion or to circularity.[46]

Pure rolling motion can be defined instead as a motion of an object composed of a rotation and a translation occurring at equal speed, hence the cycloid can be described as a curve traced by a point on a circle that moves with rotational velocity

---

[41]The group of manuscripts 38_{11}, 38_{12}, 38_{13} is also discussed in Knobloch (2006).

[42]See Hofmann (1975, p. 247). The excerpts made by Leibniz of a lecture on gear wheels given by Roemer are still extant in the following manuscript: CC 1187–1188.

[43]Hofmann (1975, p. 247–248). La Hire issued a booklet on the cycloid (*De Cycloide*) in September 1676, after Leibniz had left Paris.

[44]For instance, Mersenne referred to the curve as "la figure qui descrit la boule qui roule sur un plan" Mersenne (1636), p. 120.

[45]Pascal defined this curve in the following way: "Ce n'est autre chose que le chemin que fait en l'air le clou d'une roue, quand elle roule de son mouvement ordinaire, depuis que ce clou commence à s'elever de terre, jusqu'à ce que le roulement continu de la roue l'ait rapporté à terre, après un tour entier achevé: supposant que la Roue soit un cercle parfait; le Clou un point dans sa circonférence, et la Terre parfaitement plane" (Pascal 1913, p. 7).

[46]I want to point out, in this connection, that this is not a matter of a punctiliousness regarding rigorousness of definition which results only from our retrospective view on the past. Early modern mathematicians were perfectly aware of such conceptual difficulties and no less rigorous than us. For instance, in the article *De lineae super linea incessu* (Leibniz 2011, p. 257) Leibniz discussed in detail the difficulties of identifying rolling motion and the motion of a wheel. The article was written in 1706, but as I will show below, there is an undeniable continuity between Leibniz's early reflections and his later publications on the foundations of geometry.

equal to its translational velocity. This kinematical description had been known to the mathematicians since the first half of seventeenth century. For example, it appeared as early as 1637 in a letter from Roberval to Mersenne, where the cycloid is explicitly distinguished from an ellipse or any other known curve on account of its mode of generation (Mersenne 1933, p. vol. 6, p. 175). A couple of years afterwards, Descartes and Mersenne discussed this very curve in a way that left no doubt as to its definition by the combination of simultaneous motions (Descartes 1897–1913, vol. 2, p. 313, and p. 406).

From this kinematic definition another definition can be deduced that does not make direct appeal to the speeds of the motions: the cycloid is the locus of the point on a circumference of a moving circle, such that, when the circle rotates around its centre of an arc GI, each point on it also covers a distance IB, equal to the length of the arc (Fig. 6.7). Eventually, when the circle completes a full rotation, each of its points has covered a distance equal to the length of the circumference.

The search for a mathematically correct description of the cycloid had an immediate relevance for the practice of early modern geometers. In fact, the cycloid was for them an interesting object both in geometry and in physics. During the seventeenth Century this curve had mesmerized the attention of the mathematical community and the learned public on two occasions: from the mid 1630s of the seventeenth Century to the 1640s, when the investigations of Roberval, Descartes and Torricelli took place, and by the beginning of the 1660s, when Pascal stirred the public interest by publishing a series of challenging question about several properties of the curve.[47] It resulted that a proper definition of that curve would reasonably lead to an easier discovery of its geometrical properties, such as its tangents, its area and its center of gravity. This is outstanding, for example, for the problem of constructing its tangents, which was immediately found by Roberval once the kinematic generation of the curve was clear (see Fig. 6.8).

Leibniz's interest in the cycloid was also motivated by the "brilliant uses" of the curve, namely its application to physics and chronometry (cf. AVII3, 38_{12}, p. 486). Leibniz's judgement matches well with the view of contemporary mathematicians,

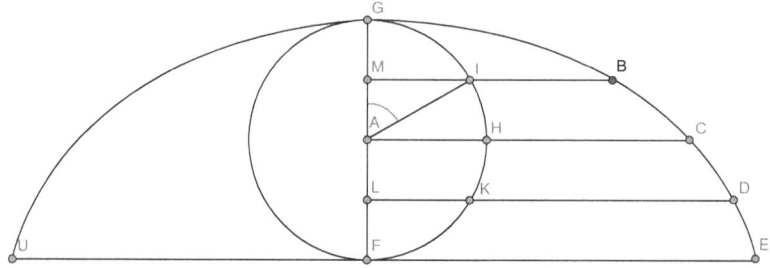

Fig. 6.7 The cycloid. $GI = IB, GH = HC, GK = KD, GF = FE$

---

[47]Whitman (1943), p. 312, 314.

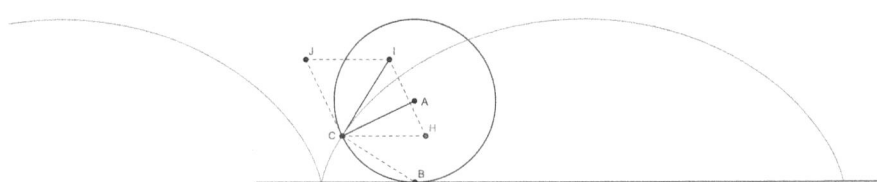

Fig. 6.8 The tangent to a cycloid constructed according to Roberval's method. Roberval considered the tangent CJ to the moving circle at C ($\overline{CJ}$ being the vector representing the rotational velocity) and a horizontal segment CH ($\overline{CH}$ is the horizontal translational velocity). By definition, both velocities are equal, so that $CJ = CH$. The tangent to the cycloid is the diagonal of the parallelogram $CJIH$

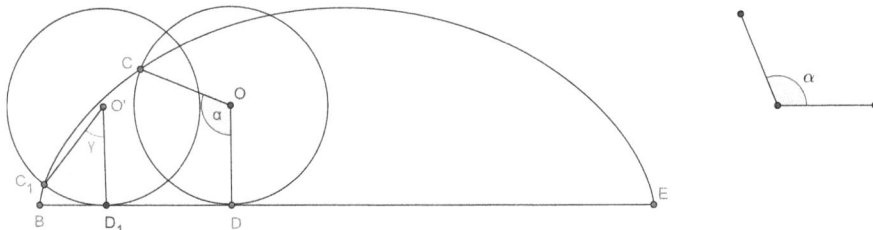

Fig. 6.9 We want to divide the angle α into an equal number of parts, e.g. trisect it, using a cycloid. On a given circle with center O and radius $OD = r$, let us construct an arc $CD = \alpha r$. Let us then roll the circle over the straight line BE, so that point C on the circumference will trace an arc of cycloid CB. By construction, $CD = BD = \alpha r$. Using elementary methods, we can trisect the segment BD. Let BD_1 be the segment measuring one third of BD. Therefore, the length of BD_1 is also equal to one third of the arclength CD. In order to construct an arc measuring one third of CD, and thus solve the trisection problem, let us imagine that the circle with centre at O rolls on BE until the point of contact between the circle and line BE coincides with D_1. When this happens, the circle will have cut the cycloid at a point C_1. By construction, $C_1D_1 = BD_1 = \frac{1}{3}CD$. Hence the angle $\angle D_1O'C_1$ is one third of α

as Huygens' opinion in the *Horologium Oscillatorium* confirms. In this treatise, the cycloid is praised and considered: "very useful for investigations in astronomy and the art of navigation." (Huygens 1888–1950, vol. 18 p. 86). Huygens had in mind particularly the isochronism of the cycloid. This is the fact that an object sliding along a cycloidal slope takes the same amount of time to reach the lowest point, no matter where it started its descent. According to Huygens, this property, discussed in the *Horologium*, might be applied to the realization of accurate and stable clocks and, in this way, to solve related practical problems such as the determination of the longitude of a ship at sea.

Moreover, to Leibniz the cycloid had also a major importance in pure mathematics. Among the mathematical problems that are solved via this curve, in fact, there is the division of an angle into an equal number of parts. This result stems directly from the geometrical definition of the curve. Let us suppose a cycloid UGE given (as in Fig. 6.9) generated by a circle of unitary radius, and let us divide the arc GI into a number n of equal parts, for example $n = 3$. Then, Because the motion generating the cycloid also rectifies the rolling circle, the segment GI will rectify the corresponding circular arc DG. At this point, it is elementary to construct with ruler and compass a segment equal to $\frac{GI}{3}$. Using the cycloid, an arc equal to the segment $\frac{GI}{3}$ can be constructed too as shown in the figure, and the original problem is then solved.

The problem of angular division could seem a mere recreational problem today but, at least during his Parisian years, Leibniz attributed a crucial value to it. Indeed, together with the analogous problem of inserting an arbitrary number of integer or irrational mean proportionals between to given segments, which can be solved in its generality by using a logarithmic curve, the division of the angle structured the domain of transcendental geometry, namely the geometry beyond the limits set by Descartes:

> In Geometry, there are two subjects difficult to handle: the ratio and the angle, and the section of the angle stands on a par with the section of the ratio, or Logarithm. Hence the trisection of the angle is a solid problem, just like the trisection of the ratio; and the division of the angle or the ratio in five parts is a supersolid problem, and goes beyond the conic loci.[48]

In this context, the cycloid and the logarithmic curve have a symmetric role, because they are chosen by Leibniz in order to systematize the domain of problems which, by their nature, cannot be associated to any finite polynomial equation.

All in all, it is not surprising that Leibniz pondered over the geometrical nature of these lines, and specifically of the cycloid. As mentioned above, Descartes discussed rolling curves in his correspondence with Mersenne and Fermat, and rejected them as mechanical.[49]

According to Descartes' standard of exactness, in fact, the cycloid did not qualify for geometry. Firstly, its kinematical definition was not conducive to any construction in geometry, because it involved a coordination of two motions which is not feasible by the sole use of linkages. Moreover, even when the curve was described without involving kinematic elements, its construction depended on the possibility of comparing arc-lengths and segments, which is an operation certainly not allowed in Cartesian geometry.

The mechanical nature of the cycloid was also emphasised by the similarity between the genetic definition of the cycloid and the traditional description of the Archimedean spiral, raised in the mathematical literature of the time.[50]

---

[48]"Duo sunt in Geometria difficilia tractatu, ratio et angulus; ac sectio anguli pariter ac rationis sive Logarithmi. Anguli enim trisectio problema solidum est, prorsus ac trisectio rationis; et sectio anguli vel rationis in quinque partes problema est sursolidum, et ultra locum conicum excurrit" (AVII6, 51, p. 555).

[49]See Descartes (1897–1913), vol. 2, p. 313, and p. 406. See also Van Schooten's opinion: "De supra dicta linea AFE notandum, eam duobus motibus describi, inter se distinctis; recto nempe, quo circulus $ABCD$ defertur ab A ad E, et circulari, quo puncto in ejus circumferentia A (quod Trochoidem describit) rotatur circa centrum, dum movetur per lineam rectam ipsi AE aequalem & parallelam" Descartes (1659–1661), p. 264.

[50]For instance, we read in Lalouvère's treatise on the cycloid: "Maluimus enim per duorum lationem quam per unius tantum, huius genesim figurae explicare, non solum quia ita res planius

This survey shows that there were sound reasons to range the cycloid with mechanical curves with respect to the demarcation set in Cartesian geometry. In spite of this fact, the wealth of mathematical discoveries related to a single, non-geometrical object like the cycloid must have raised concerns: how could it be that so many geometrical results were proved about a curve that could not even be considered geometrical? These concerns were indeed tackled by Leibniz in his notes from 1674 and 1675.

Between October 1674 and the beginning of 1675, Leibniz also studied other kinds of rolling curves that he called "parabolic trochoids". This family of curves is described by a point on the axis of a given parabola when it rolls along one of its tangents (Fig. 6.10). Leibniz devoted at least five manuscripts to the study of this curve: AVII3, 38_{11}- 38_{12}, 38_{13}, Cc 827 and Cc 831 (now in AVII7, 44), and took over once more the study of the cycloid in January 1675 (cf. in particular, AVII5, 23, 26, and the manuscripts: AVII7, 44, 49_1, 49_2).

Just like the cycloid, the rolling (*provolutio* in Leibniz's terminology) of the curve can be defined kinematically, by specifying that it is composed of a parabola rotating around a chosen point on the axis, while it translates along the given tangent, both motions occurring with equal speed.

A non-kinematical description of the curve can be given too, although I could not find it in the sources. For this, let us mark, on the parabola in Fig. 6.10, a point F on the axis (we have chosen the focus, but any other point would do), a point P on the parabola, and let us trace the tangent PA to the parabola. Then we can define

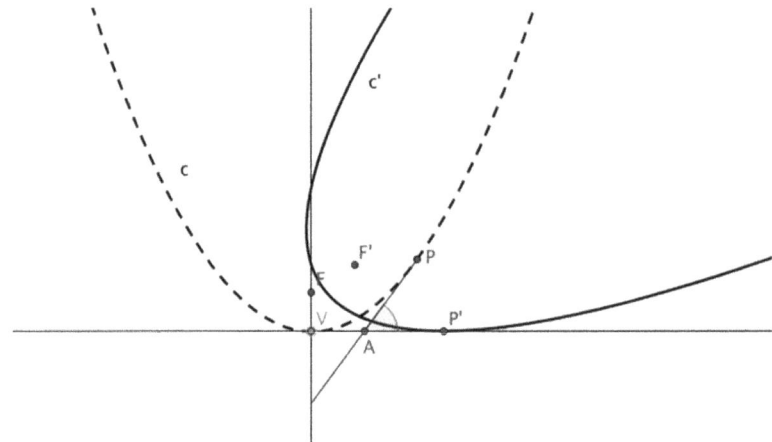

Fig. 6.10 The parabolic trochoid: $VP = VP'$

---

procedere nobis visa est; sed quia etiam Archimedes spiralis lineae generationem per motum rectae et per puncti in illa recta lationem tradidit" Lalouvere (1660, p. 3). It should be stressed that, unlike the case of the spiral, the motions which produce a cycloid stand in a simple ratio, since they are equal.

the parabolic trochoid as the curve traced by the point F as the parabola moves in such a way that, when its axis rotates clockwise around F of an angle equal to the slope of the tangent AP, the parabola has translated along the tangent to the vertex of a distance equal to the arc VP (Fig. 6.10).

Leibniz's interest in the description of the cycloid and the parabolic trochoid was primarily related to the applications of both curves in mathematical problem-solving (cf. AVII3, 38_{12}, p. 486) But in what did the usefulness of these curves consist? In the first instance, it is already implicit in their definition: since during the rolling motion the circle or the parabola are always tangent to the basis of their rolling at different points, their respective arcs are rectified along the basis.

The curve traced by a point on the axis of a parabola which rolls upon a straight line leads to similar applications. Manuscripts AVIII, 3, 38_{11}, 38_{12}, and 38_{13}, together with the tracts Cc 827 and Cc 831 (AVII7, 44) (both from December 1674), contain Leibniz's attempts to use this trochoid as a geometrical device to solve an important computational problem: the construction of logarithms.[51]

It must be recalled that the term "logarithm" does not characterize the same notion for us as it did for early modern geometers. According to a standard definition given in H. Briggs' *Arithmetica logarithmica* (1624), for example, logarithms were understood as: "numbers with constant differences matched with numbers in continued proportion".[52]

Logarithms appeared primarily in connection with the matching of arithmetical and geometrical progressions: the numbers in arithmetical progression were in fact called "logarithms" of the corresponding numbers in geometrical progression.[53]

As a consequence, logarithms were defined discretely and could be either represented by tabulating their values for each number or, using a geometrical description, by pairing segments in geometrical progression with segments in arithmetical progression in order to obtain a net of points belonging to a logarithmic (or exponential) curve. The net could be made more and more dense by subdividing the given intervals, but one could not provide any point on the logarithmic curve, unless one possessed a general method for computing the logarithm of any given

---

[51]"Trochoeidis parabolicae ope credo effici potest, ut qualibet data figura alia describatur quae sit in ratione logarithmorum ejus" (Cc 827), "Quemadmodum ergo constat per volutionem circuli Geometrice describi posse lineam quadratrici aequipollentem; ita ego reperi per parabolae provolutionem describi posse Curvam eosdem cum logarithmica usus habituram. Itaque Trochoeides circularis est figura Angulorum Geometrica, et Trochoeides parabolica est figura Rationum Geometrica. Illam Mersenno debemus, hac ego primus credo usus sum." (Cc 831). See also 38_{13}, p. 504.

[52]Burn (2001, p. 4). The same definition can be found in Pardies' *Eléments de Géometrie* (1671), read by Leibniz in Paris.

[53]The computational advantage introduced by logarithms consisted in replacing the operation of multiplication of two numbers with that of the addition of the corresponding logarithms in the table. This was, as Whiteside notes: "A cherished ideal when there were no automatic computing techniques at more than the most elementary level" (Whiteside 1967, p. 216).

quantity. Hence the logarithmic curve could only be described by successive approximations, as Leibniz was keen to stress in the manuscript we are examining.[54]

Thus, in Leibniz's intentions, the use of the parabolic trochoid as a geometrical method for constructing the logarithm of any desired quantity was to fully replace in usefulness the logarithmic curve, for which a construction via a continuous motion was still lacking at the time.[55]

With hindsight, we can recognise that the mathematical connection between the parabolic trochoid and the logarithms is the fact that the focus of a rolling parabola traces a catenary, namely the transcendental curve described by the shape of a hanging chain,[56] whose equation can be expressed, through an adequate choice of the coordinate system, as:

$$y = \frac{e^x + e^{-x}}{2}.$$

Therefore, if a catenary is given, the logarithms of as many quantities as desired can be easily constructed using only ruler and compass, and a logarithmic curve can be constructed geometrically too.[57]

In the text of 1674, Leibniz did not recognise that the focus of a rolling parabola described a catenary (in fact it is not known whether he ever made such a discovery. See Probst 2001, p. 1036–1037). Nevertheless, Leibniz was able to connect the parabolic trochoid to the logarithms thanks to two fundamental discoveries of the "Archimedean mathematics" from the middle of seventeenth century.

The first discovery concerns the equivalence between the problem of rectifying a parabolic arc and that of squaring a corresponding sector of a hyperbola. The rectification of the parabola had been a topic widely researched around the middle of seventeenth century, with its share of controversies.[58]

---

[54]Leibniz's knowledge of the logarithmic curve came from James Gregory's *Geometriae Pars Universalis* (especially the preface, where the logarithmic curve is mentioned), and from Gaston Pardies' treatise *Elemens de Geometrie*, published in 1671. Both references are made in the manuscript AVII3, 38_{12}, pp. 484ff., 492.

[55]Cf. for example: AVII3, 38_{11}, p. 481: "descripta trochoeide parabolae habebitur constructio logarithmorum geometrica et sectio rationis in data ratione"; and AVII3, 38_{12}, p. 485: "Ecce ergo lineam geometricam quae id omne praestat quod a logarithmica exacte si possibile esset, descripta posset expectari". The construction of a logarithmic curve by a continuous motion was attempted however in 1675: see Knobloch (2006), p. 118–119.

[56]For a mathematical proof, see: Texeira (1908–1915, p. 227); and the more recent: Argarwal (2010).

[57]Leibniz employed the catenary for this very purpose in a text published in the *Acta Eruditorum* in 1691 (*De linea in quam flexile se pondere proprio curvat, ejusque usu insigni ad inveniendas quotcumque medias proportionales et logarithmos*, in Leibniz (2011), p. 75), where he proved synthetically how this curve could be employed as a practical instrument to calculate logarithms. (see Bläsjo 2016; Raugh, Probst, 2019).

[58]See Yoder (1988), especially chapter 7.

According to the manuscripts in our possession, Leibniz first came to know about the rectification of the parabola through Huygens' account in his *Horologium Oscillatorium*.[59]

However, when discussing the parabolic trochoid, Leibniz had more likely in mind an analogous theorem suggested by Van Heurat's tract on the rectification of curves, appended to the second edition of Descartes' *Geometria* (Descartes 1659–1661, vol. I, p. 517. See Yoder 1988, p. 125–126; Panza 2005, p. 119–132.).

Van Heuraet's theorem, according to Leibniz's interpretation (for instance, in AVII5, 16, p. 138), can be stated as follows. Let a curve OGL be given, with axis OE, let GL be an infinitesimal part of the curve, and AG and BL two ordinates. GWB will be an infinitesimal triangle, in Leibniz's terminology, the "*triangulum characteristicum*" of the curve. With this curve we associate a companion curve $\lambda\gamma$, as in Fig. 6.11, on which we construct the characteristic triangle $\gamma\delta\lambda$. Moreover, the curve must be constructed so as to obey the following constraint: the rectangle bounded by $\lambda\gamma$ and by a given segment C (arbitrarily chosen) is equal to the rectangle with sides BL and $AB = GW$. Van Heuraet's theorem states that the area of the figure $OABLG$ is equal to the area of the rectangle whose sides are C and the arc $\lambda\gamma$ rectified.

Leibniz interpreted this theorem analytically, as we can read in 38_{12}, and proved that if the given curve is an equilateral hyperbola OGL, with equation: $y^2 = a^2 + x^2$, its companion curve is a parabola with equation: $y = \frac{x^2}{2a}$. Therefore, as a special case of Van Heuraet's theorem, the rectangle formed by the parabolic arc λO rectified and by a chosen segment $C = a$ is equal to the region $OABLG$ of an equilateral hyperbola of equation: $y^2 = a^2 + x^2$ (AVII3, 38_{12}, p. 491).

The second fundamental discovery involving the hyperbola and certainly known to the young Leibniz is the following: if points are marked, in geometric progression, along one asymptote of an equilateral hyperbola, and lines are drawn to these points parallel to the other asymptote, then the areas of the hyperbolic sectors (namely the sectors formed by the given asymptote, a portion of the curve and two successive parallels) are equal (Fig. 6.12). Therefore the sequence of the areas forms an arithmetic progression corresponding to the geometric progression formed by the sequence of their basis. As a consequence, the areas of these sectors are logarithms of the respective basis.

The relation between logarithms and hyperbolic areas appeared for the first time in the lengthy treatise *Opus Geometricum Quadraturae Circuli et Sectionum Coni* (1647) of Gregoire de St. Vincent, and was then developed by A. de Sarasa (in

---

[59]Huygens (1888–1950), vol. XVIII, p. 219–220. See AVII4, 2, for Leibniz's own annotations. In his *Horologium Oscillatorium*, Huygens enunciated this theorem in a synthetic form, omitting the proof. However, a proof of this statement can be given, on the basis of Huygens' own remarks in a manuscript note from October 1657, using analytic geometry (see Hofmann 1975, p. 106–107).

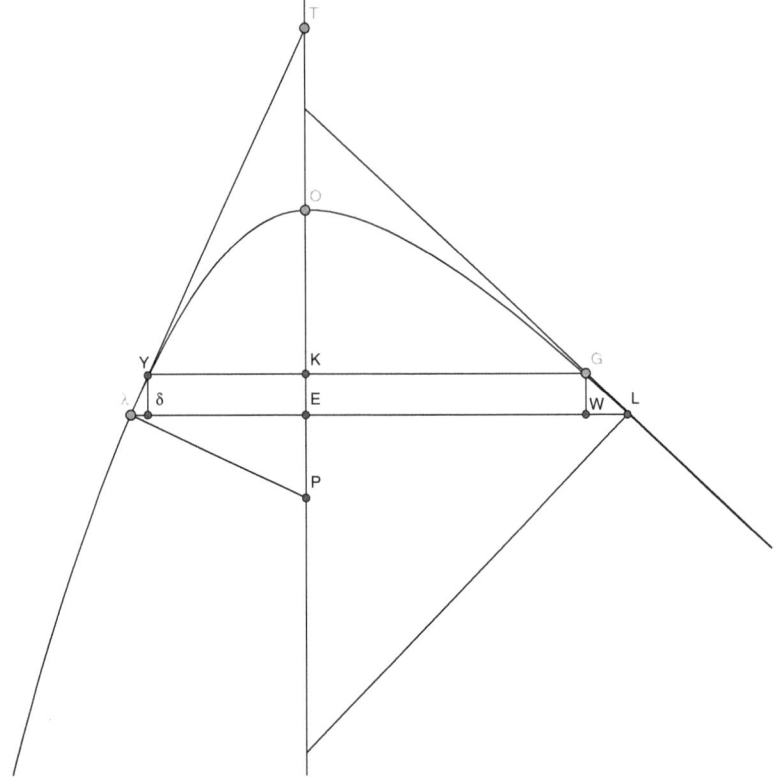

Fig. 6.11 Leibniz's analytical interpretation of Van Heuraet's theorem. Let $GW = 1$, $TE = x$, $EL = y$ and let another, arbitrary quantity a be given. The hyperbola OGL has equation: $y^2 = a^2 + x^2$. A companion curve $\lambda \gamma O$ is associated with the hyperbola, such that $\lambda \gamma . a = EL.GW$, $\lambda \gamma$ being an infinitesimal arc of the companion curve. If we translate this equivalence analytically, we shall have: $(\lambda \gamma)a = \sqrt{a^2 + x^2}$. Applying Pythagoras's theorem to the infinitesimal triangle $\gamma \delta \lambda$, there results: $\lambda \delta = \sqrt{\frac{a^2 + x^2}{a^2} - 1} = \frac{x}{a}$. Since $\lambda \delta$ is an infinitesimal portion of λE, the length of the whole segment λE will be: $\Sigma_{\lambda \delta} \frac{x}{a} = \frac{x^2}{2a}$. Hence, $\lambda E = v = \frac{x^2}{2a}$

the *Solutio problematis a R. P. Marino Mersenno Minimo propositi*, 1649) and, successively, by N. Mercator in his *Logarithmotechnia* (1667).[60]

Both results are beautifully mastered in Book III of Huygens' *Horologium*. In particular, Huygens suggested that the interlocking between the rectification of the parabola, the squaring of the hyperbola and the logarithms could be exploited to

---

[60]For Sarasa, see Burn (2001). Mercator's work, in particular, had a deep influence on Leibniz's early mathematical studies, as can be seen already from Leibniz's own notes to this work. In a forthcoming book on Leibniz's arithmetical quadrature of the circle I shall discuss at greater length the relations between Mercator's and Leibniz's techniques of integration based on infinite series.

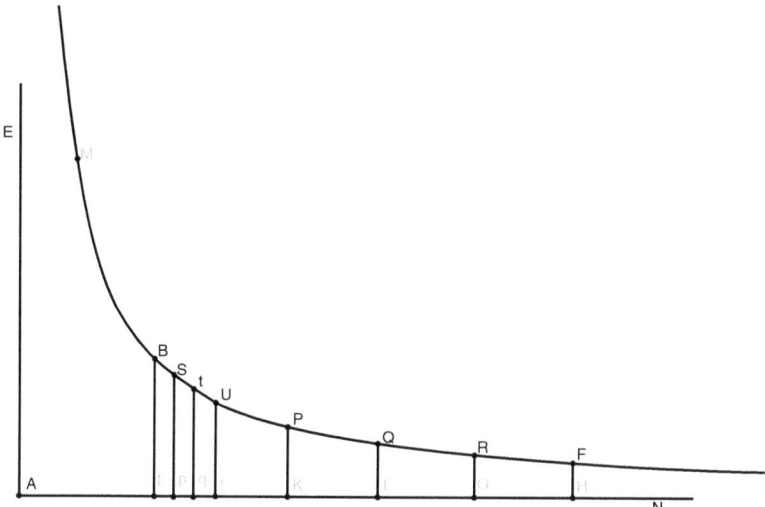

Fig. 6.12 In the equilateral hyperbola with centre A, asymptotes AE, AN, $AI = BI = 1$, the hyperbolic areas IU, IP, IQ, IR, IF form an arithmetical progression, while the base segments Ir, IK, IL, IO, IH form a geometric progression

solve the rectification of a parabolic arc numerically. Relying on the parabolic trochoid instead, Leibniz was possibly convinced that he could follow the inverse path to this and use the rectification property of the trochoid in order to give a geometric representations of logarithms, bypassing any references to areas.[61]

Probably with these results in mind, Leibniz set out to show, in the manuscript AVII3 38_{13} (in all likeliness written immediately after the 38_{12}) that the abscissas of any point of a parabolic trochoid depends on the arc length of a segment of the rolling parabola, hence the curve can be usefully employed for the geometrical construction of logarithms. A reconstruction of the original argument can be found in Probst (2001), on which I shall rely here.

Let us consider the following case (Fig. 6.13): a parabola with vertex L and focus A rolls without slipping on its horizontal tangent RL from left to right, until it touches it at a point F. By definition of the rolling motion, the segment LF is equal to the arc LF_1. At the end of its rolling, the parabola will have vertex in D, axis DB, and focus in B. The curve traced by the moving focus from A to B is an arc of parabolic trochoid.

---

[61] We can interpret in this sense the following passage: "San Vincentius revocavit logarithmos ad spatia hyperbolica (...) Heuratius hyperbolam ad curvam parabolicam; ego curvae sectiones in plano conservo ope trochoidis parabolicae. Ergo iunctis omnibus illis inventionibus inter se inventa est repraesentatio logarithmorum geometrica in plano, ope trochoeidis parabolicae." (AVII3, 39, p. 559).

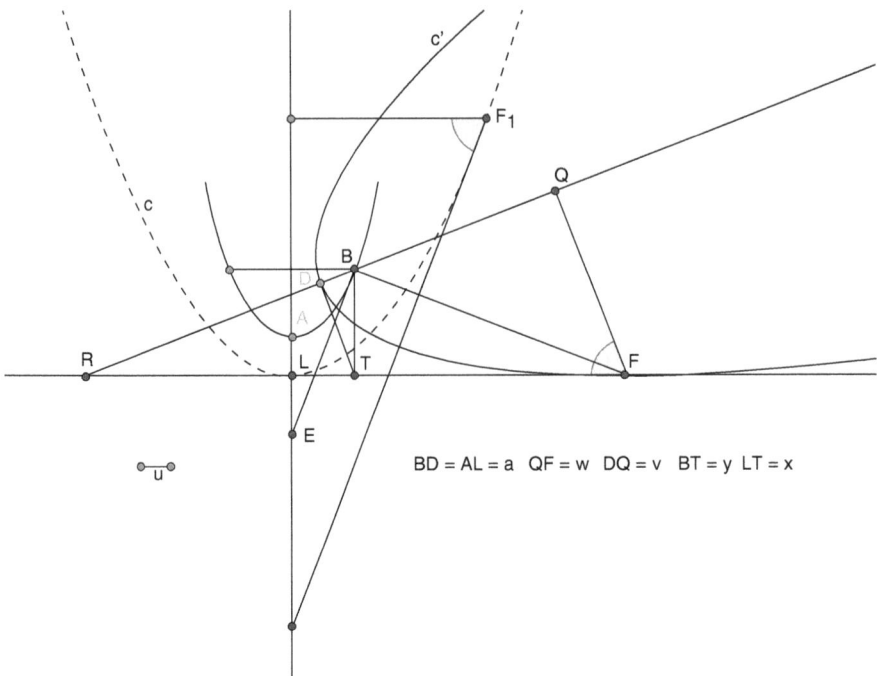

Fig. 6.13 The parabola c rolls to the right until it reaches the position of a parabola c', with vertex in D, focus B, and passing through point F. Let us trace the ordinate QF to the parabola c', and the tangent DT to its vertex D. Once an arbitrary unity u has been fixed, all the other segments can be named as shown in the figure. We want to calculate the abscissa $x = LT$ and the ordinate $y = BT$ of the point B, belonging to the parabolic trochoid AB. The parabola c' will have the Cartesian equation: $4aw = v^2$. Using the similarity between triangles RDT and BDT, we can write: $DB : DT = DT : RD$. Since RF is tangent to the parabola DF, $RD = DQ$. Moreover, DT is perpendicular to RQ, hence: $DT^2 = DB.RD$. Applying Pythagoras's theorem, we obtain: $DT^2 + DB^2 = BT^2$, hence: $y = \sqrt{av + a^2}$. The abscissa $LT = LF - TF = x$ depends on the rectification of a parabolic segment instead, since LF is equal, by construction, to the arc DF. TF can be easily calculated applying Pythagoras' theorem: $TF^2 = BF^2 - BT^2 = (v - a)^2 + 4av - av - a^2 = v^2 + av$. In short, we shall have (with s_{DF} indicating the arc length of the arc DF): $x = LF - TF = s_{DF} - \sqrt{v^2 + av}$. (Figure from Probst 2001)

Following Leibniz's manuscript 38_{13} (in particular, p. 498), we can determine the equation of the given trochoid by calculating the abscissa and the ordinate of a point B on the curve with respect to the axis RF and LA. With reference to Fig. 6.13, the ordinate $BT = y$ will be:

$$BT = y = \sqrt{av + a^2},$$

while the abscissa $LT = LF - TF = x$ depends on the rectification of a parabolic segment, since LF is equal, by construction, to the arc DF, or LF_1. A little algebra shows (with s_{DF} indicating the arc length of the arc DF) that:

Fig. 6.14 Construction of the
parabolic trochoid. From
AVII3, $38_1 1$, p. 482

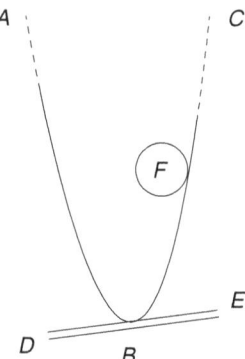

$$LF - TF = x = s_{DF} - \sqrt{v^2 + av}$$

The length of the arc DF is equal, on the basis of Van Heuraet's theorem discussed above, to the area of a corresponding sector of an equilateral hyperbola with equation: $w = \frac{\sqrt{4a^2 + v^2}}{2a}$.

Since the area of an hyperbolic sector is connected to the logarithmic function, Leibniz was able to finally prove the connection between the curve traced by the focus of a rolling parabola and the logarithms. To my knowledge, Leibniz did not provide, in the Parisian manuscripts, any description of a method to actually compute logarithms using the parabolic trochoid (such a method would be described later, in a series of paper on the catenary. See: Bläsjo 2016).

It was, nevertheless a necessary condition, in order to be able to use the cycloid and the parabolic trochoid in geometry, to give a construction of these curves complying with exactness. Leibniz must have faced the following alternative: either giving a kinematic construction, which involved the (mechanical) operation of controlling the speeds of motions, or a non-kinematic definition, which involved the rectification of the arcs. Whereas the first mode was simply not feasible in geometry, since it would violate the requirement of unicity of motion, the latter mode clearly involved a circularity: in order to construct rolling curves in this way, in fact, one would have already to possess a method to measure circular and parabolic arcs.

Probably finding himself caught in into this dilemma, Leibniz set out to study alternative constructions for both the parabolic trochoid and the cycloid in manuscripts 38_{11} and 38_{13}. For example, a detailed description of several apparata for the tracing of the parabolic trochoid can be found in manuscript AVII3, 38_{11} (p. 481–483), from October 1674.[62]

Lebniz began by illustrating the following device (Fig. 6.14) . Let a heavy metallic lamina be cut in the shape of a parabola, and let it be inserted between two

---

[62] See Knobloch (2006, p. 116). I could not find any other extant example in the literature, so we must conclude that Leibniz was the first to devise this construction.

joined rods that form a sort of track ("*duos asseres coniugatos*"). Let the apparatus thus formed be fixed upon a flat surface like a wall. Then let a heavy globe be placed inside the concave part of the lamina, so that it can roll downward along this concavity when the rods are inclined. The parabolic lamina will roll along the track in the direction determined by the inclination of the rods. If we then fix a pencil at a point on the axis of the parabolic shaped lamina, for instance at its focus, a segment of parabolic trochoid will be traced on the wall while the parabola rolls.

Such a mixture of physical and mathematical considerations, although surprising for a reader of today, is quite common in Leibniz's mathematical thought and, more generally, in the way mathematicians of the time approached the problem of describing new curves. We can think of other examples like the way Huygens introduced evolutes with a direct reference to the motion of strings (see the previous section), or such a well-studied case as that of the tractrix (Huygens devoted many pages to find out a physical way to eliminate lateral movement by physical means, and thus obtain a geometrical construction of the curve. An exhaustive study is in Bos 1988).

This being said, it seems to me that the function of the apparatus in Fig. 6.14 is related solely to the problem of finding a geometrically legitimate method for the construction of the trochoid, and not finding a more precise, although approximate description of the curve for the sake of its practical applications.

Let us observe, in the first instance, that this device would merely reproduce the motion of a parabolic "wheel", if it were not for the ball rolling along the parabolic lamina. I think that the ball has a simple, and in principle crucial role in the functioning of the device: if the rolling motion of the parabola and the rolling of the sphere inside it start at the same time, and we are somehow granted that all the speeds of the motions involved are equal, then F reaches B, the vertex of the parabola, when the parabola has covered, along the basis DE, a distance exactly equal to the arc BF. Therefore, the rolling ball could be a simple method for keeping track of the distance travelled by the parabola along the basis. In this way, Leibniz may have thought that the rectification property of the trochoid could be used without necessarily presupposing it for its generation.

Leibniz was however dissatisfied with this construction, since it still relied on two motions that must be set equal in order to have a perfect rolling. The construction was therefore too mechanical for him.[63] In manuscript 38_{11} Leibniz also toyed with the idea of changing the initial configuration. For example, he proposed that one could rotate and translate the plane (or table) underlying the parabola, leaving the latter fixed. The trochoid could therefore be traced by keeping the tracing pin at one chosen point (for instance, the focus) while the configuration moves.[64]

---

[63] AVII3, 38_{11}, p. 482: "Non nisi mechanica ope vectis moveri possit, sed tunc aliud malum. Quod scilicet saepe circa punctum fixum rotabitur, nec progredietur."

[64] "Eadem curva NB. describetur etiam si zona ponatur immobilis, sed DE circa ipsam moveantur nunc cum pariete ipsis asseribus DE affixo . . . " (ibid).

Letting the table move while the parabola remains fixed does not eliminate the two distinct motions, but this idea brought Leibniz to his final assessment of the problem.

In fact, in tract 38_{11} (p.483) and in tract 38_{13} (p. 504), Leibniz eventually proposed "an elegant remedy" (*remedium satis elegans*) in order to trace the parabolic trochoid as exactly as a Cartesian curve.[65]

Starting from the sketchy description found in the manuscripts (especially manuscript 38_{11}), I propose here the reconstruction of two devices for the tracing of the cycloid and the parabolic trochoid, which effectively lead to a construction of both curves by unique and continuous motion, and therefore to the exact solution of the rectification of their arcs. I argue that with these machines in mind, Leibniz conclusively asserted that the parabolic trochoid and the cycloid were geometrical curves.

Let us start from the parabolic trochoid. Leibniz had the fundamental idea of introducing, in the device of Fig. 6.14, a string connecting points D and B and, revolving around the parabolic arc BC, joining C to E as well. The new apparatus is represented in Fig. 6.15. The string is maintained in tension by two weights positioned at D and E. The apparatus for the parabolic trochoid becomes, modified in this way, a sort of lever, whose fulcrum is in B.

We can now imagine that the parabolic lamina (indicated by the letter c in Fig. 6.15) has a groove along which the lever DBE can move, maintaining itself always tangent to the parabola. Meanwhile, the parabola lies on a plane, which is free to translate and rotate, while the parabola is kept fixed. We must also assume

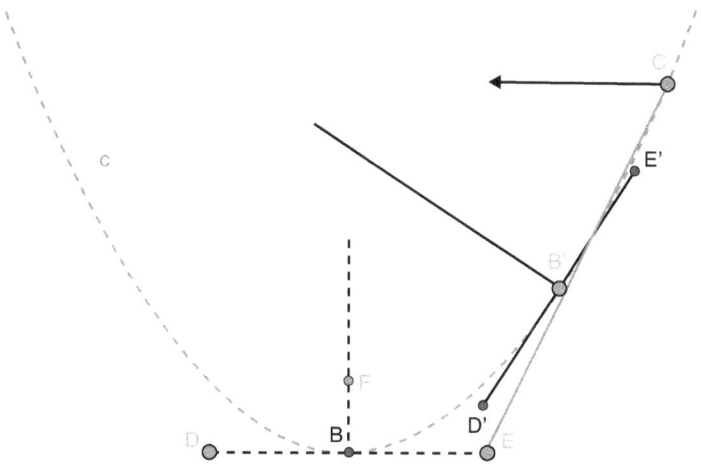

Fig. 6.15 Construction of the parabolic trochoid by strings

---

[65]"nam linea ista tam exacte describi poterit, quam parabola secundi generis a Cartesio proposita" (AVII 38_{12}, p. 484)

that the string revolving around DBE is tied to the underlying moving plane, so that the plane is translated by simply pulling the string from point C, and it is rotated while the lever moves along the parabolic lamina.

If we now imagine pulling the free end of the string attached to C in the direction indicated by the arrow in the figure, the string wrapping the parabolic lamina will drag the lever DBE to move along the parabola, in a counterclockwise sense. Let us imagine, for instance, that we pull the string until point B reaches B'. While DBE moves around the parabola, the whole plane under it rotates, counterclockwise, up to the point of forming an angle equal to the slope of $D'B'E'$, with respect to its initial position.

Meanwhile the string drags the underlying plane to translate from right to left for a distance equal to the length of the arc BB', namely the portion of string that has been "pulled" in order to bring B to B'.

As a result, the overall motion will be equivalent to a rotation of the parabola up to an angle equal an angle equal to the slope of the tangent $D'B'E'$, while it translates across a distance equal to the arc length BB': hence, according to the definition, the motion of the plane will be equivalent to the rolling of the parabola along the tangent to its vertex B. Thus, a parabolic trochoid can be described by inserting a fixed pen at the focus of the parabola. The pen will remain unmoved, the curve will be traced on the moving plane.

Leibniz suggested that a similar mechanism could be used for the cycloid as well, that can thus be traced exactly, leaving the tracing pin unmoved ("Eadem methodo et cycloeides optime describetur, paries vero vel tabula in qua designanda est curva ipsi BE affixa cum eo movebitur, stylus vero erit immobilis", 38[11], p. 483).

In order to obtain such a construction, it is sufficient to insert a circle tangent to DBE at the place of the parabola, around which a string must be wrapped (Fig. 6.16). When the string is pulled from point C, it also moves the underlying plane, imparting a rotation of an arc BB', and a translation equal to the arc length

Fig. 6.16 A construction of the cycloid by strings

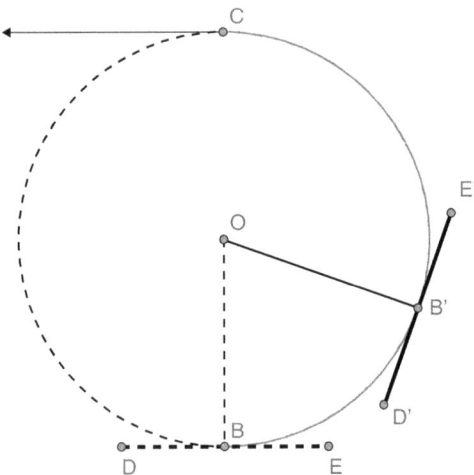

BB', so that a pencil fixed at B, which remains unmoved, will trace a cycloid according to its definition.

Hence, by introducing a construction via strings, Leibniz was able finally to attain a description of the parabolic trochoid through one continuous motion: the unicity is given by the mere dragging of one end of the string, which can be viewed as a principal motion, unaffected and unaltered by the speeds of the composing motions. Moreover, this construction does not presuppose the actual measurement of arcs, since the operation of rectification is exactly performed by the unwinding of the string, that moves the plane. Finally, let us note that, although Leibniz referred to his machines in terms of strings, weights and arms, and other concrete elements, no physical properties of the strings or of the components of these devices determine the output: the shape of the curve is fixed by the geometric properties of these configurations alone.

6.4   The Parabolic Trochoid and the Cycloid as Geometrical Curves

The geometrical nature of rolling curves is further asserted by Leibniz in the manuscript 38_{12} that immediately follows the construction of these curves expounded in the previous section:

> If everything which is exact is geometrical, it must be admitted that these constructions will be exact: indeed this line [a parabolic trochoid] can be described as exactly as the parabola of the second kind proposed by Descartes. In this it differs greatly from the spirals, which one cannot describe exactly, because of two motions, one circular and the other along the radius, whose proportion should be certain, but which cannot be given by our powers. To the point that these spirals may not be described but by divine art, by the aid of an intellect whose distinct thoughts (*cogitationes distinctae*) occur in a time less than any given time; which I suppose is not even true of angels. The contrary happens for our line [i.e. the parabolic trochoid]: this is described by a continuous motion, and it is not found point by point.[66]

A few lines below, Leibniz remarked that: "geometrical lines are those understood to be described by one *tractus*" (AVII3, 38_{12}, p. 486: "Geometricae sunt, quae uno tractu descriptae intelliguntur"), employing the same word, *tractus*, that would be later used to qualify the construction of Bertet's curve (see previous section).

---

[66]Si id omne geometricum est, quod exactum est, fatendum est has constructiones fore exactas: linam linea ista tam exacte describi poterit, quam parabola secundi generis a Cartesio proposita. In quo longe differt a spiralibus quas exacte non describas, ob duos motus, a se invicem independentes, alterum in circulo, alterum in radio, quorum certa debet proportio, quam dare non est in nostra potestate, adeo ut spirales illae non nisi divina arte describi possint, ope intelligentiae, cuius cogitationes distinctae fiant in tempore minore quolibet dato; quod nec de angelis verum puto. Quod secus est in linea nostra: Eadem linea continuo describitur motu; non per puncta invenitur" AVII3, 38_{12}, p. 484.

In the following month, Leibniz declared the geometrical character of the cycloid using a similar argument:

> I cannot see what what would cause one to refrain from calling the cycloid geometrical, since it can be described exactly, by one continuous motion and extremely simply.[67]

In both excerpts, the admission that "exactness implies geometricity" (...*id omne geometricum est, quod exactum est...*) and the characterization of exact constructions ("uno tractu" ..."cum uno motu continuo eoque admodum simplici") closely remind us of Descartes' standards of geometricity (see also Knobloch 2006, p. 115). However, Leibniz adhered only to the geometrical criterion based on the unicity and continuity of the tracing motion.

The same reasoning allows us to understand Leibniz's rejection from geometry of the Archimedean spiral, which stands as an explicit reference to the first few pages of book two of the *Géométrie*. In fact Leibniz clearly points out, in the aforementioned passage, that the main ground for considering the Archimedean spiral ungeometrical are that the curve is described by two motions, and that these motions required to be synchronized in a ratio unknown to us.[68]

By the end of 1674, Leibniz was also able to connect the study of this curve to his advances in the arithmetical quadrature of the circle. Indeed, since the geometrical description of the Archimedean spiral depended on the exact determination of the ratio between the unitary radius and the circumference (namely π), Leibniz could conclude that the latter problem depended, in its turn, on the exact computation of the area of the circle, which he had discovered to be equal to the infinite series:

$$1 - \frac{1}{3} + \frac{1}{5} - \frac{1}{7} \ldots = \sum_{n=0}^{n=\infty} \frac{-1^n}{2n+1}.$$

This result was certainly known to him by the autumn of 1673 (Probst 2006, p. 813). Therefore, whereas Descartes' considerations on the non-measurability of the ratio between the circumference and the diameter were possibly based on a traditional denial that curvilinear and rectilinear magnitudes were comparable at all, Leibniz was able to delve into the question by deploying more sound mathematical reasons.

---

[67] AVII5, 26, p. 202: "...Nam Cycloidem exempli gratia non video quid prohibeat appellari Geometricam, cum uno motu continuo eoque admodum simplici exacte describi possit ...". This manuscript dates from January 1675.

[68] For a similar discussion concerning the quadratrix of the ancients see the unpublished *De descriptionibus curvarum*, Cc2 831.

Even though his opinion on the squarability of the circle remained uncertain between 1674–76,[69] Leibniz certainly recognized both the technical and the conceptual issues involved in finding the sum of the above series.

The arithmetical representation of the series for the area of the circle made it possible to know this measure only through finite approximations (albeit to an infinite degree of accuracy) and only by a symbolical, non-intuitive knowledge. Leibniz shared the Cartesian ideal of the superiority of intuitive knowledge, the echo of which may possibly be heard in the above passage. An intuitive knowledge of the sum of the series was required in order to describe an Archimedean spiral in a geometrical way, but this would be possible only by imagining each term of the infinite series, and by grasping such a sum without the intermediary of symbolic representations. Leibniz believed that such a possibility was precluded for the human intellect.[70]

For these reasons, the description of the spiral, at least according to the methods known so far, could not stand on a par with the description of geometrical curves obtained by one continuous motion. At most, as Leibniz would declare in a text from 1676 (*Praefatio opusculi de quadratura circuli arithmetica*, AVII6, 19), a spiral may be traced approximately by the aid of a "material circle":

> This line [i.e. an Archimedean spiral] is not in our power ... nor it can be constructed by us, without a material circle, in such a way that the radius around a centre and the pin along the radius move with equal or proportional velocity.[71]

Although this construction by a "material circle" is not detailed here, Leibniz might have in mind simple mechanisms for the tracing of the Archimedean spiral that were well-known in the seventeenth century. An example is a device presented by Daniel Schwenter, in his *Geometriae Practicae novae libri IV* published in 1625 (see Crippa 2014, p. 232), made by a string wrapped around a pencil, which has been fixed at the centre of a disc. While the string is unwrapped by pulling one of its ends, the pencil rotates making the string rotate too while it is pulled.

---

[69] See AVII6, 11, p. 111. More optimistically, Huygens remarked, in November 1674: "Car le Cercle, suivant vostre invention estant a son quarré circonscrit comme la suite infinie de fractions $1 - \frac{1}{3} + \frac{1}{5} - \frac{1}{7} + \frac{1}{11}$ etc. à l'unité, il ne paroistra pas impossible de donner la somme de cette progression ni par consequent la quadrature du cercle, apres que vous aurez fait voir que vous avez determinè les sommes de plusieurs autres progressions qui semblent de mesme nature. Mais quand mesme l'impossibilitè seroit insurmontable dans celle dont il s'agit, vous ne laisserez pas d'avoir trouvè une proprietè du cercle tres remarquable, ce qui sera celebre a jamais parmi les geometres.", in Huygens (1888–1950), vol. 7, p. 393–394. Leibniz's subsequent research on the nature of π, has been investigated in Arthur (1999), who argues that by 1676 Leibniz defended the position that the number π was not algebraic (i.e. a rational or a surd number).

[70] See also the interesting tract *Demonstratio propositionum primarum*, AVI, 2, p. 481: "Quemadmodum enim nemo computare posset, praesertim numeros ingentes, sine nominibus vel signis numerabilis, loco numeri enim deberet sibi distincte imaginari omnes in eo comprehensas unitates".

[71] AVII6, 19, p. 170–171: Talis autem linea non est in potestate, neque enim (sine circulo materiali) effici hactenus a nobis potest, ut aequali aut proportionali velocitate moveantur semper radius circa centrum et stylus in radio.

In such a construction for the Archimedean spiral, the rectilinear and circular motions are not eliminated but physically connected via a thread or string. In spite of this fact, the precise drawing of an Archimedean spiral depends on the ability of the constructor, who must pay attention that both motions are uniform and coordinated. Moreover, not knowing the exact ratio in which the motions have to be set (i.e., not knowing the exact value of π) implied that any constructions performed *in concreto* could only approximate the curve, and not offer its exact description.[72]

In another interesting passage from the manuscript 38_{12}, Leibniz referred to "mechanical curves" as those traced by a "guiding intelligence" (*intelligentia directrice*, 38_{12}, p. 486). We can compare this idea with the construction of the spiral given above, which shows how the more attentive the work of the constructor, the more accurate will be the resulting curve. But such idea of the aim of mechanics as the search for precision in the concrete world is already evoked in Book II of Descartes' *Géométrie*, where mechanics is opposed to geometry as a science that searches for a different canon of exactness ("iustesse"), namely the exactness of manual creations ("la iustesse des ouurages qui sortent de la main") as opposed to the exactness of reasoning that is pursued in geometry.[73] Hence curves traced by approximate constructions belong properly to mechanics, whereas geometry is concerned with the exactness of theoretical constructions, to be achieved, for Descartes as well as for Leibniz, by a single and continuous motion.

If the distinction between geometrical and mechanical curves offered in the *Géométrie* captured well the nature of the Archimedean spiral, it failed, in Leibniz's opinion, on account of the nature of rolling curves. The reason is now clear in the light of the constructions by string illustrated in the previous sections.

By introducing a string, in fact, Leibniz was able to generate rolling curves in a geometrical way, that is, by a single and continuous motion analogous to the motion employed for the construction of evolutes. In the light of these constructions, the inclusion of the Archimedean spiral, the cycloid and the parabolic trochoid under the same class was seen by Leibniz as a categorical error, on a par with the ancient error of including the conchoid in the same class as that of the spiral and the quadratrix:

> I confess that if we must report the definition of geometry given to us by Descartes, ours will not be such. In the same way he [Descartes] correctly accuses the ancients, who excluded

---

[72]Leibniz was arguably unaware, while writing these notes, of another machine for the construction of the spiral that is to be found among Huygens' notes (Cf. Huygens 1888–1950, vol. 11, p. 216; Bos 2001, p. 348). This machine involved indeed a couple of cylindrical wooden disks, one above the other, on the top of which a ruler was also placed. A tracing pin was contrived to move along a ruler while the latter rotates. In this way, the machine reproduces the genetical description of an Archimedean spiral, but it effectively works by imparting a single motion, and can be so conceived that the output does not depend on the knowledge of the ratio between the speeds of the component motions (see Panza 2011, p. 82). It is possible that Leibniz could consider the Archimedean spiral as geometrical, had he known this construction. As we shall see below, in his later papers he indeed ranged the spiral among geometrical curves. According to Bos (quoted above), Descartes might have rejected this construction as mechanical, since it involved the bending of a straight line into a curve. However there is no strong textual evidence that Descartes knew this machine either.

[73]Descartes (1952), p. 43. For the original, see Descartes (1897–1913), vol. 6, p. 389.

the conic sections, or certainly those loci called "linear", from the number of geometrical lines; so he must be blamed in return, because, having restrained the name of "geometrical" to the analytic ones, he deprived science of a necessary aid. I think he had hidden reasons, so that he could boast of having found the method and the tangents of every geometrical curve. I wish he could show in what respect the description of the evolute of the circle, or of our trochoid is not exact. Indeed, it is certain that they are as exactly described as any geometric curve equally composite. He reproaches the ancients with having – even though they were aware of the imperfection of the helices and the spirals – nevertheless mixed these up with the conchoids and the cissoids. Thus I reproach him, because he confused the trochoids and evolutes with helices and spirals. He objects to the lines that he calls non-geometrical, saying that they depend on two different motions, whose proportion cannot be exactly maintained. This is indeed correct for the helices and the spirals: but this is not correctly predicated for the trochoids described according to our manner and for the evolutes.[74]

It should be noted that Leibniz stressed the geometrical nature of the trochoids *described by his method*. Later on, in a manuscript from January 1675 he would give the following general definition of rolling curves:

> I call 'trochoid' any line described by the continual application (*applicatione continua*) of the path of a rigid curve on another line (and I am not here discussing the motion of a wheel moving on a plane or surface).[75]

These words become clear if we consider that Leibniz was envisioning the generation of the trochoids using strings discovered at the end of 1674 and suggested in the manuscript 38_{11}. The rolling wheel ("motu rotae ....excurrente") is not invoked in such a construction, as is stressed in the passage above. On the contrary, the description occurs via a "continual application" of a curve (a circle, or a parabola) on a straight line: with the expression "*application continua*", Leibniz was presumably referring to the unwinding of a string that allows one to construct a trochoid by a single and continuous motion, according to the protocol I have examined in the previous section.

---

[74]"Fateor si ferenda est definitio geometricarum quam dedit nobis Cartesius, nostra talis non erit: Sed quemadmodum ille veteres iure culpat, quod a geometricarum numero conicas, aut certe quos vocabant lineares locos, exclussissent; ita; ille rursus culpandus est, quod geometricarum nomine ad analyticas coarctato; scientiam auxilio necessario privat; causam credo habens unicam in arcanis, ut scilicet iactare posset, omnium curvarum geometricarum a se methodum tangentesque traditas. Sane quo ille argumento utitur in veteres, eo ego in illum: quicquid exactum, id inquit geometricum est. Recte. Velim ergo ostendat, in quo descriptio evolutae circularis, vel trochoeidum nostrarum exacta non sit. Sane tam exacte descriptas esse certum est quam ullam geometricam aeque compositam. Culpat veteres, quod agnita helicum et spiralium imperfectione conchoeides quoque et cissoeides cum ipsis confusas exclussissent. Ita ego illum, quod cum helicibus et spiralibus, trochoeides et evolutas confudit. Obicit lineis quas non geometricas vocat, quod a duobus diversis motibus dependeant, quorum proportionem exacte servare non liceat. Recte illud quidem in helices vel spirales: at non recte dicitur in trochoeides nostra methodo descriptas et evolutas" (AVII 38_{12}, p. 485).

[75]AVII5, 23, p. 192: "Trochoeidem voco quaecunque applicatione continua lineae cujusdam curvae rigidiae ad aliam lineam (: neque enim hic de motu rotae per planum superficiemve excurrente loquor :), describitur".

However, the admission of the "continual application" of a curve onto a straight line constituted a violation of another constraint set in Descartes' geometry, as Leibniz was well aware. As clearly stated in the passage from 38_{12} his definition of geometry differed from the Cartesian one, despite some similarities. The crucial difference seems to lie, in the light of the manuscripts analyzed in this work, in the different use of strings in the two geometries.

A major consequence, which is already evident from the example of the involutes, is that constructibility by one continuous motion may not be represented analytically, via finite algebraic equations. The cycloid is a case in point of this phenomenon:

> I call those figures 'analytical', in which the relation of the ordinate to the abscissa can be explained by an equation (...) I prefer to call Analytical those figures which others, after Descartes, have called Geometrical. I don't see, for example, what forbids one to call the cycloid geometrical, since it can be described exactly by one continual motion; it is moreover extremely simple, and has wonderful properties. But I deny it is Analytical, because the relation between ordinates and abscissas cannot be explained by any equation.[76]

Meanwhile, together with the distinction between "geometrical" and "analytical" curves, Leibniz had also introduced the term "transcendental" to denote such geometrical, non-analytical curves.[77]

The impossibility of expressing by finite polynomial equations the loci of these rolling curves could be derived as an immediate consequence of the non-algebraic rectifiability of their corresponding arcs. As shown in Fig. 6.17 (AVII5, 23, p. 193), Leibniz argued that the abscissa of an arbitrary point (A) on the cycloid $A(A)$, generated by the rolling of the circumference AGF, depends on the length of the arc GF of the generating circle, and concluded, on the basis of the impossibility of the algebraic rectification of a circular arc, that the relations between the abscissas and the ordinates of a cycloid cannot be expressed by a finite polynomial equation (AVII5, 26, p. 202ff.). The case of the parabolic trochoid is analogous with respect to the quadrature of the hyperbola, as has been discussed above.

A proof that neither the rectification of a circular nor that of a parabolic arc (or of a corresponding hyperbolic sector) can be effected by algebraic methods was given by Leibniz in his treatise on the arithmetical quadrature of the circle, completed in 1676 (AVII6, n. 51, prop. LI). Although both impossibility proofs were given later than the period we are considering here, it can plausibly be argued that Leibniz had thought of them already between the end of 1674 and the beginning of 1675 (Cf., for instance: AVII3, 39, p. 572; see also AVII6, 8).

---

[76]"Figuras Analyticas appello, in quibus relatio ordinatae ad abscissam aequatione explicari potest (...) Figuras malim vocare analyticas, quas alii post Cartesium Geometricas....", AVII5, 26, p. 202. It should be pointed out that by "figure" (*figura*) Leibniz here means a curve, not the area delimited by a curve (Knobloch 2015, p. 91).

[77]The cycloid is explicitly considered by Leibniz as a transcendental curve, for instance in: AVII3, 38(12), p. 492; AVII3, 39, p. 555; AVII6, p. 566.

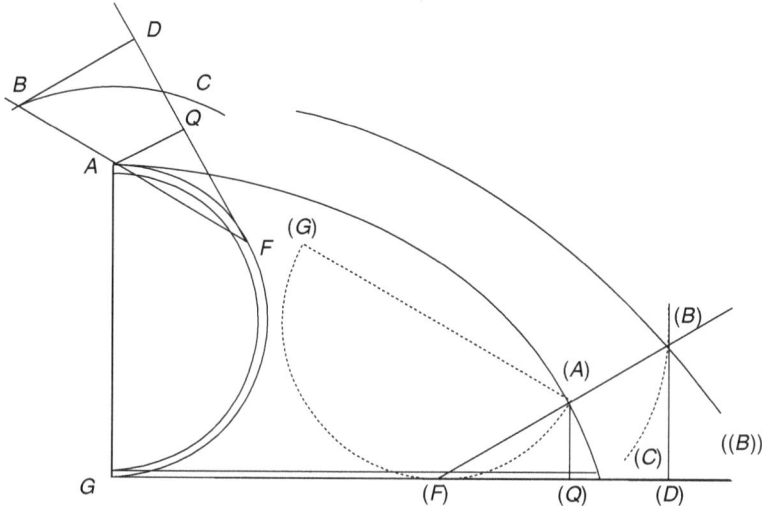

Fig. 6.17 Figure from: AVII5, 23, p. 193. The cycloid studied in Leibniz's manuscripts

6.5 Conclusions

The research presented in this article endorses the thesis, already advanced in other studies (see, in particular, Knobloch 2006; Probst 2012) that the young Leibniz challenged Descartes' ideal of exactness in a veritable struggle to redesign the limits of geometry. In this endeavour, Leibniz was motivated by two fundamental convictions: firstly, he believed that Descartes' method of analysis was unjustly restricted to "rectilinear problems". Secondly, he objected that in Descartes' geometry exactness had been unjustly restricted to motions implemented by geometrical linkages, and geometrical curves identified with analytic or algebraic ones.

Leibniz argued that legitimate curves are those that can be constructed by one continuous motion, regardless of changes in speed. This was his early ideal of exactness and geometrical rigour, rooted in the constructional criteria explained in Descartes' *Géométrie*. On the basis of this criterion, Leibniz identified three classes of geometrical, yet non-algebraic (transcendental) curves: involutes (and in particular the involute of the circle), the curves of Bertet, and the trochoids. With these examples in mind, he went on to refute, starting from a draft to Oldenburg from March 1675, the Cartesian equivalence between geometricity and analyticity. In particular, in order to defend the geometrical nature of trochoids, Leibniz had to overtly challenge the Cartesian opinion that these were truly mechanical curves on a par with the Archimedean spiral.

Leibniz was able successfully to show that the above mentioned curves had a right to stand among curves constructible by one continuous motion, provided the use of strings that can bend, superpose to arcs and return to their original shape (keeping their length invariant) was admitted in geometry. Postulating such an operation was an explicit violation of an injunction set by Descartes in his *Géométrie*: the use of strings is therefore the true moment of rupture between Leibniz's and Descartes' conceptions of exactness in geometrical constructions. Likewise, the use of strings that can bend and adapt to curves, can be taken as the *proprium* of synthetic constructions in Archimedean geometry.[78]

It should also be pointed out that between 1674 and 1676 Leibniz still countenanced certain curves as mechanical, like the Archimedean spiral and the quadratrix of the ancients, perhaps because the only known methods for their construction violated the one-motion criterion. But as E. Knobloch aptly noted: "For Leibniz the boundary between geometric and nongeometric lines is not fixed once and for all. It can happen that a nongeometrical line becomes geometrical when a way of describing it is found (for example, the logarithmic curve) and that a nonanalytical line becomes analytical (...) In other words, Descartes adhered to a mathematically fixed, closed, static realm of geometry, while Leibniz adhered to a mathematically changing, open, dynamical realm of geometry in which the classification of lines depends on our current knowledge" (Knobloch 2006, p. 118).

This consideration matches quite well the textual evidence at our disposal. In fact, in a letter to Tschinrhaus written in 1678, i.e. few years later than the time period we are considering in this article, Leibniz returned to the theme of the classification of curves:

I have already spoken to you once about my method, through which one can describe geometrically the logarithmic and other transcendental curves, like the quadratrix and the other that seemed mechanical to Descartes, since he did not know that they could be described by means of certain rules, through a continuous motion dependent on a unique one. I deem this geometrical description of the transcendental lines to count among the highest of my discoveries. Indeed it really broadens the walls of Geometry immensely. Actually, as Descartes shows that curves of higher degrees must be received in Geometry, because they can be described by one tracing (*tractus*), through the sole motion of rulers depending on a unique movement (...) so I shall show that transcendental curves (...) can be described by a wholly similar rule of motion, namely only by movable rulers that are constrained one to another according to a precise rule.[79]

It seems that Leibniz was here expanding into a research that could "broaden the walls of Geometry immensely" the very project expounded 3 years before to

---

[78] This point may also contribute to explaining the central role of rectification problems with respect to quadrature problems in the early calculus (On this point, see Bläsjo 2012). Introducing strings that can adapt to curves and then become straight again is tantamount to introducing a geometrical operation of rectification.

[79] AIII, 2, 171, p. 427: "Jam olim tibi de methodo mea qua Geometrice describi potest logarithmica aliaeque lineae transcendentes, ut quadratrix et aliae quae Cartesio Mechanicae videntur, quia eas per regulas quasdam motu continuo ab uno pendente describi posse nesciebat. Haec descriptio linearum transcendentium Geometrica inter potissima mea inventa habeo. Vere enim Geometriae pomoeria in immensum amplificat. Ut enim Cartesius ostendit curvas altiorum graduum in Geometria recipiendas quia uno tractu per solarum regularum motum ab uno pendente describi possunt (...) ita ego ostendam curvas transcendentes (...) posse describi simili plane motus ratione, solis regulis mobilis sese certa ratione ducentibus."

Oldenburg. In fact, he still anchored the standard of exactness in the criterion of unicity and continuity of the motion. Let us note, in particular, that Leibniz explicitly mentioned the quadratrix and other lines "that seemed mechanical to Descartes", which must include the Archimedean spiral as well. We can thus claim that, by 1678, Leibniz further broadened the territory of geometry, adhering to the same view on geometrical exactness.

The same concept of geometrical exactness in terms of constructions by a single and continuous motion reappeared years later, in the *De geometria recondita et analysi indivisibilium et infinitorum* an article first published in the *Acta Eruditorum* in June 1686 (Cf. Leibniz 1989, p. 127, 2011, p. 49). We read there:

> It is at least indispensable to admit in geometry the curves that can only allow their construction [namely, the solution of determinate, transcendental problems like the quadrature of the circle]. Now, these curves can be exactly constructed by one continuous motion, as is well shown by the Cycloid and other similar figures, so that we must not judge them Mechanical, but Geometrical, especially because the resources they offer leave several miles behind them the curves of ordinary Geometry, with the exception of the circle and the straight line, and they possess very extraordinary properties that directly contain geometrical results. This is the reason why the error committed by Descartes in excluding them from Geometry was as serious as the one committed by the Ancients, who rejected as ungeometrical certain solid or linear loci.[80]

The similarity between this text and the manuscript notes from 1674 and 1675 is remarkable: the cycloid is judged geometrical because it is a useful curve and because it is generated by a continuous motion, hence Descartes fell into the same blunder as the ancients by excluding this curve from geometry. Of course, although there is no mention of the parabolic trochoid in the 1686 article, exactly the same argument would hold for it too.[81]

---

[80]"...ideo necesse utique est, eas quoque lineas recipi in Geometriam, per quales solas construi possunt; et cum eae exacte continuoque motu describi possint, ut de cycloide et similibus patet, revera censendas esse non Mechanicas sed Geometricas; praesertim cum utilitate sua lineas communis Geometriae (si rectam circulumque exceperis) multis parasangis post se relinquant, et maximi momenti proprietates habeant, quae prorsus Geometricarum demonstrationum sunt capaces. Non minor ergo Cartesii Geometria eas excludentis, quam veterum lapsus fuit, qui loca solida aut linearia tanquam minus Geometrica rejiciebant." (Leibniz 2011, p. 51, 1989; Knobloch 2015, p. 95). Similar consideration on the nature of the cycloid appear, a few years later, in a letter to Huygens: "Lorsqu'on demande si cette construction est Geometrique [Leibniz refers to a device for tractional motion just illustrated] il faut convenir de la definition. Selon mon langage je dirois qu'elle l'est. Aussi crois je que la description de la cycloide ou de vos lignes faites par l'evolution, est Geometrique. Et je ne vois pas, pourquoy on restreint les lignes Geometriques à celles dont l'equation est Algebrique" (AIII5, 191, p. 648).

[81]Note that Leibniz did not employ, in the aforementioned text, the syntagm *uno continuo motu*, but simply *continuo motu*. However, this difference seems to be rather a matter of emphasis than of substance: in both cases, Leibniz conceived the cycloid described in terms of one continuous motion instead of two, unlike Descartes.

In the *Supplementum geometriae dimensoriae* ("A supplement to mensural geometry") from September 1693.[82] Leibniz discussed the nature of the cycloid, the spiral and the quadratrix, and held these curves to be geometrical because they "have very useful properties, and are applicable to transcendental quantities" (Knobloch 2006, p. 119).

He also added, concerning these curves:

> In order to construct transcendental quantities an application or adaptation of curves to straight lines has hitherto been used, as occurs in the description of the cycloid or in the case of the unwinding of a thread, or leaf, tied up with a line or surface (...) Should someone wish to describe geometrically the spiral of Archimedes or the quadratrix of the ancients, that is, by means of an exact, continuous motion, he can easily do that by a certain adaptation of a straight line to a curve, such that the rectilinear motion is adapted to the circular.[83]

It is tempting, and indeed not unreasonable, to read behind the geometrical description of the cycloid by means of a thread (or string) a reference to the construction of the cycloid sketched in the manuscripts from 1674. Moreover, the above passage suggests, in line with the letter to Tschirnhaus from 1678, that Leibniz had devised in the meantime a construction of other "mechanical" curves, obtained through the application, or adaptation of curves to straight lines, hence through the use of strings.[84]

The key word here is indeed the term: "*applicatio*", a term employed in 1675, as we have commented in the previous section, for a construction of the cycloid which does not involve wheels. Our analysis has revealed that such a construction could be effected by the unwinding of a string around a circle: hence the "applicatio" could likely refer to the operation of rectifying a curve by superposing a string upon it.

---

[82]The whole title reads: "Supplementum geometriae dimensoriae seu generalissima omnium tetragonismorum effectio per motum: similiterque multiplex constructio lineae ex data tangentium conditione", e.g.: "A supplement to mensural geometry, or the most general solution of all quadratures by motion: and similarly the numerous constructions of a curve from a given condition of its tangents", in Leibniz (2011), p. 133. For a French translation, see Leibniz (1989). Important studies have been devoted to this paper. See for instance: Bos (1988), Tournes (2009, chap. 1), and the recent Bläsjo (2015).

[83]"Ita ad construendas quantitates Transcendentes hactenus adhibita est applicatio, seu admensuratio curvarum ad rectas, uti fit in descriptione cycloeidis, aut evolutione fili vel folii lineae vel superficiei circumligati (...) Quin & si quis spiralem Archimedis, aut quadratricem veterum Geometrice (hoc est motu continuo exacto) describere velit, hoc facile praestabit quadam rectae ad curvam admensuratione, ut motus rectus circulari attemperetur" (Leibniz 2011, p. 134). English translation in Knobloch (2006), p. 119.

[84]Such constructions have not been found, so far. One hypothesis is that Leibniz was thinking of an apparatus similar to the one illustrated by Huygens, mentioned above (Huygens 1888–1950, vol. 11, p. 216; Bos 2001, p. 348). Another promising place to explore is a manuscript from 1692 in which Leibniz discussed the Archimedean spiral: LBr 79 Bl. 150. This discussion probably occurred within an exchange with Bodenhausen in 1691–1692 (see AIII5, 64).

The importance of this mathematical operation is also emphasized by the fact that Leibniz compared, in the *Supplementum geometriae dimensoriae*, the role of the application or adaptation of curves to straight lines for the construction of transcendental quantities to the use of linkages for the construction of algebraic quantities. Thus, just like the construction of curves by linkages constituted the synthetic part of the Cartesian analysis based on finitary objects (polynomial equations), the construction of curves by strings constituted the synthetic part of the Leibnizian analysis based on infinitary objects.

In the same article, Leibniz developed the idea of using strings in geometry through the introduction of tractional motion (generalized to other curves than the tractrix), and through a sketchy description of an "integraph", a device for solving any quadrature problem by means of tractional motion (this instrument made it possible to solve inverse-tangent problems, i.e. differential equations, geometrically). By means of tractional motion, and through the integraph which constituted its implementation, Leibniz was also highly confident that geometry could also be brought to completion. However, this topic will take us well beyond the boundaries set at the beginning of this article, and requires further study of published and unpublished manuscripts.[85]

For the time being, even acknowledging that the exploration of Leibniz's concepts of exactness and geometricity requires more textual study, we can claim that the fundamental insight into geometrical exactness that Leibniz acquired during his Paris period shaped his subsequent conception of what should count as geometry. The fundamental criterion that determined the geometrical nature of a curve was its constructibility via a single and continuous motion. This criterion could be realized in various ways: through the ruler and the compass (Euclidean geometry), or through geometric linkages (Cartesian geometry), or through the use of strings applied to curves, for instance in the case of other transcendental curves.

Leibniz inaugurated in Paris a programme intended to give a geometrical foundation to his infinitesimal analysis by extending the limits of Euclidean and Cartesian geometrical constructions, while maintaining the fundamental ideal of exactness that underscored these latter.[86]

This foundational endeavour was not an easy nor an obvious one. It was probably a struggle that occupied almost 20 years of Leibniz's mathematical career. In this article, I confined myself to describing its beginnings.

---

[85]On this, see Bos (1988), Tournes (2009), and Bläsjo (2015). For a mathematical study of tractional motion, see: Milici (2015).

[86]This conclusion seems to agree with the conclusion in Bläsjo (2015). According to Bläsjo, this motive also guided Leibniz's effort in his 1693 article: "In this way he [Leibniz] enlarged the domain of constructible curves vastly beyond the algebraic curves admitted by Descartes, while still adhering very strictly to Descartes's requirement of single-motion tracing and to the Euclidean-Cartesian construction framework generally" (Bläsjo 2015, p. 49).

Bibliography

Acerbi, Fabio. 2007. *Tutte le opere: testo greco a fronte*. Milano: Bompiani.

Arana, Andrew. 2016. Imagination in mathematics. In *The Routledge Handbook of Philosophy of Imagination*, ed. Amy Kind, 463–477. Abingdon: Routledge/Taylor & Francis Group.

Archimedes. 1881. *Archimedis Opera Omnia cum commentariis Eutociis*, ed. Johan Ludvig Heiberg. Leibniz: Teubner.

Argarwal, Anurag, and James Marengo. 2010. The locus of the focus of a rolling parabola. *College Mathematics Journal* 41(2): 129–133.

Arthur, Richard. 1999. The transcendentality of π (pi) and Leibniz's philosophy of mathematics. *Proceedings of the Canadian Society for History and Philosophy of Mathematics* 12: 13–19.

Baron, Margareth. *The Origins of Infinitesimal Calculus*. New York: Dover.

Beeley, Philip. 2008. *De Abstracto et Concreto*: Rationalism and empirical science in Leibniz. In *Leibniz: What Kind of Rationalist?* ed. Marcelo Dascal, 85–98. New York/Heidelberg/Berlin: Springer.

Belaval, Yvon. 1968. *Leibniz critique de Descartes*. Paris: Gallimard.

Bläsjo, Viktor. 2012. The rectification of quadratures as a central foundational problem for the early Leibnizian calculus. *Historia Mathematica* 39: 405–431.

Bläsjo, Viktor. 2015. The myth of Leibniz's proof of the fundamental theorem of calculus. *Nieuw Archief voor Wiskunde, 5* 16(1): 46–50.

Bläsjo, Viktor. 2016. How to find the logarithm of any number using nothing but a piece of string. *The College Mathematics Journal* 47(2): 95–100.

Bos, Henk. 1974. Differentials, higher-order differentials and the derivative in the Leibnizian calculus. *Archive for the History of Exact Sciences* 14: 1–90.

Bos, Henk. 1980. Huygens and mathematics. In *Studies on Christiaan Huygens: Invited Papers from the Symposium on the Life and Work of Christiaan Huygens*, ed. Henk Bos et al., 126–146. Lisse: Swets and Zeitlinger.

Bos, Henk. 1981. On the representation of curves in Descartes' *Géométrie*. *Archive for the History of Exact Sciences* 24: 295–338.

Bos, Henk. 1988. Tractional motion and the legitimation of transcendental curves. *Centaurus* 31: 9–32.

Bos, Henk. 2001. *Redefining Geometrical Exactness: Descartes' Transformation of the Early Modern Concept of Construction*. New York/Berlin/Hidelberg: Springer.

Boutroux, Pierre. 1920. *L'idéal scientifique des mathématiques*, Paris: Alcan.

Breger, Herbert. 1986. Leibniz' Einführung des Transzendenten. *Studia Leibnitiana* 14: 119–132.

Breger, Herbert. 2008. The art of mathematical rationality. In *Leibniz: What Kind of Rationalist?* ed. Marcelo Dascal, 141–152. New York/Heidelberg/Berlin: Springer.

Burn, Robert P. 2001. Alphonse Antonio de Sarasa and Logarithms. *Historia Mathematica* 28(1): 1–17.

Child, James M., trans. 1920. *The Early Mathematical Manuscripts of Leibniz*. Chicago: Open Court.

Crippa, Davide. 2014. *Impossibility Results: From Geometry to Analysis*. Université Paris Diderot (Paris 7): Doctoral Dissertation.

Descartes, René. 1659–1661. *Renati Descartes Geometria Editio Secunda*, 2 vols, ed. Frans van Schooten et al. Amsterdam: Apud Ludovicum et Danielem Elzevirios.

Descartes, René. 1897–1913. *Oeuvres de Descartes*, 12 vols, ed. Charles Adam and Paul Tannery. Paris: Cerf.

Detlefsen, Michael. 2005. Formalism. In *The Oxford Handbook of Philosophy of Mathematics and Logic*. Oxford: Oxford University Press.

Fabri, Honoré. *Opusculum Geometricum de Linea Sinuum et Cycloeidis*. Roma: Corbelletti.

Giacardi, Livia. 1995. Newton e Leibniz e il teorema fondamentale del calcolo integrale. Aspetti geometrici e aspetti algoritmici. In *Geometria, Flussioni e Differenziali*, ed. Marco Panza and Clara S. Roero, 289–328. Napoli: Istituto Italiano per gli Studi Filosofici di Napoli.

Harari, Orna. 2008. Proclus' account of explanatory demonstrations in mathematics and its context. *Archiv für Geschichte der Philosophie* 90(2): 137–164.

Heinekamp, Albert, ed. 1986. *300 Jahre Nova Methodus von G. W. Leibniz (1684–1984): Symposion der Leibniz-Gesellschaft im Congresscentrum "Leewenhorst" in Nordwijkerhout ...August 1984.*. Studia Leibnitiana. Sonderhefte: Stuttgart.

Hofmann, Joseph. 1975. *Leibniz in Paris (1672–1676). His growth to mathematical maturity.* Trans. A. Prag and D.T. Whiteside. Cambridge: Cambridge University Press.

Huygens, Christiaan. 1888–1950. *Oeuvres complètes publiées par la Société hollandaise des sciences*, vol. 22, ed. B. de Haan. The Hague: M. Nijhoff.

Itard, Jean. 1975. La lettre de Roberval à Torricelli d'Octobre 1643. *Revue d'histoire des sciences pures et appliquées* 28(2): 113–124.

Itard, Jean. 1984. *Essais d'histoire des mathématiques*. Paris: Blanchard.

Knobloch, Eberhardt. 2006. Beyond cartesian limits: Leibniz's passage from algebraic to 'Transcendental' mathematics. *Historia Mathematica* 33: 113–131.

Knobloch, Eberhardt. 2015. Analyticité, equipollence et théorie des courses chez Leibniz. In *G. W. Leibniz, Interrelations Between Mathematics and Philosophy*, ed. Norma Goethe, Philip Beeley, and David Rabouin, 89–110. Dordrecht/Heidelberg/New York/London: Springer.

Laloubere, Antoine de. 1660. *Veterum Geometria promota in septem de Cycloide Libris*. Toulouse: Colmiers.

Leibniz, G. Wilhelm. 1923–. *Leibniz: Sämtliche Schriften und Briefe: Akademie Ausgabe*. Reihe I–VIII. Akademie der Wissenschaften zu Göttingen and Berlin-Brandenburgische Akademie der Wissenschaften: Akademie-Verlag.

Leibniz, G. Wilhelm. 1989. *La Naissance du Calcul Différentiel. Introduction, Traductions et Notes par Marc Parmentier*, ed. Marc Parmeniter. Paris: Vrin.

Leibniz, G. Wilhelm. 2011. *Gottfried Wilhelm Leibniz: Die mathematischen Zeitschriftenartikel*, trans and ed. H.-Jürgen Hess and Malte-Ludof babin. Hildesheim: Georg Olms Verlag.

Mahnke, Dietrich. 1926. Neue Einblicke in die Entdeckungsgeschichte der höheren Analysis. *Abhandlungen der Preussischen Akademie der Wissenschaften, Physikalisch-Mathematische Klasse* 1: 1–64.

Mancosu, Paolo, 1999. *Philosophy of Mathematics and Mathematical Practice in XVII Century*. Oxford: Oxford University Press.

Mancosu, Paolo. 2007. Cartesian mathematics. In *A Companion to Descartes*, ed. Janet Broughton and John Carrero, 103–123. New York: Wiley-Blackwell.

Mancosu, Paolo, and Arana Andrew. Descartes and the cylindrical helix. *Historia Mathematica* 37(3): 403–427.

Mersenne, Marin. 1636. *L'harmonie universelle, contenant la théorie et la pratique de la musique*. Paris: chez Sebastien Cramoisy.

Mersenne, Marin. 1933–1988. *La correspondance du P. Mersenne religieux minime*, 18 vols, ed. C. de Waard. Paris: G. Beauchesne; then Presses universitaires de France; then Ed. du Centre national de la recherche scientifique.

Milici, Pietro. 2015. A geometrical constructive approach to infinitesimal analysis: Epistemological potential and boundaries of tractional motion. In *From Logic to Practice: Italian Studies in the Philosophy of Mathematics*, ed. Gabriele Lolli, Marco Panza, and Giorgio Venturi, 3–21. Dordrecht/Heidelberg/New York/London: Springer.

Molland, A. George. 1976. Shifting the foundations: Descartes' transformation of ancient geometry. *Historia Mathematica* 3: 21–49.

Panza, Marco. 2005. *Newton et les origines de l'analyse*. Paris: Blanchard.

Panza, Marco. 2011. Rethinking geometrical exactness. *Historia Mathematica* 38: 42–95.

Pascal, Blaise. 1913. *Les grands écrivains de France. Oeuvres de Blaise Pascal.*, vol. VIII, ed. L. Brunschvicg, P. Boutroux, and M. Gazier. Paris: Hachette.

Pappus. 1876–1878. *Collectionis quae supersunt*, ed. F. Hultsch. Berlin: Weidman.

Probst, Siegmund. 2001. Die Trochoide des Brennpunkts der Parabel. Leibniz' unbewusste erste Konstruktion der Kettenlinie. In *Nihil sine Ratione. Mensch, Natur und Technik im Wirken von*

G. W. Leibniz, vol. 3, ed. Poser Hans, Asmuth Christoph, Goldenbaum Ursula, and Li Wenchao, 1035–1038. Berlin: Leibniz Gesellschaft.

Probst, Siegmund. 2006. Zur Datierung von Leibniz' Entdeckung der Kreisreihe. In *Einheit in der Vielheit. VIII. Internationaler Leibniz-Kongress*, ed. Breger Herbert, Hernst Jürgen, and Erdner Sven. 813–817. Hannover: Gottfried-Wilhelm-Leibniz-Gesellschaft.

Probst, Siegmund. 2012. Leibniz und die Cartesische Geometrie (1673–1676). In *Zeitläufe der Mathematik*, ed. H. Fischer and S. Deschauer, 149–158. Augsburg: Dr. Erwin Rauner Verlag.

Rabouin, David. 2009. *Mathesis Universalis. L'idée de "mathématique universelle" d'Aristote à Descartes*. Paris: PUF.

Raugh, Mike, Siegmund Probst. 2019. The Leibniz catenary and approximation of e - An analysis of his unpublished calculations. *Historia Mathematica*. https://doi.org/10.1016/j.hm.2019.06.001. Available online 25 June 2019.

Roero, Clara S. 1995. Sul retaggio della tradizione geometrica nel calcolo infinitesimale Leibniziano. In *Geometria, Flussioni e Differenziali*, ed. Marco Panza and Clara S. Roero, 353–395. Napoli: Istituto Italiano per gli Studi Filosofici di Napoli.

Smith, David E., and Marcia Latham trans. 1952. *The Geometry of René Descartes*. New York: Dover.

Texeira, F. Gomez. 1908–1915. *Traité des courbes spéciales remarquables planes et gauches. tr. de l'Espagnol*. Coimbre: Imprimerie de l'université.

Thiele, Rüdiger. 2009. Über das Wirken Leonhard Eulers als Wissensvermittler. *Berichte und Abhandlungen der Berlin-Brandenburgische Akademie der Wissenschaften*. https://edoc.bbaw.de/frontdoor/index/index/docId/17. Accessed 16 May 2016.

Tournès, Dominique. 2009. *La Construction Tractionnelle des Equations Différentielles*. Paris: Albert Blanchard.

Vuillemin, Jules. 1962. *La Philosophie de l'Algèbre*, Paris: PUF.

Whiteside, Derek, T. 1961. Patterns of mathematical thought in later XVII century mathematics. *Archive for the History of Exact Sciences* 1: 179–388.

Whitman, E.A. 1943. Some historical notes on the cycloid. *The American Mathematical Monthly, 50* 5: 309–315.

Yoder, Joella. 1988. *Unrolling Time: Christiaan Huygens and the Mathematization of Nature*. Cambridge: Cambridge University Press.

Chapter 7
Leibniz and the Calculus of Variations

Jürgen Jost

Abstract The brachistochrone problem of Johann Bernoulli is considered as the origin of the calculus of variations. The solutions presented by Johann and Jacob Bernoulli and by Newton and Leibniz were all different and highly original. Leibniz' solution has received less attention than those of the Bernoullis, but I show here that his abstract idea was also general and powerful enough for a general theory, although the history of mathematics took a different path. In fact, his approach quite naturally emerges from his earlier treatment of the refraction of light by his then new calculus, i.e., his derivation of Fermat's principle. I then analyze the development of his conceptions about the speed of light from that treatment through his work on the brachistochrone problem to his Nouveaux essais from 1706. From the work of the Bernoullis and Leibniz on variational problems, also an analogy between mechanical and optical problems emerged, and this naturally leads to Leibniz' considerations on the physical concept of action and extremal principles. In contrast to later formulations of such a principle by Maupertuis and Euler, Leibniz devoted much effort to deriving more abstract principles based on considerations of symmetry and determination, as analyzed in De Risi (Geometry and monadology: Leibniz's analysis situs and philosophy of space. Birkhäuser, Basel/Boston, 2007). Some of his corresponding ideas look surprisingly modern, for instance in the light of Feynman's path integral approach to quantum mechanics. Leibniz' ideas are put into the perspective of modern science in Jost (Leibniz und die moderne Naturwissenschaft. Springer, 2019).

J. Jost (✉)
Max Planck Institute for Mathematics in the Sciences, Leipzig, Germany
e-mail: jost@mis.mpg.de

© Springer Nature Switzerland AG 2019
V. De Risi (eds.), *Leibniz and the Structure of Sciences*, Boston Studies in
the Philosophy and History of Science 337,
https://doi.org/10.1007/978-3-030-25572-5_7

253

7.1 Introduction

This essay is decidedly ahistorical. It examines some ideas of Leibniz from the perspective of modern mathematics and physics. Today, we can see connections between various aspects of Leibniz' thinking that were not evident in his days, and the question is to what extent he did perceive, imagine, suspect, surmise or divine them. I am not in a position to answer this question. I can only provide certain glimpses. But I believe that the question is worth to be further pursued. Of course, I shall also touch several more concrete questions that have been discussed and analyzed extensively in the literature on Leibniz. In some cases, the answers are still controversial. Again, I cannot answer them, but only provide a different perspective.

   Although the contribution of Leibniz towards the calculus of variations analyzed in this paper is concerned with a problem, that of the brachistochrone, that had been posed as a mechanical one, there are obvious connections to geometric optics, both in the contemporary discussion and in Leibniz' own writings. Leibniz' theory of light, and, in particular, its relation with the contemporary research of Huygens and Rømer seems not to have been studied extensively. The basic reference Sabra (1981) about theories of light in the seventeenth century concentrates on other actors and says very little about Leibniz. McDonough (2009) discusses Leibniz' treatment of the laws of optics in the light of his teleological principles, and, more generally, his ideas about the calculus of variations and the principle of least action can also be seen from the perspective of that principle. As we shall see, some enigmas remain.

   I am much indebted to Vincenzo de Risi for several discussions, insightful comments and crucial references. I also thank a referee for his helpful comments.

7.2 The Calculus of Variations

Proceeding ahistorically, I shall start with the contemporary theory. The calculus of variations is a well established and active field of modern mathematics, see for instance Butazzzo et al. (1998), Jost and Li-Jost (1998), and for an advanced treatment with a historical perspective, Giaquinta and Hildebrandt (1996). Within that theory, solving a variational problem consists of three steps:

1. Set up the problem in integral form, that is, as the integral over a Lagrangian containing the unknown function.
2. Derive the Euler-Lagrange equations.
3. Solve these equations.

The difficulty of step 1 varies with the problem at hand. In some sense, it is more a question of mathematical modelling than of mathematics itself. In particular, most important variational problems arose from physical applications, rather than from mathematical considerations. Step 2, in contrast, is straightforward. This is the main achievement of Euler and Lagrange. For the story to be presented below, however, this was the most difficult part, and at the same that of most retrospective interest. The protagonists of the story, the competing brothers Johann and Jacob Bernoulli, Leibniz and Newton, developed ideas that were seminal for the later development of the theory when they grappled with step 2 for the concrete problem of the brachistochrone. Step 3, in contrast, for most variational problems is the most difficult one. In fact, often the Euler-Lagrange equations cannot be solved explicitly. Therefore, Hilbert and many other mathematicians of the twentieth century developed more abstract methods that guaranteed the existence of a solution, without identifying it explicitly. The approximate computation of such a solution then became one of the key challenges of numerical analysis. Curiously, however, for the brachistochrone problem that we shall discuss, step 3 was the easiest one, because the protagonists knew its solution, the cycloid, already. In fact, the cycloid played a crucial role in Huygens' work on pendulum clocks. Huygens had found that when a pendulum swings along a cycloid, its period of oscillation remains constant, independently of its amplitude, see for instance Bell (1947). Therefore, the cycloid was also known as the isochrone.

Let us consider some examples (see Jost and Li-Jost 1998):

- We want to minimize the arc length of the curve $(x(t), y(t)) : [a, b] \to \mathbb{R}^2$ in the plane among all curves with prescribed boundary values at $t = a$ and $t = b$. This leads to the variational problem

$$\int_a^b \sqrt{\left(\frac{dx}{dt}\right)^2 + \left(\frac{dy}{dt}\right)^2}\, dt \to \min. \qquad (7.1)$$

The solution is the straight line between the prescribed boundary points, which can be parametrized so as to satisfy the differential equation $x''(t) \equiv y''(t) \equiv 0$ where we use the abbreviations $x'(t) = \frac{dx}{dt}$ and $x''(t) = \frac{d^2x}{dt^2}$ (which, of course, are not Leibnizian). The problem can be generalized for finding shortest curves on surfaces. For convex surfaces, this problem was already studied by Leibniz and Johann Bernoulli. It was solved in general form by Euler (1728–1732/1952) and Clairaut (1733).

- A more general problem is

$$\int_a^b \frac{\sqrt{\left(\frac{dx}{dt}\right)^2 + \left(\frac{dy}{dt}\right)^2}}{\gamma(x(t), y(t))}\, dt \to \min, \qquad (7.2)$$

where $\gamma : \mathbb{R}^2 \to \mathbb{R}$ is a given positive function. This variational problem arises from Fermat's principle that a light ray chooses the path of shortest time among all possible paths. If the speed of light in a given medium is $\gamma(x, y)$, we obtain the variational problem (7.2).

- Historically, the calculus of variations started with a problem that formally is a special case of (7.2), although the original interpretation was different. This is the so-called brachistochrone problem posed by Johann Bernoulli and which we shall discuss below in more historical detail. In fact, Johann Bernoulli realized the connection of the brachistochrone problem with Fermat's principle (see for instance the discussion in Schramm (1984)). The problems asks to connect two points (x_a, y_a) and (x_b, y_b) in $\mathbb{R}^2$ by such a curve that a particle obeying Galileo's law of gravity and moving without friction travels the distance between those points in the shortest possible time. Here, x denotes the height, that is, the vertical distance, whereas y is the horizontal distance. After falling the height x, the particle has speed $(2gx)^{\frac{1}{2}}$ where g is the constant of gravitational acceleration. The time the particle needs to traverse the path $(x(t), y(t))$ then is

$$\int_a^b \sqrt{\frac{\left(\frac{dx}{dt}\right)^2 + \left(\frac{dy}{dt}\right)^2}{2gx(t)}} dt \to \min. \tag{7.3}$$

In these examples, the parametrization by $t \in \mathbb{R}$ is arbitrary, and, if the curves are monotonic, we can choose either x or y as the parameter. When we choose the height x as the parameter in (7.3), this becomes

$$I(y) = \int_a^b \sqrt{\frac{1 + \left(\frac{dy}{dx}\right)^2}{2gx}} dx \to \min. \tag{7.4}$$

We use Euler's abbreviation $p = \frac{dy}{dx}$ and $F_p = \frac{\partial F(x,y,p)}{\partial p}$, with $F(x, y, p) = \sqrt{\frac{1+p^2}{2gx}}$ for the integrand of the variational problem (7.4). As will be explained in a moment, a solution has to satisfy the Euler-Lagrange equation

$$\frac{d}{dx} F_p = \frac{d}{dx} \frac{1}{\sqrt{2gx}} \frac{p}{\sqrt{1 + p^2}} = 0. \tag{7.5}$$

That is, F_p has to be constant. Denoting that constant by c, the solution is

$$p = \sqrt{\frac{2cgx}{1 - 2cgx}}, \tag{7.6}$$

which defines a cycloid, the curve generated by a point fixed on a circle when the latter rolls along a line.

The last example already exhibited equations, (7.5) and (7.6), that a minimum of the variational problem should satisfy. How are these equations, corresponding to step 2 in the above scheme, obtained? Here, we shall not describe the derivation by Lagrange (1760–1761/1894) that now is universally employed (see Carathéodory 1935/82; Butazzzo et al. 1998; Jost and Li-Jost 1998) nor the original derivation of these equations for general variational problems by Euler (1744/1952), but rather show how Leibniz' method for the brachistochrone problem can be used for their derivation.[1]

---

[1] In Gerhardt's edition Leibniz (1855/1971) of Leibniz' mathematical works, Leibniz' derivation of the solution is given as a "Beilage" (pp. 290–295) to a letter written to Johann Bernoulli (letter XXIX, pp. 284–290), which, as Gerhardt explains on p. 117 of Leibniz (1855/1971), was found among Leibniz' manuscripts.

We consider a variational problem of the form

$$\int_a^b F(x, y(x), \frac{dy}{dx}(x))dx \rightarrow \min \tag{7.7}$$

among all curves $y(x)$ satisfying suitable boundary conditions $y(a) = y_a$, $y(b) = y_b$. An easy, but fundamental insight is that if a specific curve $y^*(x)$ is optimal, then so is any portion of it. That is, whenever $a \leq x_0 < x_1 \leq b$, then the curve $y^*(x)$ restricted to the interval $x_0 \leq x \leq x_1$ minimizes

$$\int_{x_0}^{x_1} F(x, y(x), \frac{dy}{dx}(x))dx \rightarrow \min. \tag{7.8}$$

Thus, global optimality implies local optimality (this is known as Euler's Lemma), and the rules of Leibniz' calculus can then turn that into a condition for infinitesimal optimality.[2]

In abstract form, Leibniz' idea (Leibniz 1855/1971, pp. 290–295, Beilage for XXIX) then is the following. Given two points (x_0, y_0), (x_1, y_1) on the solution curve, with $x_1 - x_0 =: 2\varepsilon$, take some $x_h = hx_0 + (1-h)x_1$ for $0 < h < 1$ between x_0 and x_1 and seek the value y_h such that the point (x_h, y_h) minimizes

$$F(x_0, y_0, \frac{y_h - y_0}{2h\varepsilon}) + F(x_h, y_h, \frac{y_1 - y_h}{2(1-h)\varepsilon}). \tag{7.9}$$

See the following diagram

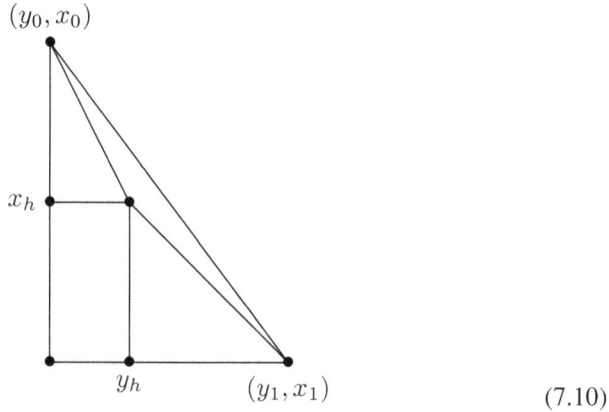

$$\tag{7.10}$$

[2]Leibniz expresses the transition to the limit as follows:"Quod si jam concipiamus viam polygonam facillimam ita continuari, ut constet ex angulis numero infinitis, qui incident in horizontales infinitesime distantes seu vicinissimas, habebimus Lineam facillimi descensus" (Leibniz 1855/1971, p. 292), that is, the brachistochrone, or as Leibniz calls it, the tachystoptote.

Note that, as in Leibniz' scheme, the independent variable x here is the vertical coordinate, and the dependent y the horizontal one.

Thus, we try to approximate a piece of the solution curve not just coarsely by the straight line between by (x_0, y_0) and (x_h, y_h), but in a more refined manner by the broken arc from (x_0, y_0) to (x_1, y_1) via (x_h, y_h) and seek the best value of y_h for such an approximation, in the sense that the approximated action be minimized.

The computations to follow are not Leibniz'. Their purpose is to explore his ansatz in general form and to show that this easily implies the Euler-Lagrange equation. Leibniz himself treated the problem of the brachistochrone, but then pointed out that his method also applies to other variational problems, like that of the catenary.

Differentiating (7.9) w.r.t. y_h leads to the necessary condition

$$\frac{1}{2h\varepsilon}F_p(x_0, y_0, \frac{y_h - y_0}{2h\varepsilon}) - \frac{1}{2(1-h)\varepsilon}F_p(x_h, y_h, \frac{y_1 - y_h}{2(1-h)\varepsilon}) + F_y(x_h, y_h, \frac{y_1 - y_h}{2(1-h)\varepsilon}) = 0.$$
(7.11)

We put $h = \frac{1}{2}$, that is, $x_1 - x_h = x_h - x_0 = \varepsilon$.

In order to handle the difference of derivatives w.r.t. p, we approximate for small ε (see for instance Jost (2005) for the calculus background)

$$\frac{1}{\varepsilon}F_p(x_0, y_0, \frac{y_h - y_0}{\varepsilon}) - \frac{1}{\varepsilon}F_p(x_h, y_h, \frac{y_1 - y_h}{\varepsilon})$$

$$\approx F_{pp}(x_0, y_0, \frac{dy}{dx}(x_0))\frac{y_h - y_0}{\varepsilon^2} + F_{pp}(x_h, y_h, \frac{dy}{dx}(x_h))\frac{y_h - y_1}{\varepsilon^2}$$

$$- F_{py}(x_0, y_0, \frac{dy}{dx}(x_0))\frac{y_h - y_0}{\varepsilon} - F_{px}(x_0, y_0, \frac{dy}{dx}(x_0))$$

$$\approx -F_{pp}(x_0, y_0, \frac{dy}{dx}(x_0))\frac{d^2 y}{dx^2}(x_0) - F_{py}(x_0, y_0, \frac{dy}{dx}(x_0))\frac{dy}{dx}(x_0) - F_{px}(x_0, y_0, \frac{dy}{dx}(x_0)).$$
(7.12)

Inserting this into (7.11) and letting $\varepsilon \to 0$ leads to the necessary condition for a minimizer $y(x)$ of the variational integral (7.7)

$$F_{pp}(x_0, y_0, \frac{dy}{dx}(x_0))\frac{d^2 y}{dx^2}(x_0) + F_{py}(x_0, y_0, \frac{dy}{dx}(x_0))\frac{dy}{dx}(x_0) + F_{px}(x_0, y_0, \frac{dy}{dx}(x_0))$$

$$= F_y(x_0, y_0, \frac{dy}{dx}(x_0)).$$
(7.13)

In more condensed form, with an obvious notation, this can also be written as

$$F_{pp}y'' + F_{py}y' + F_{px} = F_y.$$
(7.14)

This is the famous Euler-Lagrange equation. As already mentioned, it was first derived in general form by Euler (1744/1952) by a method different from that presented here; Euler's method can be seen as a generalization of that used by Jacob Bernoulli for the brachistochrone problem. Later, Lagrange (1760–1761/1894) found another derivation which was then subsequently adopted, because it is more elegant.

Thus, we have shown that Leibniz' idea naturally leads to a derivation of the Euler-Lagrange equation, alternative to that of Jacob Bernoulli, and almost 50 years before it was derived in general form by Euler.

Of course, in the problem (7.4) at hand, $F_y = 0$, and therefore, the computations of Leibniz are simpler, and as in (7.5), (7.6), we arrive at

$$F_p \equiv \text{const.} \tag{7.15}$$

So much for the abstract setting. Before turning to the circumstances inducing Leibniz to develop the idea just presented, we need to recall some aspects of the discussion of the propagation of light in the seventeenth century.

7.3 Light

The competing and evolving theories of light in the seventeenth century are treated in Sabra (1981). For our purposes, only certain details are relevant.

Fermat invoked a principle of least time to derive Descartes' law for the refraction of light at the interface between two optical media of different densities (Fermat 1891–1922). The history is somewhat curious. First, it was the general opinion in Leibniz' time that the law had not been discovered by Descartes, but rather by Snellius, from whom Descartes had learned it. That was pointed out by C.Huygens, and Leibniz of course knows that (see e.g. Leibniz (1682/1985) from 1682, where he refers to Spleiss as the origin of the hypothesis that Descartes had learned the law of refraction from Snell, or Leibniz (1880/2008), p. 318 or p. 448). Moreover, Descartes assumed that light travels faster in a denser medium. Actually, at certain points, Descartes had assumed that light travels with infinite velocity. The speed of light will remain important when the role of Leibniz will be discussed, but here, I shall not enter into Descartes' physical theories. Fermat, in contrast, assumed that light travels more slowly in a denser medium. Nevertheless, from his principle that light follows the path of least time between two points, he could derive the law of Snellius-Descartes that had been based on the opposite assumption about the relative speed of light in different media. Finally, the teleological principle on which Fermat had based his reasoning was vehemently rejected by the Cartesians, and Fermat could not prevail in the ensueing discussion.

One of the first applications of Leibniz' calculus was his derivation of the refraction law from Fermat's principle of least time, see Leibniz (1682/1985).[3] In fact, curiously, he did not invoke a principle of least time, but rather one of least resistance. In the introduction of Leibniz (1682/1985), he states "*Lumen a puncto radiante ad punctum illustrandum pervenit via omnium facillima*" (emphasis in the original), and further down, he specifies this by "Jam in diversis mediis viae difficultates sunt in composita ratione & longitudinis viarum, & resistentiae mediorum", that is, by the products of the length of the path and the resistance along it in the various media.[4]

Here is a diagram that is a simplification of Leibniz' original figure, intended to bring out the similarity with (7.10) most clearly. AB indicates the interface between the two media; the denser medium, that is, the medium with higher resistance, is below, the less dense one above this line. One seeks the path of least resistance connecting C and G. Leibniz then finds that path by minimizing the total resistance w.r.t. the position of the point D where the light passes into the denser medium. The solution is easily obtained by calculus.

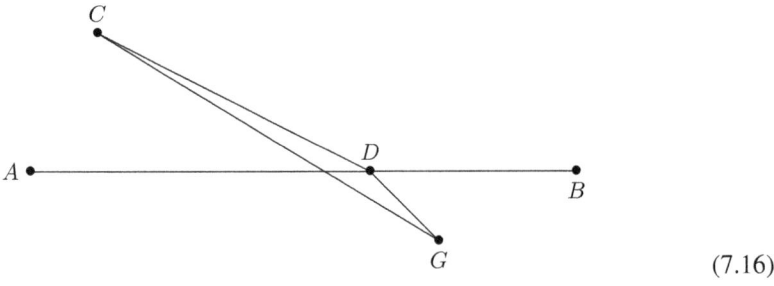

(7.16)

We can now see that Leibniz' solution of the brachistochrone problem in some sense represents a generalization of his treatment of Fermat's principle. We had already pointed out above that the brachistochrone problem (7.3) can be seen as a special case of (7.2). In turn, (7.2) is a generalization of the problem of light refraction. It can be seen as the problem of the propagation of light in a continuously varying medium, as opposed to refraction at an interface between two different media. Therefore, it seems natural, at least in hindsight, to transform Leibniz' scheme for deriving the result of Snell and Fermat into an infinitesimal version appropriate for continuously varying media. The point (x_h, y_h) in the first diagram simply takes the role of D in the current one. It is not clear to me, however, to what extent he already saw and exploited this analogy when he solved the brachistochrone problem. His method for solving this problem, however, can be

---

[3]I quote the mathematical texts of Leibniz mostly from Leibniz (2011) where the original bibliographical references can be found. Of course, for most of his mathematical writings, the standard reference is Leibniz (1855/1971).

[4]There are also several manuscripts of Leibniz (1906/1995) where he develops his ideas about refraction.

interpreted in that way. In any case, Johann Bernoulli, with whom Leibniz discussed the brachistochrone problem after they both had solved it, explicitly pointed out that analogy, and in Leibniz (1697), Leibniz also discusses it.

We return to the propagation of light. After thus easily having obtained Snell's result on the basis of a principle similar to Fermat's, but with the superior computational tool of his calculus, he first criticizes Descartes' reasoning.[5]

But then something strange happens. He proposes to consider a pointlike particle instead of the light ray. Leibniz reasons that since in the denser medium, it will push more particles more lightly, it will therefore move faster in that medium. From this, he concludes that Descartes had been right with his assumption that light travels faster in a denser medium, and in fact, that the velocity is inversely proportional to the resistance. Fermat's contrasting assumption – which is of course the correct one – that light travels more slowly in the denser medium is not discussed. In particular, while Leibniz like Fermat formulates a teleological principle, his principle, in contrast to Fermat's, is not that of least time. Leibniz' reasoning sounds somewhat strange, almost as if he did not really mean to make such claims about the actual speed of light in different media, but rather only that when the analogy with the motion of the pointlike particle were taken serious, Descartes' claims would follow.

As we shall see below in Sect. 7.5, in a treatise written around 1696/7, however, Leibniz speaks of a principle of least (or, more precisely, stationary) time for the paths of reflected and refracted light, although the more important and general principle becomes that of the most determined path.

The paper Leibniz (1682/1985) that we have just discussed appeared in the 1682 volume of the Acta eruditorum. Huygens then worked on the propagation of light; he presented his theory to the French Academy of Sciences in 1679, and his monumental treatise Huygens (1690/1912) appeared in 1690 (see e.g. Bell 1947). For our story, the following facts are relevant. Huygens proposed that light does not propagate like a particle, but rather as a wave. (Newton subsequently pointed out that light cannot propagate as a longitudinal wave, and since nobody at that time could imagine transversal waves for the propagation of light, this led to the rejection of Huygens' theory, until it was finally confirmed in the nineteenth century by Fresnel

---

[5]Descartes had invoked the analogy of a ball losing speed when going through a medium with more friction. Leibniz criticizes this with the argument that a light particle regains its status when returning to the original medium: "cum tamen radius lucis ex medio magis resistente in medium minus resistens, primo simile, rursus ingressus, priorem statum recuperet, & posito duorum mediorum similium, primi & ultimi, superficies (illius emittentem, hujus recipientem) esse planas parallelas, Directionem recipiat per refractionem posteriorem illi parallelam, quam habuit ante priorem." (Leibniz (1682/1985), cited after the text in the supplement of Leibniz (2011), p. 40). In an undated manuscript, Leibniz (1880/2008), pp. 304–309, about Descartes, Leibniz repeats the arguement against Descartes that light does not lose force by passing through a rare medium, because such a passage does not affect the angle of refraction:"Car s'il est vray que l'air à cause de sa flexibilité fait perdre une partie de la force comme le tapis celle du globule qui court là dessus, cette force perdue ne sera point rendue lorsque le rayon sort de l'air et retourne dans l'eau. Cependant nous voyons que le rayon y reprend la premiere inclinaison.", Leibniz (1880/2008), p. 308.

and Young.) Huygens had concluded already in 1673 that light travels more slowly in a denser medium, in contrast to Newton, who, like Descartes and Leibniz (at least in his 1682 paper), claimed the opposite.[6] In any case, Galileo in his *Discorsi* had already raised the question of the speed of light, and Huygens clearly stated that the speed of light has to be finite, whereas Descartes, Leibniz and Newton had been ambiguous about that issue, notwithstanding their reasonings about the refraction of light that we have described. Of course, in his *Optics* of 1704 (Newton 1704), Newton developed his corpuscular theory of light; he considered light as consisting of perfectly elastic, rigid and weightless particles moving at finite speed. (In principle, Newton could argue that the speed of light is infinite in the vacuum,[7] but since Descartes and Leibniz had excluded the possibility of a vacuum, they could not appeal to a behavior of light that was different in the atmosphere and in interstellar space.) When Ole Rømer in 1676 deduced from observations of the Jupiter moons that the speed of light is indeed finite and came up with an estimate of that speed, Huygens took that as a confirmation of his theory. As far as I know, in contrast to his friend and mentor Huygens with whom he was in regular contact, Leibniz did not pay much attention to Rømer's discovery, and I don't know of any reference in Leibniz' writings.[8] He was clearly aware, however, of the philosophical importance of the fact that the speed of light is finite. In the *Nouveaux essais*, Livre II, Chap. IX (p. 122 in Leibniz (1882/1978)) (which were composed much later, around 1706), he writes "nous ne voyons que l'image proprement, et nous ne sommes affectés que par les rayons. Et puisque les rayons de la lumiere ont besoin de temps (quelque petit qu'il soit), il est possible que l'objet soit detruit dans cet intervalle, et ne subsiste plus quand le rayon arrive à l'oeil, et ce qui n'est plus, ne sauroit estre l'objet present de la veue." An immediate interaction across spatial distances, be it by light or some other physical transmission, would have been better for his philosophical system, as will be discussed in Jost (2019).

7.4   The Origins of the Calculus of Variations

The history of the calculus of variations is a well documented subject. We may refer to Carathéodory (1937), Goldstine (1980), Thiele (2007), and Freguglia and Giaquinta (2016) where many further references can be found. We do not wish to

---

[6]In a draft written in or before 1695, Leibniz (1880/2008), p. 472, Leibniz prefers Huygens' theory of light: "j'avoue que ce que M.Hugens nous a donné sur la production de la lumiere et de la refraction paroisse plus vraisemblable que tout ce qu'on en a donné jusqu'icy." In the letters he exchanged with Huygens himself, however, he is less committal and discusses the relation between light, gravity and magnetism, see for instance Leibniz (1849/1971), p. 182ff.

[7]It is, however, outside the scope of this contribution to investigate Newton's views on the speed of light.

[8]Although it was mentioned by Huygens in his correspondence with Leibniz, see for instance Leibniz (1849/1971), p. 176, in a letter from 1694.

repeat that here, but only present those aspects that are relevant for Leibniz' role. Nevertheless, the aforementioned works should be consulted for the documentation of the story unfolded below.

Although some variational problems were already discussed in Hellenistic times, the story relevant for us begins with Fermat and his principle of least time that we have discussed in the previous section.

The story continues when Johann Bernoulli, in order to challenge his elder brother Jacob, in 1696 posed the problem of finding the curve of quickest descent, that is, the curve on which a massive particle under Galileo's law of gravity travels from one point to another in shortest time. The second point is not only lower than the first, but also horizontally displaced, as in the diagram (7.10). While Leibniz proposed to call such a curve a *tachystoptote* (swiftest fall), Bernoulli prevailed with his name *brachystochrone*[9] (shortest time). As it turns out, the solution is a cycloid, a curve that had already been intensively studied, in particular by Huygens. Among other properties, Huygens had found that it is an isochrone, meaning that regardless of at which point of the curve a mass that initially is at rest starts to fall, it will always arrive at the same time at the bottom of the curve. Leibniz had derived the differential equation for the brachistochrone,

$$\frac{dy}{dx} = \sqrt{\frac{kx}{1 - kx}}, \tag{7.17}$$

for a constant k, that is, (7.6), in 1686.

The challenge of Johann Bernoulli was not only met by his brother Jacob, but also by Leibniz and Newton. In fact, this represents a singularly fortunate situation in the history of mathematics. Four of the greatest mathematicians of all times competed on a specific problem. Naturally, this led to profound ideas that were pivotal for many subsequent mathematical developments.

Newton solved the problem by appealing to the property of the cycloid discovered by Huygens, that it is the isochrone, as explained above; thus, he only had to verify that the curve he knew, the cycloid, possesses the required properties for the problem at hand, and, in particular, that for any two points as specified in the problem, a suitable cycloid always exists. The solutions of Jacob and Johann, while different, were both very ingenious. They both appeared in the May 1697 volume of the Acta Eruditorum,[10] together with that of Newton (which he had already previously published anonymously) and solutions submitted by L'Hôpital[11] and

---

[9]Johann Bernoulli wrote *brachystochrone* instead of *brachistochrone*, the spelling now usually adopted. In fact, the Greek root is $\beta\rho\alpha\chi\acute{\upsilon}\varsigma$, *short*, although its superlative is usually written as $\beta\rho\acute{\alpha}\chi\iota\sigma\tau o\varsigma$, *shortest*.

[10]See also the German translations by P.Stäckel (1894/1976).

[11]Which was incorrect, see the analysis in Peiffer (1989).

Tschirnhaus, while Leibniz only wrote the introductory article (Leibniz 1697).[12] Carathéodory (1937) was particularly impressed by Johann's solution which not only verified that the cycloid is an extremum of the problem, but indeed a minimum, anticipating some aspects of the field theoretic approach to the calculus of variations developed in the nineteenth century by Weierstraß and others. We refer to the detailed discussion by Thiele (2007). Jacob's solution contained important ideas upon which Euler could build when he created the field of the calculus of the variations as an independent mathematical discipline in Euler (1744/1952) in 1744.[13] Leibniz' solution has received less attention than those of the Bernoullis. Above, I have already described the key idea of Leibniz and how this can be naturally generalized. With that reasoning, Leibniz derived the differential equation (7.17) that a solution needs to satisfy, and as mentioned, he knew already in 1686 that the solution of that equation has to be a cycloid.[14]

7.5   The Principle of Least Action

As already discussed, when presenting his solution of the brachistochrone problem, Johann Bernoulli also pointed out the analogy between optical and mechanical problems. But that aspect cannot have escaped Leibniz either, given the analogy between his solutions of the Fermat problem and the brachistochrone problem. In fact, as explained, the latter can naturally be considered as a generalization of the former. In particular, I am wondering whether in 1696, after having solved the brachistochrone problem within a few days, he still maintained his views about the speed of light in different media as expressed in his 1682 paper. In his *Tentamen Anagogicum* in Leibniz (1890/2008), pp. 270–279,[15] he again discusses the reflection and refraction of light. In general, he speaks about the easiest, and not

---

[12]See Goldstine (1980) and Freguglia and Giaquinta (2016) for details about the solutions of Newton and Leibniz.

[13]I am working on a commented German edition of this fundamental text.

[14]Johann Bernoulli had sent the letter ot Leibniz in which he stated the brachystochrone problem on June 9, 1696, and the latter, in spite of ill health, replied to Bernoulli already on June 16. In his letter, Leibniz stated the differential equation, and in his manuscript (pp. 291–295 in Leibniz (1855/1971)), he derived the differential equation (7.17) for the solution, however, apparently without noticing that the solution is the cycloid (whose differential equation Leibniz himself had earlier derived), but only stating that he had already solved the problem in the past ("jam olim"). That the solution is the cycloid is pointed out by Johann Bernoulli, see Leibniz (1855/1971), p. 299, l.10-15, and Leibniz acknowledges this, ibid., p. 310, "Tu longius progressus cycloidem ipsam esse pulchre reperisti."

[15]In Leibniz (1890/2008), it is undated, but Gerhardt suggests that it might have been written between 1690 and 1695. Since, however, Leibniz in this text speaks about the curve of swiftest descent ("la ligne de la plus courte descente entre deux points donnés", p. 272), that is, the brachistochrone, it might have been written somewhat later, after 1696/7 when Leibniz and the Bernoullis had analyzed that problem.

the swiftest path, but when he discusses reflection in more detail, he speaks about paths of shortest (or longest) time, and he also includes refraction. The precise quote is "la regle du chemin singulier ou plus determiné en longueur du temps, a lieu generalement dans le rayon direct et rompu soit par reflexion *ou par refraction*, soit à l'egard des plans ou des surfaces courbes, concaves ou convexes, sans qu'on distingue dans cette determination le temps le plus long ou le plus court." (loc.cit., p. 277; my emphasis). The overarching principle, however, is that of the most determined path ("la regle du chemin singulier our plus determiné en longueur du temps", ibid.). For the geometric aspects of that principle, we refer to De Risi (2007).

We have seen above that global optimality implies local and hence infinitesimal optimality. The converse is not necessarily true. The shortest curves on a sphere are portions of great circles. But when a great circle passes through two antipodal points, that is, becomes longer than half an equator, it ceases to be the shortest connections between its end points. While the general theory behind this phenomenon, the theory of conjugate points, was only developed in the nineteenth century by Jacobi, this example was known to Leibniz. While such a portion of a great circle is not exactly the longest connection between its endpoints – there are arbitrarily long curves joining them –, it is at least the longest among all circular arcs. In that sense, we may say that the path now maximizes instead of minimizes the corresponding variational integral.[16]

In particular, as this example shows, infinitesimal optimality does not necessarily imply global optimality. But Leibniz found another condition characterizing such paths, be they minimizing or maximizing (in the sense just described). They are the most determined paths. This can be explained by the following easy example.

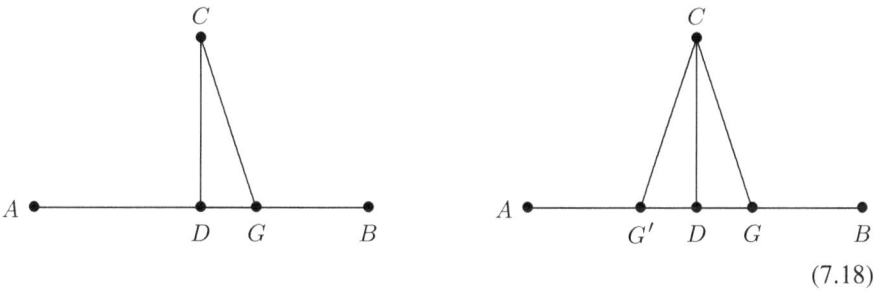

$$(7.18)$$

We seek the optimal connection from the point C to the line AB. When we take the endpoint G, the line CG has a "twin", that from C to G'. D is the only point for which the straight line from C to it does not possess such a twin. Therefore, Leibniz reasoned, it is the most determinate among all lines from C to AB. One could then argue that the two lines CG and CG' are symmetric fluctuations about the line CD,

---

[16]A mathematically more correct way to describe this is that such a circular that is longer than half an equator is a critical point of the variational integral, but not a minimum. In particular, it satisfies the corresponding Euler-Lagrange equation (for a geodesic on the sphere in modern terminology).

and because of this symmetry, their contributions cancel. The only remaining line whose contribution does not cancel, because it does not possess a symmetric twin, is the straight line from C to D. This does not imply that the realized curve is a minimum. It only implies an infinitesimal symmetry. Leibniz was aware of that.

The relationship between the various characterizations or definitions of a straight line, being the shortest, the straightest, or the most determined, was also a recurring theme in Leibniz' considerations of the foundations of geometry, and this has been analyzed in depth by V. de Risi in (2007), to which we may refer at this point.

The path integral approach of Richard Feynman to quantum mechanics depends on a least, or again, more precisely, on a stationary action principle, see Feymnan (1989), Vol.I, §26, and Vol.II, §19. Feynman's path integral is an integral over all possible lines from C to AB, with each line exponentially weighted by its length. Thus, we are no longer integrating over a single curve at a time, as in (7.8), but take an integral of integrals of the form

$$\int e^{\frac{\sqrt{-1}}{\hbar} \int F(x,y(x),y'(x))dx} dy(x) \qquad (7.19)$$

where the outer integral is over all possible curves $x \mapsto y(x)$.[17] The inner integral $\int F(x, y(x), y'(x))dx$ as before is the action of a particular curve. $\hbar$ is Planck's constant. Quantum mechanically, all paths are possible, and not only those of minimal action. The smaller the action, however, the exponentially smaller the fluctuations caused by the corresponding path in the action integral, and when one formally lets Planck's constant tend to zero, the path integral concentrates on the paths of least, or more precisely, stationary action, as the different phases of other paths will more and more cancel each other. Therefore, Leibniz' most determinate curve gives the dominant contribution to the path integral.

Thus, Leibniz was in the position to solve variational problems, and on the basis of his considerations about natural philosophy, it cannot have escaped him that there is some general principle behind those and other variational problems where one seeks a minimum (or maximum) of some variational integral. The question then is how such variational problems naturally arise from physics. In the middle of the eighteenth century, general principles of least action were formulated by Maupertuis (1748, 1751/1965) and Euler (Euler 1744/1952). Here, the variational integral to be minimized is the so-called action, that is, the integral of the energy over time, or equivalently, the momentum integrated along the path,

$$\int mv^2(t)dt = \int m\left(\frac{ds(t)}{dt}\right)^2 dt = \int m\frac{ds(t)}{dt}ds = \int mv(s)ds. \qquad (7.20)$$

---

[17]There are considerable mathematical subtleties involved with the correct definition of such a functional integral. We do not enter that issue, but refer to Jost (2009) and the references provided there.

König (1752) then presented a copy of a letter of Leibniz where the latter had already identified that action and formulated the principle. A famous controversy ensued about the authenticity of that letter (Goldenbaum 2016), and even though Kabitz (1913) in 1912 found another copy, the authenticity of the letter is still controversial, see Breger (1999) and Goldenbaum (2004). Even the recipient of the letter, if it was authentic, is disputed, whether it was Herrmann or Varignon or somebody else. On the other hand, no plausible candidate who could have forged the letter, in case it was not authentic after all, has ever been proposed (apart from König himself, but given the details of the story, it seems that he can be ruled out). Cassirer (1902/1998) considered the letter as genuine beyond any doubt, because it fits so well with Leibniz' general thinking. Couturat (1901) reaches the same conclusion and attributes the principle of least action to Leibniz and not to Maupertuis.

In any case, Leibniz had derived and analyzed the concept of the action as in (7.20) in his letters to Johann Bernoulli, which had been edited and published by the time Maupertuis developed his principle. In the letter of March 1696, he writes "Actiones motrices (aequabiles intelligo) ejusdem mobilis sunt in ratione composita effectuum immediatorum, nempe longitudinum percursarum et velocitatum. Porro longitudines (aequabiliter percursae) sunt in ratione composita temporum et velocitatum. Ergo actiones motrices sunt in ratione composita ex simplice temporum et duplicata velocitatum; adeoque iisdem temporibus vel temporum elementis, actiones motrices ejusdem mobilis sunt in ratione duplicata velocitatum, vel, si diversa sint mobilia, in ratione composita ex simplice mobilium et duplicata velocitatum." (Letter XXV in Leibniz (1855/1971), p. 259). Similarly in the letter where, among several other issues, he discusses the solution of the brachistochrone problem (Letter XXIX in Leibniz (1855/1971), p. 287), he writes "actionem ... composita ratione suorum principiorum, potentiae et temporis". He also explains why when the same result is achieved in half the time, the energy is quadrupled, but the action is only doubled (ibid., p. 286). In those letters, however, he did not claim that the action always assumes a minimum or a maximum in physical processes, although the main topic of the Letter XXIX is such an extremal problem, that of the brachistochrone. But with these aspects in such close intellectual proximity, it was only a small step, either for Leibniz himself, for Maupertuis, or for a forgerer, to the principle of least action.

The principle of least action has been extensively discussed in the literature, see for instance Yourgrau and Mandelstam (1979), Pulte (1989), Stöltzner (2012) and for teleological principles in general Schramm (1984).

7.6 Conclusion

The brachistochrone problem, originally posed by Johann Bernoulli as a challenge for his brother Jacob, is considered as the origin of the calculus of variations, an application of Leibniz' calculus to problems with infinitely many degrees of freedom. Although it was developed as a systematic theory only later by Euler, the solutions presented by Johann and Jacob Bernoulli and by Newton and Leibniz were all highly original. In hindsight, the solutions presented by the Bernoullis were seen to contain already many seminal ideas for the development of the discipline even beyond Euler. Newton's solution, while ingenious, did not lead to methodological advances, as he simply reduced the problem to Huygens' cycloid. Leibniz' solution has received less attention than those of the Bernoullis, but I show here that his abstract idea was also general and powerful enough to develop the general theory, although the history of mathematics took a different path. In fact, his approach quite naturally emerges from his earlier treatment of the refraction of light by his then new calculus, that is, his derivation of Fermat's principle. This then raises some important issues about the conceptions about the speed of light in different media by Leibniz and his contemporaries. I find that Leibniz' views have changed and developed in an interesting and in hindsight perhaps not always consistent manner in the course of his life. In particular, in his *Nouveaux essais* (see Leibniz 1858/2004), written around 1706 in response to Locke's philosophy, he was aware of the epistemological implications of the finiteness of the speed of light (see Jost (2019) for a further discussion). And, of course, the brachistochrone problem can also be seen as an instance of the principle of least action, in particular in view of the analogy between mechanical and optical problems that emerged from their formulation as variational problems. While it is still controversial whether Leibniz did explicitly formulate such a principle of least action, that principle naturally arises from his mathematical, philosophical and physical reflections. Leibniz was aware, however, in contrast to the formulation by Maupertuis, that the action is not always minimal, but might only be stationary. Therefore, he devoted much effort to deriving more abstract principles based on considerations of symmetry and determination, as analyzed in depth by De Risi (2007). Some of his corresponding ideas look surprisingly modern, for instance in the light of Feynman's path integral approach to quantum mechanics. Leibniz' ideas will be put in a systematic manner into the perspective of modern science in Jost (2019).

Bibliography

Bell, A. 1947. *Christian Huygens*. London: E. Arnold.
Breger, H. 1999. Über den von Samuel König veröffentlichten Brief zum Prinzip der kleinsten Wirkung. In *Pierre Louis Moreau de Maupertuis. Eine Bilanz nach 300 Jahren*. ed. H. Hecht, 363–381. Berlin: Berlin Verlag Arno Spitz GmbH.
Butazzo, G., M. Giaquinta, and S. Hildebrandt. 1998. *One-Dimensional Variational Problems*. Oxford: Clarendon Press.
Carathéodory, C. 1937. *The Beginning of Research in the Calculus of Variations*, Osiris, vol. 3, 224–240.
Carathéodory, C. [2]1982/1935. *Calculus of Variations and Partial Differential Equations of the First Order*. New York: Chelsea; translated from the *German, Variationsrechnung und Partielle Differentialgleichungen erster Ordnung*. Berlin: Teubner.
Cassirer, E. 1902/1998. *Leibniz' System in seinen wissenschaftlichen Grundlagen*. Hamburg: Felix Meiner.

Clairaut, A.C. 1733. *Sur quelques questions de maximis et minimis*, 186–194. Paris: Hist Acad Sci.

Couturat, L. 1901. *La logique de Leibniz, d'après des documents inédits*. Paris: Alcan.

Euler, L. 1728–1732/1952. *De linea brevissima in superficie quacunque duo quaelibet puncta iungente*. Comm Acad Sci Petropolitanae, vol. 3, 110–124; in OPERA, I, vol. XXV, ed. C. Carathéodory, 1–12. Bern.

Euler, L. 1744/1952. *Methodus Inveniendi Lineas Curvas Maximi Minimive Proprietate Gaudentes sive Solutio Problematis Isoperimetrici Latissimo Sensu Accepti*. Lausanne and Geneva: M. M. Bousquet & Socios. OPERA, I, vol. XXIV, ed. C. Carathéodory. Bern.

Fermat, P. 1891–1922. *Œuvres de Fermat*. Paris: Gauthier-Villars et fils. T. II, p. 354, p. 457, and T.I, p.170.

Feynman, R., R. Leighton, and M. Sands. [2]1989. *The Feynman Lectures on Physics*, 3 vols. Englewood Cliffs, N.J: Prentice Hall.

Freguglia, P., and M. Giaquinta. 2016. *The Early Period of the Calculus of Variations*. Cham: Birkhäuser.

Giaquinta, M., and S. Hildebrandt. 1996. *Calculus of Variations*, 2 vols. Berlin/New York: Springer.

Goldenbaum, U. 2004. *Appell an das Publikum*, 2 Bde. Berlin: Akademie-Verlag.

Goldenbaum, U. 2016. *Ein gefälschter Leibnizbrief? Plädoyer für seine Authentizität, Hefte der Leibniz-Stiftungsprofessur*, Bd.6. Hannover: Wehrhahn Verlag.

Goldstine, H. 1980. *A History of the Calculus of Variations from the 17th Through the 19th Century*. New York: Springer.

Goldstine, H.H. 1991. Introduction. In *Die Streitschriften von Jacob und Johann Bernoulli, Variationsrechnung*, bearbeitet und kommentiert von H. H. Goldstine mit historischen Anmerkungen von P. Radelet-de Grave, 1–113. Basel/Boston/Berlin: Birkhäuser.

Huygens, C. 1690/1912. *Traité de la lumière*. Leiden: Pierre vander Aa; English translation by S. Thompson, Isis 1,273. London.

Jost, J. [3]2005. *Postmodern Analysis*. Berlin: Springer.

Jost, J. 2009. *Geometry and Physics*. Berlin, Heidelberg: Springer.

Jost, J. 2019. *Leibniz und die moderne Naturwissenschaft*. Monograph, to appear in the Series *Wissenschaft und Philosophie – Science and Philosophy – Sciences et Philosophie*. Berlin, Heidelberg: Springer.

Jost, J., and X. Li-Jost. 1998. *Calculus of Variations*. Cambridge: Cambridge University Press.

Kabitz, W. 1913. Über eine in Gotha aufgefundene Abschrift des von S. König in seinem Streite mit Maupertuis und der Akademie veröffentlichten, seinerzeit für unecht erklärten Leibnizbriefes. In *Sitzungsberichte der Königlich Preussischen Akademie der Wissenschaften*, 2, 632–638. Berlin: Halbband.

Knobloch, E. 2012. *Leibniz and the brachistochrone*, Documenta Math, Extra Vol. ISMP, 15–18.

König, S. 1752. Appel au public, du jugement de l'Académie royale de Berlin, sur un fragment de lettre de Mr. de Leibnitz, cité par Mr. Koenig.

Lagrange, J.L. 1760–1761/1894. *Essai sur une nouvelle méthode pour déterminer les maxima et les minima des formules intégrales indéfinies*, OEUVRES, vol. I, 333–362. Misc Soc Tur, vol. 11, 173–195. German translation by P. Stäckel, Ostwald's Klassiker, vol. 47, 3–56, Leipzig (Reprinted Wiss.Buchges., Darmstadt, 1976) and English translation (in part) by Struik, D.J. *A Source Book in Mathematics*, 407–418.

Leibniz, G.W. 1682/1985. Unicum Opticae, Catoptricae, & Dioptricae Principium. Autore G.G. L., Acta eruditorum, 185–190; reprinted 10–14 in Leibniz, G.W., *Œuvre concernant la physique*, ed. J. Peyroux. Paris; German translation in Leibniz, G.W. (2011), 19–28, and in Leibniz, G.W., Schöpferische Vernunft, translated and ed. W. von Engelhardt, 287–298. Münster, Köln: Böhlau-Verlag.

Leibniz, G.W. 1697. *Communicatio suae solutionis problematis*. Autore G.G.L., Acta eruditorum, 201–205, reprinted in Leibniz (1858/2004), 331–336.

Leibniz, G.W. 1849/1971. *Leibnizens mathematische Schriften. Band II. Briefwechsel zwischen Leibniz, Hugens van Zulichem und dem Marquis de l'Hospital*, ed. C.I. Gerhardt, Halle, Part I, Vol. II; reprinted: Leibniz, G.W., *Mathematische Schriften 1*. Hildesheim/New York: Georg Olms.

Leibniz, G.W. 1855/1971. *Leibnizens mathematische Schriften*, ed. C.I. Gerhardt, Halle, Part I, vol. III; reprinted: Leibniz, G.W., *Mathematische Schriften 2*. Hildesheim/New York: Georg Olms.

Leibniz, G.W. 1858/2004. *Leibnizens mathematische Schriften*, ed. C.I. Gerhardt, Halle, Part II, vol. I; reprinted: Leibniz, G.W., *Mathematische Schriften 5*. Hildesheim/Zürich/New York: Georg Olms.

Leibniz, G.W. 1880/2008. *Die philosophischen Schriften von Gottfried Wilhelm Leibniz*, vol.4, ed. C.I. Gerhardt. Berlin, reprinted: Leibniz, G.W. *Die philosophischen Schriften 4*. Hildesheim/Zürich/New York: Georg Olms.

Leibniz, G.W. 1882/1978 *Die philosophischen Schriften von Gottfried Wilhelm Leibniz*, vol.5, ed. C.I. Gerhardt. Berlin, reprinted: Leibniz, G.W., *Die philosophischen Schriften 4*. Hildesheim, Zürich, New York: Georg Olms.

Leibniz, G.W. 1890/2008. *Die philosophischen Schriften von Gottfried Wilhelm Leibniz*, vol.7, ed. C.I. Gerhardt, Berlin, reprinted: Leibniz, G.W., *Die philosophischen Schriften 7*. Hildesheim/Zürich/New York: Georg Olms.

Leibniz, G.W. 1906. *Leibnizens nachgelassene Schriften physikalischen, mechanischen und technischen Inhalts*. ed. E. Gerland. Leipzig: Teubner; reprinted: Leibniz, G.W. 1995. *Nachgelassene Schriften physikalischen, mechanischen und technischen Inhalts*. Hildesheim/Zürich/New York: Georg Olms.

Leibniz, G.W. 2011. *Die mathematischen Zeitschriftenartikel*. Übersetzt und kommentiert von H.-J. Heß u. M.-L. Babin. Hildesheim/Zürich/New York: Georg Olms.

de Maupertuis, P.-L.M. 1748. Les Loix du Mouvement et Du Repos deduites d'un Principe Metaphysique. In *Mémoires de l'Académie Royale des Sciences et Belles Lettres, t. II*, 267–294. Berlin; the Paris publication is:

de Maupertuis, P.-L. M. 1751/1965. Accord de differentes loix de la nature qui avoient jusqu'ici paru incompatibles. In *Histoire de l'Académie Royale des Sciences*, 564–578. Amsterdam: Mortier; reprinted pp. 1–28 in P.-L.M. de Maupertuis, Oeuvres, IV. Hildesheim: Georg Olms.

McDonough, J. 2009. Leibniz on natural teleology and the laws of optics. *Philosophy and Phenomenological Research* 78(3): 505–544.

Newton, I. 1704. *Opticks: or, A Treatise of the Reflexions, Refractions, Inflexions and Colours of Light. Also Two Treatises of the Species and Magnitude of Curvilinear Figures*. London: S. Smith and B. Walford.

Peiffer, J. 1989. Le problème de la brachystochrone à travers les relations de Jean I Bernoulli avec L'Hôpital et Varignon. Studia Leibnitiana Sonderheft 17: 59–81.

Pulte, H. 1989. *Das Prinzip der kleinsten Wirkung und die Kraftkonzeptionen der rationalen Mechanik*. Stuttgart: Franz Steiner.

De Risi, V. 2007. *Geometry and Monadology: Leibniz's Analysis Situs and Philosophy of Space*. Basel/Boston: Birkhäuser.

Sabra, A.I. ²1981. *Theories of Light from Descartes to Newton*. Cambridge: Cambridge University Press.

Schramm, M. 1984. *Natur ohne Sinn? Das Ende des teleologischen Weltbildes*. Graz: Verlag Styria.

Stäckel, P. 1894/1976. *Variationsrechnung, Ostwald's Klassiker*, vol. 47. Leipzig: Engelmann; reprinted Darmstadt: Wiss. Buchges.

Stöltzner, M. ²2012. Das Prinzip der kleinsten Wirkung. In *Philosophie der Physik*, ed. M. Esfeld, 342–367. Frankfurt/M: Suhrkamp.

Thiele, R. 2007. *Von der Bernoullischen Brachistochrone zum Kalibrator-Konzept*. Turnhout: Brepols.

Yourgrau, W., and S. Mandelstam. 1979. Variational principles in dynamics and quantum theory. New York: Dover (reprint of the 3rd ed., 1968).

Chapter 8
Teleology and Realism in Leibniz's Philosophy of Science

Nabeel Hamid

Abstract This paper argues for an interpretation of Leibniz's claim that physics requires both mechanical and teleological principles as a view regarding the interpretation of physical theories. Granting that Leibniz's fundamental ontology remains non-physical, or mentalistic, it argues that teleological principles nevertheless ground a realist commitment about mechanical descriptions of phenomena. The empirical results of the new sciences, according to Leibniz, have genuine truth conditions: there is a fact of the matter about the regularities observed in experience. Taking this stance, however, requires bringing non-empirical reasons to bear upon mechanical causal claims. This paper first evaluates extant interpretations of Leibniz's thesis that there are two realms in physics as describing parallel, self-sufficient sets of laws. It then examines Leibniz's use of teleological principles to interpret scientific results in the context of his interventions in debates in seventeenth-century kinematic theory, and in the teaching of Copernicanism. Leibniz's use of the principle of continuity and the principle of simplicity, for instance, reveal an underlying commitment to the truth-aptness, or approximate truth-aptness, of the new natural sciences. The paper concludes with a brief remark on the relation between metaphysics, theology, and physics in Leibniz.

Abbreviations

A   *Sämtliche Schriften und Briefe*. 1923-. Edited by Berlin-Brandenburgische Akademie der Wissenschaften, and Akademie der Wissenschaften zu Göttingen. Berlin: de Gruyter. (Cited by series, volume, and page)
AG   *Philosophical Essays*. 1989. Edited and translated by Roger Ariew and Daniel Garber. Indianapolis: Hackett.
C   *Opuscules et fragments inédits de Leibniz: extraits des manuscrits de la Biblothèque Royale de Hanovre*. 1903. Edited by Louis Couturat. Paris: Felix Alcan.

N. Hamid (✉)
Concordia University, Montréal, Canada

© Springer Nature Switzerland AG 2019
V. De Risi (eds.), *Leibniz and the Structure of Sciences*, Boston Studies in the Philosophy and History of Science 337,
https://doi.org/10.1007/978-3-030-25572-5_8

GM *Leibnizens mathematische Schriften.* 1849–1863. Edited by C.I. Gerhardt. Berlin: A. Asher
     & Comp. (Cited by series, volume, and page)
GP *Die philosophischen Schriften,* 7 vols. 1875–1890. Edited by C.I. Gerhardt. Hildesheim:
     George Olms. (Cited by series, volume, and page)
L   *Philosophical Papers and Letters.* 1969. Edited and translated by Leroy E. Loemker.
     Dordrecht: Kluwer.
LS   *The Leibniz-Stahl Correspondence.* 2016. Edited and translated by Justin E.H. Smith and
     Francois Duchesneau. New Haven: Yale University Press.
LW   *Briefwechsel zwischen Leibniz und Christian Wolff.* 1860. Edited by C.I. Gerhardt. Halle:
     H.W. Schmidt.
NE   *New Essays on Human Understanding.* 1996. Edited and translated by Peter Remnant and
     Jonathan Bennett. Cambridge: Cambridge University Press.
UP   "Unicum opticae, catoptricae et dioptricae principium". 1682. *Acta eruditorum,* 1:185–190.
     Translated by Jeffrey McDonough. http://philosophy.ucsd.edu/faculty/rutherford/Leibniz/
     unitary-principle.htm.
WF   *Leibniz's 'New System' and Associated Contemporary Texts.* 1997. Edited and translated by
     R.S. Woolhouse and Richard Francks. Oxford: Clarendon Press.

8.1 Introduction

Among the signature features of Leibniz's mature philosophy is a commitment to
a harmonious, teleological structure of reality. Bucking the trend of seventeenth-
century mechanical philosophy, Leibniz attempts to reintroduce teleological notions
of ends and final causes in natural science. Indeed, beyond separating the domains
of mental and physical reasons, which are coordinated through a divinely instituted
harmony, Leibniz further contends that there are "two kingdoms even in corporeal
nature," and that mechanical laws "depend on more sublime principles" of order
and wisdom.[1] Teleological principles, thus, should be valid not just for describing
mental acts of desiring or willing, but also for physical events of impact and
collision.

How precisely Leibniz understands the co-governance of nature by mechanical
and teleological principles remains contested. It is one thing for Leibniz to recognize
final causes as governing an order of mental events parallel to that of physical
events, but an altogether different proposition to argue for teleology in the physical
domain itself. For one thing, it appears flatly to violate his own firm contention,
shared with many contemporaries, that we should explain "all the phenomena
of physics mechanically."[2] From his early conversion to the mechanistic view
of nature, through his important contributions to mathematical physics over the
next few decades, and into his later, monadological metaphysics, Leibniz remains
committed to the search for mechanical causes as the goal of science.[3] While we

---

[1]"Tentamen anagogicum" (ca. 1696), GP VII 273; L 478–9.

[2]"New System" (1695), GP IV 487; WF 22.

[3]Writing to Remond in 1714, he recounts his conversion to the new, mechanical philosophy in
the 1660s: "After finishing the *Écoles Triviales* I fell upon the moderns, and I recall walking in a

certainly cannot assume that Leibniz held fixed or unambiguous positions on many topics over the course of his life, the value he attaches to mechanical explanation remains constant. Given his firm commitment to the sufficiency of mechanism, what role does he think is left for teleology?

This essay identifies an autonomous place for classical teleological principles— roughly, variations on the old thesis that 'nature does nothing in vain'—in Leibniz's philosophy of physical science. It develops a reading of Leibniz as a scientific realist in which principles such as continuity and simplicity are indispensable. Specifically, I argue that Leibniz was committed to a thoroughgoing realism about the semantics of physical science: the claims of the new mechanical sciences of the seventeenth century, for Leibniz, have genuine truth conditions, or aspire to give true descriptions of phenomena. It is in this sense that Leibniz's philosophy of science embeds semantic realism: when properly interpreted, theoretical statements about the physical world can be said to be true or false.

To be sure, the truth-aptness of scientific claims on the reading of Leibniz developed here does not consist in metaphysical truths at the fundamental level of ontology. For, at least by the last decade of his life, if not earlier, Leibniz espouses a sophisticated form of metaphysical idealism according to which "there is nothing in things except simple substances, and in them perceptions and appetites."[4] For Leibniz, there is no non-mental reality as such. Whatever exists is either a mind (or a mind-like thing; in his later terminology, a monad), or a modification of one. On this picture, the material world of chairs and tables reduces to, or results from—to use Leibniz's technical locution—the perceptions of quasi-spiritual beings.[5] Consequently, the truth-orientation proper to physical science, for Leibniz,

---

grove on the outskirts of Leipzig called Rosental, at the age of fifteen, and deliberating whether I should keep the substantial forms. Mechanism finally prevailed and led me to apply myself to mathematics" (G III 606; L 655). In 1678, during an intense period of work in physics and optics, he writes to Hermann Conring: *"everything happens mechanically in nature,* that is, according to certain mathematical laws prescribed by God" (A II.1604; L 189). And to Burcher de Volder in 1703: "in phenomena . . . everything is explained mechanically" (GP II 250; L 529).

[4]Letter to de Volder, June 30, 1704, GP II 270; L 537.

[5]This view of the dependence of ordinary material objects on perceptions gives rise to the question of whether Leibniz can retain any place for material substance realism. The issue of the reality of material substances, however, should be kept apart from Leibniz's realism *per se.* For the ideality of Leibnizian monads does not make them unreal or illusory. On the idealist reading of Leibniz, minds are the most real beings; material bodies, meanwhile, are ontologically subordinate and thus are interpreted as having derivative reality. Readings of Leibniz as an idealist in this sense include Gueroult (1967), Adams (1994), Rutherford (1995), and De Risi (2007). Garber (1985) influentially challenged the idealist interpretation as a correct account of Leibniz's middle period, proposing instead a corporeal substance account on which bodies have reality independently of minds, and inspired others to develop broader, non-idealist readings of Leibniz. Phemister (2005), Hartz (2007), and McDonough (2016) are some of the authors to have followed Garber in defending readings of Leibniz as a material substance realist, where corporeal substances have an independent, foundational ontological status. In his 2009 *Leibniz: Body, Substance, Monad,* Garber responds to critics, but agrees that, by the *Monadology* period, Leibniz has embraced a metaphysical idealism on which minds and their experiences are the only ultimately real beings.

is ultimately demarcated from the aims of metaphysics or morals, which seek their first principles through inner experience, or reflection on the thinking self.[6] As Maria Rosa Antognazza has recently argued, "for Leibniz, physics proper is the study of natural phenomena in mathematical and mechanical terms without recourse for its explanations to metaphysical notions." At the same time, however, Leibniz recognizes that a physics unanchored in metaphysics would be seriously deficient. Antognazza continues that, the autonomy of physics "does not imply for Leibniz that physics can say on its own all that there is to be said about the natural world. Quite the opposite. Leibniz inherits from the Aristotelian tradition the view that physics needs metaphysical roots or a metaphysical grounding."[7] Leibniz's division between the metaphysical and the physical does not amount to a separation between a realm of inquiry guided by the epistemic value of truth and one guided by utility or convenience. Physical science aims at the truth as well, though in its own, properly delimited field of appearances.

In the absence of any mind-independent material reality as such, the truth or falsity of scientific claims ultimately rests in facts about the intersubjective contents of perceptions. Leibniz's realism does not aim to uphold common-sense intuitions about a reality independent of any perceiver whatsoever. But while the objects of knowledge are not mind-independent material bodies, they are also not mere illusions. For Leibniz, true concepts of physical objects possess fixed, univocal content, the explication of which constitutes the aim of science. This content consists in the descriptive terms and mathematical laws that express regularities among the objects of outer experience. Explicating this content, however, requires science to reach beyond the actual and possible observational evidence that would license belief in statements involving theoretical terms. Physical science, for Leibniz, must recognize certain non-empirical principles, those which posit unity, simplicity, and harmony in the phenomenal realm, as necessary conditions for the truth-aptness of its first-order, empirical claims. Truth conditions for theoretical statements are distinct from their verification conditions. To formulate truth conditions for an empirical claim commits us to recognizing normative epistemic principles for theory construction distinct from the facts which would confirm or disconfirm those claims. This condition on the truth-aptness of scientific claims brings with it teleological principles of harmony and orderliness in nature.[8]

---

[6] To Thomas Burnet, for example, he writes: "Locke did not well understand the origin of necessary truths, which do not depend on the senses, or on experiences, or on facts, but on the consideration of the nature of the soul, which is a being, a substance, having unity, identity, action, passion, duration, etc. We need not be astonished if these ideas and the truths which depend on them are found in us, although we need reflection to perceive them, and sometimes need experiences to elicit our reflection or attention, to make us notice of what our own nature provides us" (26 May, 1706; G III, 307–308).

[7] Antognazza (2017, 21).

[8] The realism I attribute to Leibniz is similar to Psillos' (1999, 10–13) understanding of semantic realism, with the important caveat that Leibniz rejects the realist intuition of the mind-independence of theoretical entities.

Leibniz's view that the new physics requires teleology acquires its significance in this semantic context. Mechanical principles alone, according to Leibniz, provide inadequate support for a realist interpretation of theoretical terms and laws even when restricted to the domain of phenomena. A suite of non-empirical principles—such as the principle of continuity, the principle of maxima and minima, and the principle of simplicity—enter Leibniz's scientific work to interpret the data gathered from experiments and observation in models that aim to reconstruct the true order of appearances. On Leibniz's envisioned reconciliation of efficient and final causes, teleological principles are always at work in science. They serve a variety of necessary functions: they mediate the unification of empirical laws, guide choice between competing hypotheses, and inform the classification of natural kinds. In these roles, teleological principles are not merely heuristical. They do not simply provide an easier method for deducing physical laws, or a convenient scheme for organizing experimental data, but are indispensable for interpreting empirical results as expressive of nature. In other words, on the reasonable supposition that the working scientist takes herself to be investigating the truth about nature, teleological principles inescapably figure as constraints on the semantics of any theory. Leibniz's defense of teleology in physics is motivated by a concern to articulate a realist interpretation of the new natural science of the seventeenth century.[9]

My strategy is as follows: Section Two spells out some possible meanings of Leibniz's thesis that there are two realms in physical nature. It suggests reasons why Leibniz seeks a deep unity of mechanical and teleological principles, such that each serves a distinct, necessary role in scientific explanation. Section Three discusses Leibniz's views on the aims of science in general, and identifies unification as a key virtue in his conception of explanation. Section Four elucidates the central claim of the paper by showing how teleological commitments to harmony, considered as a semantic notion of unity in diversity, figure in Leibniz's defense of heliocentrism. The essay concludes with some reflections on Leibniz's views on the relation between science and metaphysics (and theology).

8.2 Two Realms in Corporeal Nature

Leibniz's various discussions on the place of teleology in natural science do not readily present a univocal position. In "Discourse on Metaphysics" (1686), for instance, Leibniz suggests a heuristical role for teleology: "The way of final causes ... is easier [than the profounder way of efficient causes] and is often useful for understanding important and useful truths, which one would be a long

---

[9]My interpretation departs, accordingly, from François Duchesneau's, inasmuch as, on my view, Leibniz does not think that the justification of teleological principles consists in their utility or fecundity in explaining particular phenomena. I agree with Duchesneau, however, in that I see Leibniz as giving teleological principles a constitutive role in nature; see Duchesneau (1993, 260–2). I will return to Leibniz's deeper foundations for teleological reasons in the conclusion.

time seeking by the other more physical route."[10] By contrast, in notes from the late 1670s, Leibniz suggests an explanatory equivalence of efficient and final causal laws: "All natural phenomena can be explained by final causes alone, just as if there were no efficient cause; and all natural phenomena can be explained by efficient causes alone, as if there were no final [cause]."[11] A third view, expressed in the "Tentamen anagogicum" (1696), hints at distinct but complementary roles for efficient and final causes, such that, even if every natural fact were mechanically explicable, teleological reasons would still be required to ground mechanical principles: "all natural phenomena could be explained mechanically if we understood them well enough, but the principles of mechanics themselves cannot be explained geometrically, since they depend on more sublime principles which show the wisdom of the Author in the order and perfection of his work."[12] Do these remarks admit of a single, consistent position concerning the status of teleological principles?

It is likely that Leibniz entertained several positions concerning natural teleology over the course of his multi-faceted career, and I won't attempt to force a uniquely correct view onto his corpus. That he does, on occasion, grant the heuristical value of teleological reasoning based on attributions of functions and goals, whether or not they are supported by causal mechanisms, is confirmed in other texts. Commenting on Descartes' *Principles of Philosophy* in 1692, for instance, Leibniz contends that, from the mere discovery of usefulness of some phenomenon, we can confidently infer the existence of a reason for it in God's mind.[13] Certainly, such reasoning in the context of discovery is typically considered innocuous, and Leibniz offers examples of such useful, non-explanatory discoveries: "in the natural world . . . the discovery of the magnetic needle is and will be a great thing, even if its workings remain forever unexplained to us."[14] Yet, this constitutes only the weakest respect in which Leibniz recognizes the validity of teleology.

The independent sufficiency of final causal explanations motivates the second of the interpretive options above, namely, that efficient and final causes constitute parallel laws of the physical world. The view ascribes to Leibniz a thesis of explanatory overdetermination, though of a unique sort. Overdetermination here is not captured in the standard example of two stones striking a window simultaneously—that sense of causal overdetermination remains within the realm of efficient causes. Rather, the equipotency of efficient and final causes is the thesis that the breaking of the window is determined independently by two different kinds of causes, so that every physical fact can, in principle, be sufficiently explained by either efficient causal

---

[10]GP IV 448; L 317.

[11]A VI.4B 1403. The notes are titled "Definitiones cogitationesque metaphysicae" and dated by the Academy editors to 1678–80.

[12]GP VII 272; L 478.

[13]GP IV 360–1; L387.

[14]Cited in Antognazza (2017, 36n).

or final causal laws.[15] The main impetus for such a reading comes from Leibniz's work in optics. In fact, in each of the three texts cited in the opening paragraph of this section, Leibniz illustrates his defense of final causes by appeal to a derivation of the laws of reflection and refraction from an optimization principle. This result, already described by Pierre de Fermat earlier in the century, and repeated by Leibniz in his 1682 "Unicum opticae, catoptricae et dioptricae principium," demonstrates the two basic laws of optics—the law of reflection and the law of refraction—by using the principle that a ray of light always travels through the easiest path; or, stated more generally, that "nature, proposing some end to itself, chooses the optimal means."[16] Employing as a first hypothesis that "light irradiating from a point reaches an illuminated point by the easiest path," Leibniz's strategy is to use this principles alone to derive both the law of reflection—the equality of the angle of incidence and the angle of reflection—and the basic law of refraction—that the ratio of the sines of the angles of incidence and refraction is equivalent to the reciprocal of the ratio of the resistances of the media through which light passes.[17] That Leibniz was very impressed with the result is evident from the numerous occasions on which he cites it in defense of final causes.[18]

Neither the laws nor their association with teleology is original with Leibniz. Already in Hellenistic times, Hero of Alexandria had formulated the problem of finding the angles of incidence and reflection under the teleological supposition that light strives to move over the shortest possible distance.[19] The use of such principles continues to flourish in modern mathematical sciences, where they collectively comprise the variational calculus. The distinctive feature of variational principles is that they express the maximization or minimization of some physical quantity. Light rays minimize the time taken to travel between two points. The hexagonal cells of honeycombs maximize storage space per ounce of wax. The spherical shape assumed by a single drop of water minimizes the surface tension of water molecules. Such phenomena have struck some observers, from antiquity to today, to indicate an intrinsic proclivity in nature toward harmony. The strikingly contingent character of such arrangements have further suggested intentional activity that could effect such apparent coincidences through advance knowledge of the optimal end state—the minimization of propagation time for light rays, the maximization of honey stores, the conservation of molecular energy. The variational procedure inverts the style of reasoning used in classical mechanics. Whereas a typical mechanical derivation—

---

[15]McDonough (2008, 2009, 2010) defends such a reading. "Equipotency" is his term to describe the relation between laws of efficient and final causes (2008, 674).

[16]A VI.4B 1405.

[17]See McDonough (2008, 2010) for a detailed reconstruction of Leibniz's procedure. McDonough's translation of Leibniz's text is available at http://philosophy.ucsd.edu/faculty/rutherford/Leibniz/unitary-principle.htm

[18]Cf. GM VI 243, L 442; GP VII 274ff, L 480ff; LS 24–25.

[19]Lemons (1997, 13–14). See Darrigol (2012, ch1) for ancient and medieval Greek and Arabic precedents.

find the acceleration of a particle given inertial force and mass—requires knowledge of initial conditions, variational principles demand boundary conditions—find the shape of a drop of water such that surface tension is a minimum. The thesis of explanatory overdetermination proposes to treat this method of discovery as having independent sufficiency.[20]

There are multiple reasons for skepticism about a view that regards nature as lawfully overdetermined in this way. Attending to these problems motivates the third option identified above, namely, that, for Leibniz, efficient and final causes serve distinct but complementary functions in unified explanations.

In the first place, admitting two parallel laws of nature plainly violates the principle of parsimony. Despite the popular image of Leibniz as a flamboyant metaphysician, such a violation runs afoul of his express commitment to the old maxim not to multiply entities without necessity. Indeed, Leibniz frequently ties the principle of preferring the simpler hypothesis, or of maximizing effects from fewest causes, to the nature of divine wisdom. In "Discourse on Metaphysics", he writes of the divine act of creation that, "where wisdom is concerned, decrees or hypotheses are comparable to expenditures, in the degree to which they are independent of each other, for reason demands that we avoid multiplying hypotheses or principles."[21] Similarly, in "On the Radical Origination of Things" (1697) Leibniz identifies the principle that "a maximum effect should be achieved with a minimum outlay" as a principle of what he admires as a "divine mathematics or metaphysical mechanism."[22] That God should decree two sets of laws to institute the same series of phenomena plainly contradicts this axiom of his wisdom. While Leibniz's God is indeed committed to creating as much being as possible, he is not thereby committed to multiplying ways of lawfully relating essences.

Second, the empirical adequacy of Leibniz's mathematical constructions for deriving the laws of optics does not license inferences about the physical causes governing the behavior of light. By the lights of Leibniz's own commitment to mechanical explanation, the causes for physical phenomena must be sought in the sizes, shapes, and motions of bodies. In this regard, Ernst Mach's criticism of formal teleological explanations in his *Science of Mechanics* (1883) is instructive. For Mach, the value of Fermat's, Leibniz's, or Maupertuis' method consists in its "economical character" and that "it secures us a *practical mastery*" of physical phenomena even though it provides no explanatory insight.[23] The reason why teleological principles do not yield insight into physical processes, according to Mach, can be seen from a proper grasp of mechanical explanations. The observation

---

[20]The question of the value of these forms of reasoning continues to be contested in contemporary philosophy of science. For recent defenses of the sufficiency of mathematical explanations of the sort rendered through variational principles, see Ginzburg and Colyvan (2004), Baker (2009), and Lange (2013).

[21]GP IV 431; L 306.

[22]GP VII 303–4; L 487–8.

[23]Mach (1919, 341).

that many phenomena express maximal or minimal quantities owes not to the tendency of nature to seek elegance or economy, but simply to the fact that, when the least or greatest possible physical magnitude has been reached, no further change is possible. A catenary—the shape assumed by a rope or chain when hanging from its ends and carrying only its own mass—reaches the lowest point of center of gravity, not because it seeks that point, but because further descent is impossible once the chain is in that state. The explanatory work here is performed by facts about the forces acting upon the chain when it is in a certain shape, not in nature's inclinations toward certain ends. Mach explains: "The important thing, therefore, is not the maximum or the minimum, but the removal of *work*; work being the factor determinative of the alteration." Endorsing a dispassionate sobriety in scientific matters, he concludes that: "It sounds much less imposing but is much more elucidatory, much more correct and comprehensive, instead of speaking of the economical tendencies of nature, to say: 'So much and so much only occurs as in virtue of the forces and circumstances involved can occur.'"[24] It is true that the principle of least time saves the phenomena of the relations among angles resulting from reflected and refracted light, but the empirical adequacy of the principle should not be mistaken for a causal explanation of the propagation of light.

Finally, there are compelling textual reasons for why Leibniz should not wish to be saddled with a view of the corporeal realm as governed by two parallel sets of laws. Part of Leibniz's dialectical objective of undercutting the Cartesian natural philosophers' rejection of teleology is to recover grounds for assent to a divinely instituted intelligibility in nature. His strategy to this end depends on his defense of the thesis that the laws of nature are contingent rather than necessary, as Descartes had maintained. That is, Leibniz argues that the laws of collision and impact, which impressively predict physical interactions, cannot themselves be grounded in more basic geometrical facts and, consequently, well-confirmed facts such as those expressed in the principle of the conservation of kinetic energy, or in the law of the equality of full cause with the full effect, cannot be demonstrated as logical or mathematical identities.[25] What's lacking to adequately ground the new physics, Leibniz thinks, are further facts that would establish the necessitating, or law-like, character of the laws of motion. It is precisely this lawlikeness which, according to Leibniz, cannot be supplied through mechanical principles, but instead requires principles that he variously calls "metaphysical", "architectonic", or "laws of final causes".[26] But Leibniz does not thereby wish to assert the absolute priority of final

---

[24]Mach (1919, 459–60)

[25]Leibniz famously demonstrates, against Descartes, that it is not momentum (mv) but kinetic energy (mv^2), *vis viva*, that is conserved in collision events. For reasons of space, I cannot discuss any further Leibniz's reasons for holding conservation laws to be contingent, and therefore as, strictly speaking, outside a purely mathematical physics. In any case, this issue is well studied in the literature. See Garber (2009, ch.6) for a detailed account.

[26]GP VII 273; L 479.

causes over efficient causes in physics either. Rather, it is crucial to recognize the legitimate role of each kind of principle. It is worth reviewing again what he wrote in the "Tentamen":

> The true middle term for satisfying both truth and piety is this: all natural phenomena could be explained mechanically if we understood them well enough, but the principles of mechanics themselves cannot be explained geometrically, since they depend on more sublime principles which show the wisdom of the Author in the order and perfection of his work.[27]

Leibniz's goal in this conciliatory project is to emphasize the interconnection between the two kinds of principle, as opposed to securing their parallel coexistence. Knowledge of the mechanisms by which bodies operate and are produced is a central goal of physics, and it is axiomatic for him that we should seek complete mechanical descriptions of material phenomena.[28] In an earlier, programmatic text on the methods and aims of natural philosophy, the "Praefatio ad libellum elementorum physicae" (1678–79), Leibniz virtually identifies mechanical description with distinct explanation:

> [T]he way in which a body operates cannot be explained distinctly unless we explain what its parts contribute. This cannot be understood, however, unless we understand their relation to each other and to the whole in a mechanical sense, that is, their figure and position, the change of this position or motion, their magnitude, their pores, and other things of this mechanical kind, for these always vary the operation.[29]

Leibniz emphasizes here the indispensability of mechanical explanations for the operations of bodies, implying that an explanation cast solely in terms of teleological principles could not displace an explanation of bodies in terms of the law-governed motions of their parts. The claim here is not that one merely has the option of explaining bodily occurrences through laws of motion and impact. Rather, natural scientists must seek quantitative expressions of the regularities among phenomena and their material bases.

At the same time, an adequate foundation for the science of mechanics requires metaphysical principles, which Leibniz attributes to God's rational volition to bring about an orderly course of nature. But while Leibniz is predictably critical in the "Tentamen" of the materialists and the necessitarians who exclude God's purposes from nature, he is equally critical of the "zealous theologians who, shocked at the

---

[27] GP VII 272; L 478. He expresses this thought also in "Specimen dynamicum": "In my judgment the best answer, which satisfies piety and science alike, is to acknowledge that all phenomena are indeed to be explained by mechanical efficient causes but that these mechanical laws are themselves to be derived in general from higher reasons and that we thus use a higher efficient cause only to establish the general and remote principles" (GM VI 242; L 441).

[28] Leibniz insists on this virtually throughout his career. An unambiguous statement of it comes from the De Volder correspondence: "in phenomena... everything is explained mechanically" (GP II 250; L 529). It should also be borne in mind that for Leibniz the mechanistic principle encompasses the biological realm as well. In the *New Essays*, for instance, he writes: "I attribute to mechanism everything which takes place in the bodies of plants and animals except their initial formation" (NE 139).

[29] A VI.42008; L 288.

corpuscular philosophy and not content with checking its misuse, have felt obliged to maintain there are phenomena in nature which cannot be explained by mechanical principles." Leibniz worries that, by replying in a dogmatic manner to the excesses of the mechanical philosophy, the defenders of piety "injure religion in trying to render it a service."[30] Whereas the mechanical philosophers fail to recognize the deeper grounding (in reasons rather than in arbitrary volitions) required by the laws of motion, their polemical opponents fail to appreciate the harmony of nature represented in the mechanical picture, which alone is worthy of a perfectly wise creator. A consistent feature of Leibniz's defense of the new physics is his conviction that the simplicity and generality of its laws is uniquely befitting of divine wisdom. A brute appeal to divine providence, a mere assertion *that* mechanical laws require further support without any account of *how* such metaphysical grounds interact with mechanical principles, does not do justice either to the cause of faith or to that of reason. The moral Leibniz wishes to impress is that neither a purely mechanistic nor an exclusively final causal approach suffices for a proper understanding of nature. The 'middle term' must be such as to form a link between the two, so that each kind of principle could be shown to be indispensable in a full account of the world. Leibniz, ultimately, wishes to place stronger constraints on what constitutes an adequate physical explanation than either the mechanists or the theists. A parallelism of two equipotent sets of laws would weaken those constraints by allowing each method to be sufficient on its own.

I submit that, like the heuristic reading, the equipotency reading also captures only a limited application of teleology in Leibniz's scientific work. The use of optimality principles to derive the laws of reflection and refraction provides only weak reasons for the kind of general validity of teleological principles in natural science that Leibniz seems to want. Leibniz, as suggested by the passage from the "Tentamen," accords a privileged role to final causes, inasmuch as they are required for efficient causes to have explanatory force. Put another way, Leibniz's conception of the relation between final and efficient causes is one of complementarity—each dispenses a distinct, necessary function in explanation. Moreover, an independent sufficiency of mechanism and teleology does not sit comfortably with Leibniz's dialectical interest in their reconciliation. The deeper source of legitimacy for teleological principles accrues from their role in the construction and interpretation of scientific theories.

8.3 Teleology and Scientific Realism

Although Leibniz often and publicly highlights the success of the easiest path principle in optics, it does not represent his only strategy to defend final causes. He employs a suite of teleological principles in his natural philosophy, including

---

[30]GP VII 272; L 478.

conservation laws, the principle of the equality of cause and effect, and the principle of continuity, all of which he collectively labels "metaphysical" or "architectonic."[31] These principles serve a variety of functions in his scientific work besides occasionally providing an alternative method of deduction. Architectonic principles play an essential role, for instance, in the unification of empirical laws, and in guiding hypothesis choice. In such theoretical functions we see a substantive interaction of teleology and mechanism. In general, while mathematical-mechanical laws express regularities between physical quantities of bodies, teleological laws stitch those laws together to represent nature systematically. While the content of physical science consists in its descriptive concepts and laws, the intelligibility of phenomena by means of this content depends on its being interpreted under principles of order.

Leibniz's defense of teleology is further rooted in considerations about the aim of science in general. He accepts on behalf of the new mechanical philosophy a classical conception of *scientia* as truth-directedness. Physical science, specifically, seeks the kind of truth proper to phenomena, or to what is composite, namely a stable order and regularity in the succession of appearances. The value of such knowledge, Leibniz emphasizes, is not mere utility such as could be exploited for external profit even while one remains in a state of ignorance about causes. The goal of inquiry rather is the "perfection of the mind itself," or the attainment of clarity and precision in one's knowledge so that, "if someone were to discover some admirable device of nature and to learn its mode of operation, he would have achieved something great even if no application of his discovery to common life could be shown."[32] The perfection of the mind comes apart from a defense of physical science grounded in the practical advantages it affords. Leibniz thus rejects one contemporary current in the interpretation of the new science that

---

[31]Leibniz indicates the equivalence of these terms in "On the Correction of Metaphysics and the Concept of Substance" (1694), where he describes metaphysics as the "primary and architectonic discipline" (GP IV 468; L 432). In early modern philosophical usage, "architectonic" connotes, following Aristotle's usage of ἀρχιτεκτονικός in the *Politics* (III.111282a4–6), the *Nicomachean Ethics* (I.11094a1–26), and the *Physics* (II.2194a34-b8), a master science or art that supplies principles to a subordinate field in virtue of knowing the latter's goal of production. Aristotle's conception of an architectonic discipline as one that regulates and orders others, and for whose sake special disciplines are practiced, was alive in the seventeenth and eighteenth centuries. In his *Lexicon* of 1652, for example, Johann Micraelius notes Aristotle's extension of the classical, architectural sense of the term as *scientia bene aedificandi*, to the science of politics through which cities are properly ordered and governed. The idea of architectonic as the art of constructing intellectual systems becomes general in the eighteenth century, finding its most self-conscious expression in J.H. Lambert's *Anlage zur Architectonic, oder Theorie des Einfachen und des Ersten in der philosophischen und mathematischen Erkenntniß* (1771). Kant's notion of an "architectonic of pure reason" preserves this conception of a governing science of principles, which he identifies with his critique of the cognitive faculties.

[32]"Praefatio ad libellum elementorum physicae", A VI.41994; L 280.

regards at least some of its theoretical results as reflecting nothing more than the "Workmanship of Men," in Locke's famous phrase.[33]

But, further, Leibniz does not conceive science as offering signs of God's particular volitions, as, for instance, Samuel Clarke maintained. For Clarke, matter being indifferent to purposeful powers and laws, "there is no such thing as what men commonly call 'the course or nature' or the 'power of nature'." Rather, "[t]he course of nature, truly and properly speaking, is nothing else but the will of God producing certain effects in a continued, regular, constant, and uniform manner." We certainly detect regularities in natural experience. But nature itself does not bear the causes, thus the explanatory grounds, of its regular appearances, which reside instead in a radically free, divine will.[34] Against this voluntarist tradition in natural religion, Leibniz rejects attempts to locate the value of natural science in any suggestion of divine purposes as intimated by plant morphology or advantageous climate patterns. The intellectual perfection of the mind through scientific discovery brings joy or felicity, according to Leibniz, only because it provides insight into "the laws or the mechanisms of divine invention," that is, to the intelligible reasons behind natural patterns.[35] Mechanical explanations, in other words, ought to be interpreted as partial expressions of the divine intellect, rather than of God's arbitrary volitions or of human purposes in controlling nature. Leibniz rehabilitates teleological principles precisely in service of a naturalistic interpretation of the new book of nature, and against approaches that would construe its hard-won fruit either as the expression of so many miracles, or as conventions instituted by self-interested human observers.

These teleological or architectonic principles anchor physics in metaphysics. Yet, it is important to bear in mind that, for Leibniz, such anchoring is only partial, so that physics cannot be deduced *a priori* from metaphysics. Empirical physics has, as he sometimes puts it, only moral certainty insofar as it concerns the coherence of perceptions. As he writes to Foucher in 1675,

> The more consistency we see in what happens to us, it is true, the more our belief is confirmed that what appears to us is reality... This permanent consistency gives us great assurance, but after all, it will be only moral until somebody discovers a priori the origin of the world which we see and pursues the question of why things are as they appear back to its foundations in essence.[36]

The foundations of even the most well-confirmed physical laws lie in "essence," upon which even the kind of truth-orientation proper to phenomena depends. While this foundational, metaphysical project lies outside the legitimate bounds of physics, it nevertheless supplies principles which are assumed in the empirical

---

[33]Locke (1975, III.vi.37).

[34]Clarke, *A Discourse Concerning the Unchangeable Obligations of Natural Religion*, 1705 (1998, 149). Cf. Clarke's Second Reply to Leibniz, GP VII 359; L 680. Other notable figures in this tradition include John Ray and William Derham. Gascoigne (1989) speaks of a "holy alliance" between Anglican natural theology and Newtonian science at Cambridge at the turn of the century.

[35]"Praefatio ad libellum elementorum physicae", A VI.41994; L 280.

[36]A II.1B 391; L 154.

investigation of nature, insofar as it aims to establish the truth expressed in well-founded phenomena. In a later text, "On the Method of Distinguishing Real from Imaginary Phenomena," Leibniz identifies further criteria for well-founded, or real, as opposed to imaginary, or unreal, phenomena. These criteria include their congruity, complexity, coherence with past, regular phenomena, and our ability to give explanations for them. But the most important criterion, for Leibniz, is the harmony of perceptions within a perceiver's experience, as well as intersubjective agreement among different perceivers. Harmony also grounds the predictive order of perceptions, and it is highlighted as the most important mark that secures reality for phenomena: "the most powerful criterion, sufficient even by itself, is success in predicting future phenomena from past and present ones, whether that prediction is based upon a reason, upon a hypothesis that was previously successful, or upon the customary consistency of things as observed previously."[37] Robert Adams comments on this important text: "Real phenomena are those that form part of a coherent, *scientifically* adequate story that appears all or most of the time, at least in a confused way, to all or most perceivers. That is the story that would be told, or approximated, by a perfected physical science. Imaginary phenomena are those that do not fit in this story."[38] Teleological principles enter empirical physics to anchor the study of real phenomena in metaphysics by supplying assumptions of their maximal harmony and unity. They dispense their function in a variety of ways.

Teleological principles enter, for instance, as premises in arguments to unify empirical laws, thus to increase their generality. A signal instance of this use occurs in Leibniz's unification of the laws of motion and rest by applying the principle of continuity. In effect, by conceiving motion as a continuous quantity Leibniz is able to treat rest as a limit case of motion. Assuming the motion of body A to remain constant, we vary continuously the quantity of motion of a second body B as it collides with A until the motion of B approaches zero at the moment of collision. This state is defined as rest, after which the motion of B increases continuously in the opposite direction as it rebounds. Likewise, in the case when B is at rest, the motion of A is described as a continuously changing quantity that merges with the quantity of motion of B as the bodies collide. Applying the law of continuity to colliding bodies thus allows Leibniz to treat rest as "the limit of the cases of directed motion, or the common limit of linear or continuous motion, and so, as it were, a special case of both."[39] In other words, the same law that governs velocity can be shown to apply to its absence, or rest. The principle of continuity enables greater unification among laws of nature by supporting idealizations—in this case, treating rest as infinitesimal or vanishing motion—under which discrete rules receive a common analysis. Leibniz, in this way, provides deeper foundations for an assumption already present in Galileo and Descartes, that motion and rest are

---

[37] GP VII 320; L 364.

[38] Adams (1994, 257).

[39] "Specimen dynamicum", GM VI 250, L 447–8; cf. "Reply to Malebranche", G III 52–3, L 352.

simply different modes of bodies.[40] As a foundational project, Leibniz takes this kind of unification to be not simply for the sake of cognitive economy but rather a guide to truth. While separate rules might adequately subsume unknown cases to known ones to facilitate prediction, their sufficiency for explanation remains an open question. The discovery of lawful connection of the phenomena of motion and rest under a single, more general rule now serves as a constraint on future theorizing. Subsequent analysis of the special cases should not lead to rules that violate the unity (or harmony) known to hold among them. In this regard, Leibniz shares with Aristotle before him, and with more recent philosophers of science since, a conception of scientific explanation as aiming not just at predictive success but also intelligibility.[41]

Considerations of unity and harmony also figure as reasons for choice between equivalent hypotheses. One of the most important of such choices in the seventeenth century is between geocentric and heliocentric cosmological models, around which debate intensified following the formal condemnations in 1616 and 1633 of Copernicus' *De revolutionibus* and of Galileo's advocacy of heliocentrism.[42] A common defense of Copernicanism from Church censure following this storied affair exploited Galilean arguments for the relativity of motion. Just as an observer below the deck of a ship moving with uniform speed and direction would not be able to determine whether the ship was in motion or at rest by studying projectiles in her cabin, observers on Earth could never experimentally establish if the Sun or the Earth were in motion. But if true motions cannot be ascribed to bodies, neither the heliocentric nor the geocentric (nor, for that matter, Tycho Brahe's geoheliocentric) model could be shown to be the physically correct one on observational grounds. Copernicanism, therefore, should simply be treated as an alternative computational model, a mere calculational instrument that saves the astronomical data but cannot pretend to speak the truth, any more than the Tychonic model with its implausible intersection of the Sun's orbit with those of Mercury, Venus, and Mars.[43] It thus

---

[40]Galilei, *Dialogues* (1953, 20–1); Descartes, *Principles* Pt II, §27 (1982, 52).

[41]"Specimen dynamicum", GM VI 250, L 447. Compare Herbert Feigl on unification as a virtue: "The aim of scientific explanation throughout the ages has been *unification*, i.e., the comprehending of a maximum of facts and regularities in terms of a minimum of theoretical concepts and assumptions" (1970, 12).

[42]For a recent study of defenses of Copernicus and Galileo in the seventeenth century, see Finocchiaro (2009). Rather less familiar is the diversity of scholastic defenses of geocentrism which seriously engaged heliocentric arguments. See Grant (1984) for a helpful corrective.

[43]To be sure, Tycho had important defenders in the seventeenth century, such as the Jesuit astronomer Giambattista Riccioli. In his monumental *Almagestum Novum* (1651), Riccioli defended a modified version of the Tychonic system, with its stationary Earth, and two separate centers at the Sun and the Earth, against both the Copernican and Ptolemaic. Riccioli's main targets were the justification of Galileo's condemnation, and a defense of a geostatic cosmology, for which he found the Tychonic model better on astronomical grounds. Leibniz, for his part, accepts both Galilean relativity and the hypothesis of a mobile Earth. In what follows, the contrast between the geocentric (and geostatic) versus the heliocentric models will be treated at a higher degree of abstraction, and will exclude the specific differences between the Tychonic and Ptolemaic models.

could not offend against Joshua 10:12–14, because it leaves open a literal reading of Joshua's command to the Sun to stand still, according to this defense.[44] Leibniz's interventions in the Copernicanism controversies in the 1680s and 1690s, while ultimately unsuccessful from a diplomatic point of view, reveal his commitment to cosmological teleology as a presupposition of scientific realism.

8.4 Copernican Harmony as Determination

Unlike instrumentalist defenders of Copernicanism in the seventeenth century, Leibniz takes the empirical equivalence of the geocentric and heliocentric systems to invite reflection upon the criteria for ascribing true motion and rest. Taken as a description of physical reality, a (Galilean) relativistic analysis of motion in inertial frames leads to the unacceptable consequence that we can never assert whether a body is in motion or at rest. Leibniz, in fact, draws an even stronger conclusion: "if there is nothing more in motion than this reciprocal change [of mutual vicinity or position], it follows that there is no reason in nature to ascribe motion to one thing rather than to others. The consequence of this will be that there is no real motion."[45] If it were impossible in principle to determine which of two bodies was in motion at any given moment, then ascriptions of motion could never be interpreted as having objective truth conditions. In the context of celestial mechanics, one would have to conclude that there is no fact of the matter as to whether the Sun moves or the Earth. Leibniz correctly recognizes the situation as one in which observational evidence will always underdetermine planetary theory, which results in his doctrine of equivalence of hypotheses:

> As for absolute motion, nothing can determine it mathematically, since everything ends in relation. The result is always a perfect equivalence in hypotheses, as in astronomy, so that no matter how many bodies one takes, one may arbitrarily assign rest or some degree of velocity to any one of them we wish, without possibly being refuted by the phenomena of straight, circular, or composite motion.[46]

A physics restricted to mathematical-mechanical principles alone could not coherently claim to speak the truth or approximate truth about phenomena.

Rational principles enter here as normative criteria for deciding among competing theories. For Leibniz, the choice of the simpler or more intelligible theory is not merely a matter of convenience. As he writes to Huygens in 1694, given an equivalence of hypotheses, "when I assign certain motions to certain bodies, I do

---

[44] See Omodeo (2014) for a wide-ranging study of the reception of Copernicanism in the broader cultural and intellectual debates of the sixteenth and seventeenth centuries.

[45] "Animadversiones", GP IV 369; L 393. See Lodge (2003) for a fuller analysis of Leibniz's argument for the relativity of motion. Garber (2009, 106–115) details the development of Leibniz's views on motion as relational.

[46] "New System" G IV, 486–87, L 459.

not have, and cannot have, any other reason but the simplicity of the hypothesis (other things being equal) for the true one."[47] In Leibniz's use of simplicity as a criterion, we find an urgent concern to interpret the new science as advancing the truth about the physical world. Insofar as physics aims, as Leibniz insists it does, to uncover the truth about phenomena, it takes the regular motions of bodies to express a determinate, intelligible order. The Copernican attribution of rest to the Sun and motion to the Earth asserts that the actual Sun and the Earth instantiate the properties of rest and motion respectively; these are not merely convenient ways of speaking. But mathematical considerations alone are insufficient to ground such attributions, a conclusion Leibniz had already embraced in the 1670s.[48] Thus, the natural philosopher must look elsewhere to settle the matter. The principle of simplicity supplies one such consideration and serves as a criterion to adjudicate between competing astronomical systems.

As discussed earlier, in physics Leibniz does not take simplicity as a guide to absolute or metaphysical truth. As Paul Lodge observes, in the exchange with Huygens as well as in the "Specimen Dynamicum," Leibniz is careful to qualify the appeal to simplicity with locutions such as that one can "hold" (*tenir*) the simplest hypothesis for the true one, or "we speak as the situation demands [*loquimur, prout res postulat*] in whatever way provides the more fitting and simpler explanation of the phenomena."[49] Absolute truth, for Leibniz, is not a notion pertinent to phenomena. Yet, while Lodge rightly cautions readers not to take Leibniz in these passages to be speaking of metaphysical truth, it must also be recognized that Leibniz does not take the empirical equivalence of hypotheses to amount to their epistemic equivalence *tout court*. Instead, he appeals to a different semantic notion, which he often calls "intelligibility," in physical matters. Indeed, he expressly identifies the truth of physical hypotheses with their intelligibility.[50] Donald Rutherford explicates this restricted version of the principle of sufficient reason in the created world as "the principle of intelligibility [which] entails that whatever we assert of bodies, whatever laws we frame to describe their behavior, must be explainable in terms of the nature of body."[51] That is, any quality actually possessed by a body has to be such that it can be explained by appeal to concepts proper to an analysis of bodies. Even though we cannot have demonstrative proofs of claims about corporeal phenomena as rendered in the empirical sciences—that is, claims about bodies cannot be reduced to identities in a finite number of steps through the principle of contradiction—we nonetheless possess rational criteria for

---

[47]GM II 199; L 419.

[48]Garber (2009, 112).

[49]Lodge (2003, 299–300); GM II 199; L 419; GM VI 248; L 445.

[50]C 590–1; AG 91: "since . . . people do assign motion and rest to bodies, even to bodies they believe to be moved neither by a mind, nor by an internal impulse, we must look into the sense in which they do this, so that we don't judge that they have spoken falsely. And on this matter we must reply that the truth of a hypothesis is nothing but its intelligibility."

[51]Rutherford (1995, 241).

adjudicating competing accounts. For Leibniz takes the intelligibility of a physical hypothesis to consist in the degree of distinctness it achieves in parsing concepts of phenomena. Intelligibility thus appears as an in principle comparable quantity. In a suggestive note from 1700, Leibniz explicitly glosses intelligibility as a measurable, intersubjectively ascertainable notion: "reality should be evaluated according to the multitude and variety and order of things and thus, in a word, according to the quantity of intelligibility."[52]

The notion of intelligibility lies at the root of Leibniz's opposition to instrumentalist or conventionalist defenses of Copernicanism. In a piece from 1689, written as part of his ultimately unsuccessful diplomatic efforts to have the Vatican's ban on the teaching of heliocentrism rescinded, he is clear that the empirical equivalence of various astronomical hypotheses does not license an arbitrary choice among them. His argument for allowing Copernicanism to be taught appeals to its philosophical merits. For, Leibniz writes,

> the Copernican hypothesis . . . displays the harmony of things at the same time as it shows the wisdom of the creator, and since the other hypotheses are burdened with innumerable perplexities and confuse everything in astonishing ways, we must say that, just as the Ptolemaic account is the truest one in spherical astronomy, on the other hand the Copernican account is the truest theory, that is, the most intelligible theory.[53]

Here, Leibniz tellingly distinguishes the values of the empirically equivalent geocentric and heliocentric systems. The former is certainly a good model for the computational purposes of positional astronomy. Traditional astronomy relied on data interpreted on the assumption of uniformly rotating spheres above an apparent plane surface, upon which technologies such as navigation charts, mariner's astrolabes, chronometers, and quadrants depended. Leibniz is aware that the Ptolemaic model not only saves all the relevant data but also serves adequately the ends of seafaring. For a navigator, it makes little sense to abandon an intuitively plausible, empirically sufficient model, which furthermore supports a sophisticated array of instruments, almanacs, and shared practices.[54]

Yet, Leibniz declares the heliocentric theory to be the better *physical* theory on account of its greater intelligibility, and because "it displays the harmony of things." Now, Copernicus himself had highlighted *harmonia* as an important virtue of his model. Copernicus' conception of harmony has sometimes been understood in aesthetic terms. As Thomas Kuhn observes in his influential study, Copernicus recognized that "the real appeal of sun-centered astronomy was aesthetic rather

---

[52]The text is edited and translated in De Risi (2006, 58–63).

[53]C 591–2; AG 92. Bertoloni Meli (1988, 25–9) establishes the context of this piece. Leibniz reiterates this position some years later as part of a general defense of the use of teleological principles in the "New System": "it is reasonable to attribute true motions to bodies if we follow the assumption which explains the phenomena in the most intelligible way" (GP IV 487; L 459).

[54]In fact, even the construction of modern planetariums in the early twentieth century began by self-consciously emulating the Ptolemaic model in their use of gears and motors to simulate a uniformly rotating celestial sphere passing before a stationary observer. See Bigg (2017) for a case study of the construction of projection planetariums in interwar Germany.

than pragmatic ... The ear equipped to discern geometric harmony could detect a new neatness and coherence in the sun-centered astronomy of Copernicus."[55] But whether or not Copernicus' conception of harmony was aesthetic in the way it has been understood by Kuhn, among others, Leibniz seems to have drawn a different lesson from the achievement of heliocentric astronomy. For Leibniz, the greater harmony of Copernicus' model is a semantic feature, rather than an aesthetic one, or a pragmatic one. In other words, it is neither in virtue of its finer symmetry, nor in virtue of any practical success, but rather because of its greater truth-aptness that heliocentrism is preferable to geocentrism. Specifically, the harmony of Sun-centered astronomy consists in the determinacy which follows from its basic assumptions. Leibniz defines harmony in general as "diversity compensated by identity."[56] Harmony is the property of unity in a multiplicity of things, or of uniformity in a manifold, as would be instituted by a rule or law. This ontological sense of harmony is expressed more clearly in a later letter to Wolff: "order, regularity, and harmony come to the same thing. You can even say that it is the degree of essence, if essence is calculated from harmonizing properties, which give essence weight and momentum, so to speak."[57] Harmony as a property of a collection of objects can be measured or determined to be of a definite grade. More precisely, it is the property of a system, a collection of distinct objects unified under rules of order, which ground its true, structural features. The greater the order, regularity, or stability in a system of objects, the greater its "degree of essence." The harmony of a system, thus, is neither a matter of taste nor of practical utility, but pertains to its constitutive features. It is this sense of harmony that, I contend, Leibniz wants to highlight in his defense of Copernicanism. Two differences in how the heliocentric and geocentric models address astronomical phenomena illustrate this point.

The first example has to do with the problem of retrograde motion of the superior planets (Mars, Jupiter, and Saturn). At regular intervals, each of these planets appears to reverse its direction in the sky, briefly moving in the opposite direction from that of the other bodies in its system, before returning to its original direction. The Ptolemaic model accounts for these observations by placing the planet on a rotating epicycle, which itself is moved on a sphere, or, in a more minimalist model, on a circle called the "deferent." The resulting planetary path describes an

---

[55]Kuhn (1957, 172). Copernicus emphasizes harmony and symmetry in Bk 1, Ch 10 of *De revolutionibus*: "So we find underlying this ordination an admirable symmetry in the Universe, and a clear bond of harmony in the motion and magnitude of the Spheres [Invenimus igitur sub hac ordinatione admirandam mundi symmetriam, ac certum harmoniae nexum motus et magnitudinis orbium]" (translation in Kuhn 1957, 180).

[56]"Harmoniam diversitatem identitate compensatam." Letter to Arnauld, November 1671, A II.1B 279; L 149.

[57]"ordo, regularitas, harmonia eodem redeunt. Posses etiam dicere esse gradum essentiae, si essentia ex proprietatibus harmonicisi aestimetur, quae ut sic dicam faciunt essentiae pondus et momentum." 18 May, 1715, LW 172; AG 234.

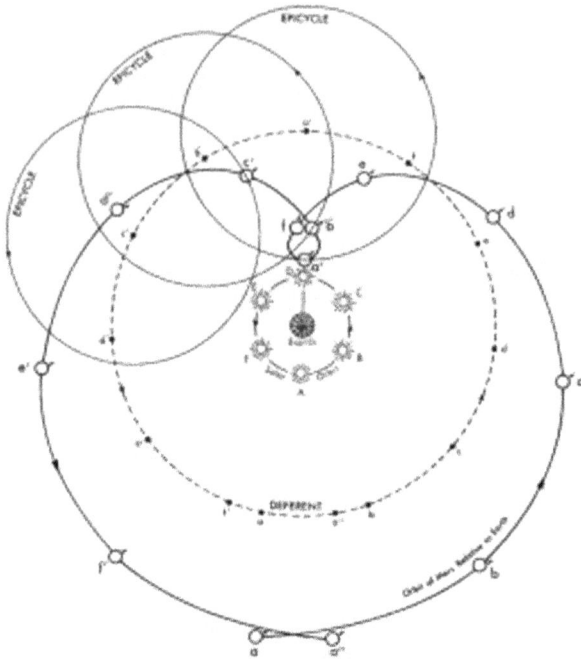

Fig. 8.1 Retrograde motion of Mars, in the Ptolemaic system

epicycloid curve, the exact proportions of which vary according to the size and speed of the epicycle. In this procedure, saving the observations from a stationary point on Earth requires separate constructions for each planet (see Fig. 8.1 for a construction showing the retrogradation of Mars). These constructions agree in the fact that the epicycles for each planet complete one revolution each year. In the Ptolemaic system, this is a brute fact, required to save the phenomena but having no physical basis.

By contrast, in the Copernican model, retrogradation of all the planets is explained as a natural consequence of the initial assumptions about the geometry of the system, namely, that the sphere of the stars is fixed, that the Earth moves around the Sun, and the Moon around the Earth. The retrograde transit of the superior planets follows from the fact that the Earth travels faster in its orbit around the Sun than do Mars, Jupiter, and Saturn. As a result, it is to be expected that, as the Earth overtakes Mars, for instance, Mars would appear to slow its eastward motion among the stars, briefly appear to move westward, and then resume its original direction (see Fig. 8.2). The fact of the annual revolution of the Earth about the sun also explains the yearly cycle assigned, in the Ptolemaic system, to the superior planetary epicycles. Further, besides furnishing a single reason for the retrogradations of the superior planets, the geometry of the model also determines precisely the size and order of the planetary orbits, once the frequency of retrograde transits is established.

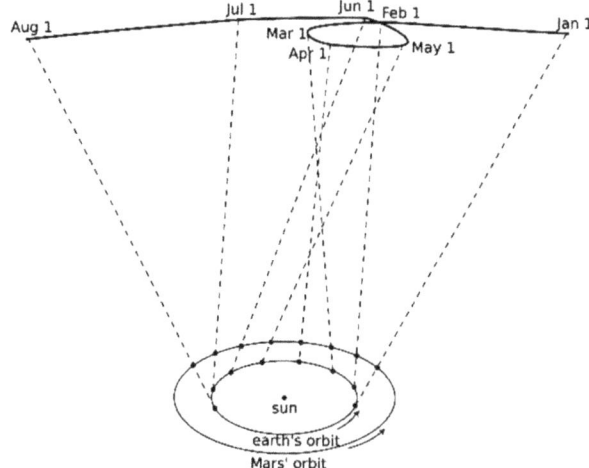

Fig. 8.2 Retrograde motion of Mars, in the Copernican system (Source: B. Crowell, 2005, www. lightandmatter.com)

Fig. 8.3 Elongations of Mercury and Venus from the Sun, in the Ptolemaic system (Source: George Benthien,https://gbenthien. net/)

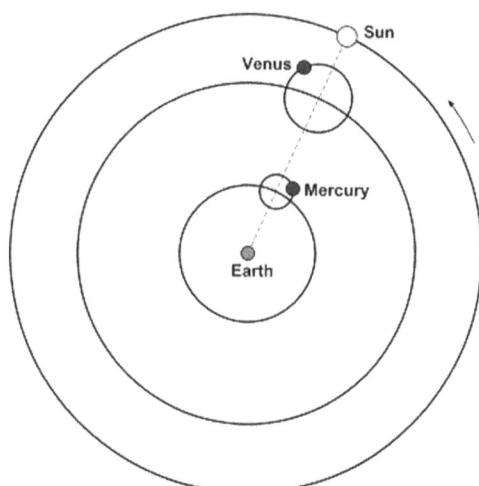

The exact order of the planets was a question on which the Ptolemaic astronomy had always remained ambivalent; Copernicus' model posits a determinate order on the basis of specific geometrical reasons.

A second astronomical problem relates to a peculiarity in the appearance of the inferior planets, namely, that these never wander very far from the Sun. The Ptolemaic solution for the apparent proximity of Mercury and Venus to the Sun was to tie the centers of the epicycles of the three bodies together so that they rotate around the Earth in sync. Differences in the sizes of Mercury's and Venus's epicycles, meanwhile, model their different relative positions in the sky (see Fig. 8.3). The construction adequately saves the phenomena only by introducing an

Fig. 8.4 Elongations of
Mercury and Venus from the
Sun, in the Copernican
system

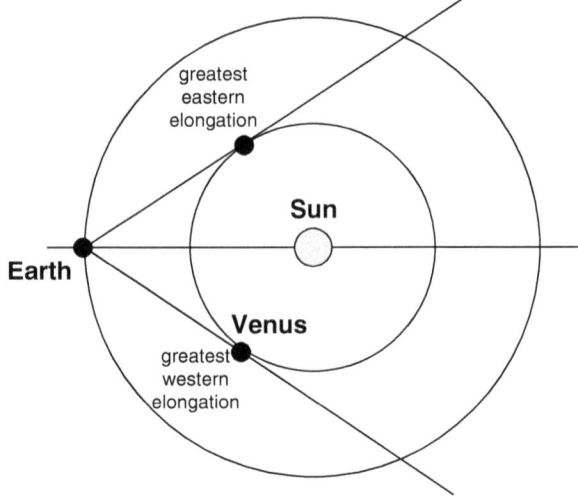

arbitrary element—a chord binding the bodies to Earth—absent elsewhere in the
system.

Sun-centered astronomy, by contrast, does not need recourse to special elements
to explain this fact. Once again, the apparent proximity of the inferior planets to
the Sun is a necessary consequence of the basic geometrical scheme. Given their
positions between the Earth and the Sun, it is to be expected that Mercury and
Venus would always appear within a narrow band from the Sun's path, because their
elongation from the Sun as observed from Earth would always be limited by their
smaller orbits relative to Earth's (see Fig. 8.4). As with the problem of retrograde
motion, these facts are also determined by a more general feature of the model,
rather than requiring a special solution.

Thus, what recommends acceptance of heliocentrism with a mobile Earth
as the better physical theory is what Leibniz calls its greater harmony, or its
determination of diverse phenomena under fewer rules. Sun-centered astronomy
reduces arbitrariness in its models, so that fewer initial assumptions are sufficient
to explain a wider range of qualitative phenomena. Neither pragmatic nor aesthetic
considerations speak especially in favor of Copernicus' scheme. On the one hand,
the old, positional astronomy remained the preferred option on practical grounds
in virtue of being embedded in cultures of material technologies and practices. On
the other, Copernicus' final scheme retained much of the ontological complexity for
which later generations would disparage Ptolemaic astronomy—for instance, while
doing away with the major epicycles with annual periods, Copernicus re-introduced
an equivalent number of smaller epicycles in saving the phenomena. Rather, it is
on theoretical grounds that heliocentrism is to be preferred, for it institutes greater
order in the phenomena, and hence contributes to the perfection of the mind.

Finally, the greater determination internal to heliocentric planetary astronomy
also makes it amenable to unification with physical science, thus to a kind of

inter-theoretic convergence that likewise supplies positive reasons of harmony. Leibniz regards the possibility of the convergence of heliocentrism with advances in kinematics as a reason for holding it to be true. In 1689, Leibniz published his response to Newton's *Principia*, the *Tentamen de motuum coelestium causis*, in which he took an important step in the direction of unifying mechanics with Kepler's elliptical model of planetary motion. Following the models of Descartes and Huygens, Leibniz hypothesizes a fluid, mechanical ether as a vehicle for the propagation of moving force by appeal to properties of shape, size, and velocity. A plenist model offers a mechanical theory by which impulsions in a physical medium could maintain the planets in harmonic circulation, as required by Kepler's laws. While Leibniz is certainly aware that experimental evidence is lacking for both the elliptical astronomy and the mechanical ether, he nonetheless hopes with this work to "come close to the true causes of celestial motions."[58] As he regards it, the new astronomy first makes possible a physical-mechanical account of celestial motion, and thus takes a step toward the ultimate end of a unified representation of nature.[59]

The upshot of this convergence of astronomical and physical theory finds its way in Leibniz's diplomatic endeavors. Touting the advantages of theoretical unity, he writes:

> For not only do the labyrinths concerning the stations and retrograde motions of the planets disappear with one mental stroke . . . but magnetic observations [of Jupiter's and Saturn's moons] are also united in a marvelous way since the Earth itself is like a magnet, not only with respect to the magnets of everyday experience, but also with respect to the heavenly bodies themselves . . . Copernicus could hardly have hoped for any greater confirmation of his view.[60]

In sum, Leibniz's advocacy of heliocentrism does not reflect an ecumenical proclivity to give Ptolemy, Copernicus, and the Church their dues, as some scholars have suggested.[61] As Leibniz discusses the matter in the *New Essays*, since the time of the initial condemnations of Galileo, when the papal authorities believed geocentrism to be in conformity with both reason and scripture, "people have become aware that reason, at least, no longer supports it."[62] By the 1690s, in fact, there is a gathering sense among European intellectuals that, despite the lack of conclusive experimental evidence, heliocentrism and the hypothesis of a mobile Earth has won the day against its detractors.[63] Leibniz can thus mount

---

[58] Bertoloni Meli (1993) gives a translation of the text in GM VI 144–161 and a commentary on its genesis and significance.

[59] In another, related text from the same year, "Tentamen de Physicis motuum coelestium Rationibus," Leibniz expresses hope for the possibility of being able to "explain the physical causes of planetary motion" (A VI.4C 2041).

[60] C 592–3; AG 93.

[61] Bertoloni Meli (1988, 27), for instance, interprets Leibniz here as an "equilibrist".

[62] NE 515.

[63] In his *Système de Philosophie*, the Cartesian Pierre Sylvain Regis, for instance, concludes his responses to the standard objections to the Earth's mobility as follows: "the hypothesis of the mobility of the Earth, which is now so common that it can be asserted that between all the various

a different response to the censure of Copernicanism than one which simply apportions equal share to the claims of reason and faith. For Leibniz, Church opposition to Copernicanism is not just dialectically inadequate, but epistemically problematic. It obstructs the dissemination of a superior physical theory, one that makes stronger claims to assent on account of its greater determinacy. In contrast to instrumentalist apologies for Copernicanism, Leibniz's intervention in the debate underscores his commitment to taking the claims of the best scientific theories at face value. As in the case of explanatory unification, his case for a realist interpretation of Copernicanism requires positing a criterion of truth that cannot be specified entirely in terms of observational evidence. Truth (or intelligibility) conditions for theoretical statements come apart from verification conditions.

8.5   Teleology and the New Science

I have highlighted an epistemological role for teleological principles in Leibniz's physics as stemming from a concern to uphold scientific realism. While Leibniz deploys teleological reasoning in a variety of contexts, the valuable lesson is not the one conveyed in his claim that non-causal principles yield complete and adequate alternative explanations of physical phenomena. Teleological principles do not underwrite a separate series of laws parallel to those of mechanical physics, as Leibniz's optical writings at first suggest. Rather, the deep lesson is that non-causal considerations—teleological, metaphysical, or architectonic, in Leibniz's various locutions—are implicated in any representation of nature that purports to track the truth or approximate truth about its appearances. Principles of harmony or intelligibility are cognitively significant insofar as a realistic attitude about our knowledge-making practices brings with it a commitment to determinate reasons underwriting our claims about nature, or, conversely, to excluding the possibility that our investigations might be in vain. Classical teleological principles of nature's intrinsic orderliness get redeployed here as coherence-making features of our epistemic practices. A sentiment akin to Leibniz's insistence that, "besides purely mathematical principles subject to the imagination, there must be admitted certain metaphysical principles,"[64] has found echoes in recent work on the limits of empiricist philosophies of science. Metaphysical reasons, as a number of recent authors have observed, are always at work in scientific practice to give precise meanings to theories and models. Criticizing a false dichotomy between metaphysics and empirical science, Anjan Chakravartty writes that,

---

opinions which are found in Astronomy, it has more supporters, not only than any other, but even than all others together" (1691, 226).

[64]GM VI 242; L 441.

it is not metaphysics that empiricists should oppose, but degrees of metaphysical speculation that fall outside the bounds of what they judge to be appropriate to the forms of empirical inquiry that most interest them, in accordance with their epistemic and other values. Indeed, were they to oppose metaphysics *simpliciter*, it appears that they would be guilty of pragmatic incoherence, since they themselves generally rely on some such speculations in one form or another.[65]

It is logically possible, of course, for the world to be a dappled mosaic, or for nature to be fundamentally unknowable, so that any amount of metaphysical speculation would be equally in vain. But then the burden of proof falls on the dogmatic proponents of a disorderly nature to substantiate their claim, or on the skeptics who would cast doubt on the possibility of knowledge. For Leibniz, the epistemic attitude proper to the scientific enterprise, supported by the gathering successes of seventeenth-century research, is the optimistic one.

To be sure, there is a crucial theological dimension to Leibnizian optimism, which undoubtedly weighs heavily in his thinking and cannot be ignored. In brief, his thesis that the human mind is made in the divine image supplies the ultimate guarantee that human inquirers could glimpse some of the truth about God's creation. A wise, benevolent God would not only construct the order of nature in such way as to exclude fundamental indeterminacies, but would also ensure that his moral subjects have the capacity to grasp enough of that order to appreciate his wisdom.[66] Leibniz's anchoring of the truths of nature in God's intellect is importantly different from competing, voluntarist theologies in his time, and directly relevant to his realism about physics. The Newtonian strategy, or even that of some Cartesians, to integrate the new physics in a theistic worldview, as Leibniz never tires of complaining, turns nature into a realm of perpetual miracles in the form of divine interventions to preserve order. In treating nature as essentially indifferent to form and order and, therefore, in need of periodic divine assistance for its regularities, such approaches evacuate the natural world of any intrinsic facts of the matter about why certain laws hold rather than others, or why certain powers are suitable means for specific effects. For the Cartesian occasionalists and the Anglican natural theologians, every empirical discovery indicates an arbitrary divine volition, but not a fact about nature itself.

On Leibniz's view of God's relation to the world of experience, by contrast, empirical investigation ought to be regarded as a source of insight into reasons of natural order. Indeed, when properly conjoined with metaphysical principles,

---

[65]Chakravartty (2007, 206). Expressing a similar thought, William Bechtel (1986, 40) casts the rationalist element in science as teleological, inasmuch as it conveys the explanatory aims of science: "All science is implicitly teleological, insofar as it adopts a semantics for its models—it doesn't let every activity or process in a target system enter into its model, but picks those it deems as causally relevant to explain the phenomenon."

[66]See Jolley (2005) for a systematic interpretation of Leibniz centered on the thesis of human subjects as mirrors of God.

empirical inquiry should lead to the same reasons which guided God in the creation. Even though the contents of perception are not the fundamental constituents of reality (that title is reserved for Leibniz's monads), a rational divinity would ensure that a definite, intelligible order appears to created minds. Leibniz sums up the epistemological upshot of his image of God thesis in a letter to Damaris Masham:

> [S]ince our understanding comes from God, and should be considered a ray of that Sun, we should conclude that what best conforms with our understanding (when we proceed methodically, and in accordance with the nature of the understanding itself) will conform with the divine wisdom; and that by following that method, we are following the procedure which God has given us.[67]

In understanding the order of nature, however partially, we access the same reasons that inclined divine wisdom to create the best of all possible worlds. Teleological principles underlie Leibniz's reconstruction of the new scientific worldview: when properly understood, the new science offers a vision of nature as a fully intelligible domain, requiring no divine assistance beyond its initial formation.

Acknowledgments I would like to thank Karen Detlefsen, Gary Hatfield, Vincenzo De Risi, and three referees for Springer for invaluable comments on earlier drafts of this paper. Versions of this paper were presented at workshops at Princeton University, University of Pennsylvania, and University of Turku, and I thank those audiences for their feedback.

Bibliography

Adams, Robert Merrihew. 1994. *Leibniz: Determinist, Theist, Idealist.* New York: Oxford University Press.

Antognazza, Maria Rosa. 2017. Philosophy and Science in Leibniz. In *Tercentenary Essays on the Philosophy and Science of Leibniz*, ed. Lloyd Strickland, Erik Vynckier, and Julia Weckend, 19–46. Cham: Palgrave Macmillan.

Aristotle. 1984. The Complete Works of Aristotle. Edited by Jonathan Barnes. Princeton: Princeton University Press.

Baker, Alan. 2009. Mathematical Explanation in Science. *British Journal for the Philosophy of Science* 60: 611–633.

Bechtel, William. 1986. Teleological Functional Analyses and the Hierarchical Organization of Nature. In *Current Issues in Teleology*, ed. Nicholas Rescher, 26–48. Lanham: University Press of America.

Bertoloni Meli, Domenico. 1988. Leibniz on the Censorship of the Copernican System. *Studia Leibnitiana* 20 (1): 19–42.

———. 1993. *Equivalence and Priority: Newton versus Leibniz.* Oxford: Oxford University Press.

Bigg, Charlotte. 2017. The View from Here, There and Nowhere? Situating the Observer in the Planetarium and in the Solar System. *Early Popular Visual Culture* 15 (2): 204–226.

Buchdahl, Gerd. 1969. *Metaphysics and the Philosophy of Science.* Cambridge: MIT Press.

Chakravartty, Anjan. 2007. Six Degrees of Speculation: Metaphysics in Empirical Contexts. In *Images of Empiricism*, ed. Bradley Monton, 183–208. Oxford: Oxford University Press.

---

[67]GP III 353; WF 211.

Clarke, Samuel. 1998. In *A Demonstration of the Being and Attributes of God and Other Writings*, ed. Ezio Vailati. Cambridge: Cambridge University Press.

Darrigol, Olivier. 2012. *A History of Optics*. Oxford: Clarendon.

De Risi, Vincenzo. 2006. Leibniz around 1700: Three Texts on Metaphysics. *The Leibniz Review* 16: 55–69.

———. 2007. *Geometry and Monadology: Leibniz's Analysis Situs and Philosophy of Space*. Basel: Birkhäuser.

Descartes, René. 1982. *Principles of Philosophy*. Translated by Valentine Rodger Miller and Reese P. Miller. Dordrecht: Kluwer.

Duchesneau, François. 1993. *Leibniz et La Méthode de La Science*. Paris: Presses Universitaires de France.

Feigl, Herbert. 1970. The 'Orthodox' View of Theories: Remarks in Defense as Well as Critique. In *Minnesota Studies in the Philosophy of Science, Volume IV*, ed. M. Radner and S. Winokur. Minneapolis: University of Minnesota Press.

Finocchiaro, Maurice A. 2009. *Defending Copernicus and Galileo*. Dordrecht: Springer.

Galilei, Galileo. 1953. *Dialogues Concerning the Two Chief World Systems*. Berkeley/Los Angeles: University of California Press.

Garber, Daniel. 1985. Leibniz and the Foundations of Physics: The Middle Years. In *The Natural Philosophy of Leibniz*, ed. Kathleen Okruhlik and James Robert Brown, 27–130. Dordrecht: Reidel.

———. 2009. *Leibniz: Body, Substance, Monad*. Oxford: Oxford University Press.

Gascoigne, John. 1989. *Cambridge in the Age of the Enlightenment*. Cambridge: Cambridge University Press.

Ginzburg, Lev, and Mark Colyvan. 2004. *Ecological Orbits: How Planets Move and Populations Grow*. Oxford: Oxford University Press.

Grant, Edward. 1984. In Defense of the Earth's Centrality and Immobility: Scholastic Reaction to Copernicanism in the Seventeenth Century. *Transactions of the American Philosophical Society* 74 (4): 1–69.

Guéroult, Martial. 1967. *Leibniz: Dynamique et Métaphysique*. Paris: Aubiers-Montaigne.

Hartz, Glenn A. 2007. *Leibniz's Final System: Monads, Matter, and Animals*. Oxford: Routledge.

Jolley, Nicholas. 2005. *Leibniz*. New York: Routledge.

Kuhn, Thomas. 1957. *The Copernican Revolution*. Cambridge, MA: Harvard University Press.

Lange, Marc. 2013. What Makes a Scientific Explanation Distinctively Mathematical? *British Journal for the Philosophy of Science* 64 (3): 485–511.

Lemons, Don S. 1997. *Perfect Form*. Princeton: Princeton University Press.

Locke, John. 1975. *An Essay Concerning Human Understanding*. Edited by Peter Nidditch. Oxford: Clarendon. (cited by book, part, and paragraph number).

Lodge, Paul. 2003. Leibniz on Relativity and the Motion of Bodies. *Philosophical Topics* 31: 277–308.

Mach, Ernst. 1919. *The Science of Mechanics*. 4th ed. Translated by Thomas J. McCormack. Chicago: Open Court.

McDonough, Jeffrey. 2008. Leibniz's Two Realms Revisited. *Nous* 42 (4): 673–696.

———. 2009. Leibniz on Natural Teleology and the Laws of Optics. *Philosophy and Phenomenological Research* 78 (3): 505–544.

———. 2010. Leibniz's Optics and Contingency in Nature. *Perspectives on Science* 18 (4): 432–455.

———. 2016. Leibniz and the Foundations of Physics: The Later Years. *Philosophical Review* 125 (1): 1–34.

McRae, Robert. 1976. *Leibniz: Perception, Apperception, and Thought*. Toronto: University of Toronto Press.

Micraelius, Johann. 1653. *Lexicon philosophicum terminorum philosophis usitatorum*. Jena: Jeremiah Mamphras.

Okruhlik, Kathleen. 1985. The Status of Scientific Laws in the Leibnizian System. In *The Natural Philosophy of Leibniz*, ed. Kathleen Okruhlik and James Robert Brown, 183–206. Dordrecht: Reidel.

Omodeo, Pietro Daniel. 2014. *Copernicus in the Cultural Debates of the Renaissance*. Leiden: Brill.

Phemister, Pauline. 2005. *Leibniz and the Natural World: Activity, Passivity, and Corporeal Substances*. Dordrecht: Springer.

Psillos, Stathis. 1999. *Scientific Realism: How Science Tracks Truth*. London/New York: Routledge.

Regis, Pierre Sylvain. 1691. *Système de Philosophie*. Vol. 3. Lyon: Anisson, Posuel, & Rigaud.

Rutherford, Donald. 1995. *Leibniz and the Rational Order of Nature*. Cambridge: Cambridge University Press.

Lightning Source UK Ltd.
Milton Keynes UK
UKHW021844050321
379874UK00003B/445